PHARMACEUTICAL PRODUCT DEVELOPMENT

PHARMACEUTICAL PRODUCT DEVELOPMENT

Insights into Pharmaceutical Processes,
Management and Regulatory Affairs

Edited by
Vandana B. Patravale
John I. Disouza
Maharukh T. Rustomjee

CRC Press
Taylor & Francis Group
Boca Raton London New York

CRC Press is an imprint of the
Taylor & Francis Group, an **informa** business

CRC Press
Taylor & Francis Group
6000 Broken Sound Parkway NW, Suite 300
Boca Raton, FL 33487-2742

First issued in paperback 2021

Version Date: 20160113

ISBN 13: 978-1-03-223801-2 (pbk)
ISBN 13: 978-1-4987-3077-8 (hbk)

Contents

Foreword

The pharmaceutical industry is among the most research-intensive and highly regulated industries propelled towards conducting exciting research in various therapeutic areas using conventional technologies as well as newer technologies and approaches with application of biotechnology, molecular biology and genetics to address unmet medical needs towards achieving novel therapies. However, the cost of research is very high and increasing with time, and the gestation period from research to market launch is long. Hence, getting a respectable return on investment is very challenging. Any small error or oversight in development or change in regulations can cost dearly with respect to time and finances. Since the pharmaceutical industry is looked upon as a 'zero defects' industry, achieving efficiency in the product development process is the key to maximising gains and mobilising resources for fabricating the next drug product. The pharmaceutical industry demands consistent production of quality pharmaceuticals with overall quality assurance and continuous quality improvement. The quality of pharmaceutical products and suitability of drugs for their intended use are determined by

- Their efficacy weighed against safety as endorsed and judged by regulators
- Their compliance to specifications developed or appropriate pharmacopoeial regulations pertaining to identity, purity and other characteristics

Developing pharmaceutical products in a timely manner and ensuring quality is a complex process that requires a systematic, scientific approach. Information about the properties of drug substances and excipients and the interaction between components of the pharmaceutical dosage forms are obtained in the early stages of research and development. This knowledge is then applied in different ways, including heuristics, decision trees, correlations and first-principle models utilising quality-by-design principles. These decisions help define the preferred route of administration and the choice of excipients. Unit operations and equipment needed to manufacture the product can then be developed. Computer-based techniques are extensively used to assist formulation scientists in managing the vast information, capturing the knowledge and providing intelligent decision making that supports pharmaceutical product development. Regulators and the pharmaceutical industry together are looking at more holistic approaches to improve processes to bring new products to the market with accelerated product development and consistent and reliable assurance of the highest level of quality. The main focus is on product life cycle management, a business transformation approach to manage products and related information across the industry. This is challenging because of the complex value chain and business processes required in a highly regulated environment. Being able to collect and analyse knowledge relating to all aspects of safety, efficacy and pharmaceutical development, including quality of a drug and drug product, allows pharmaceutical companies to quickly address industry challenges and provide solutions for unmet medical needs for our ultimate customers, patients. Successful commercialisation of the drug thus requires a scientific review of pertinent development data that provides the necessary information to assure that decisions are made regarding the potential in-licensing of a drug. The present book aims to provide a single comprehensive compilation of pharmaceutical product development and management covering all the steps from the very beginning of product conception to the final packaged form that enters the market in the form of a reference book that can equip the pharmaceutical formulation scientist with up-to-date, complete knowledge of drug product development. This book provides an extensive description of the entire set of processes a potential pharmaceutical product has to go through in order to qualify as a quality product to appear in the market, and dovetails pharmaceutical product management with core formulation development and large-scale manufacturing. Application of core science principles for product development is thoroughly discussed in conjunction with the latest approaches involving design of experiment and quality by design with a comprehensive illustration based on practical case studies of several dosage forms.

Praveen Tyle, PhD
President and Chief Executive Officer
Osmotica Pharmaceutical

Preface

The field of pharmaceutical product development has undergone a significant change over the years, from an early phase of adopting mostly empirical and unsubstantiated approaches, to a focus on a more structured process that contributes to the essential and allied processes involved in the creation of a finished product. Today, the product development process is largely inclusive of aspects related to intellectual property rights, design and strategy applying principles of quality by design (QbD), risk management and use of statistical design of experiment (DoE) to enhance the traditional product development stages, including preformulation, formulation development, scale-up and packaging. This book covers all of these aspects using a holistic approach, conceiving the final product from its genesis to market entry. Case studies discussing the application of these concepts for each type of dosage form have been presented for a better understanding of the application of these concepts. Introduction to recent quality improvement trends such as *quality metrics* and *continuous manufacturing* has also been provided for reflection and contemplation. This book seeks to disseminate the knowledge for pharmaceutical product development in an easy-to-read mode with simplified theories, case studies and guidelines for both working professionals in the pharmaceutical industry and students. While the book presents concepts in their entirety, it mainly identifies and presents realistic, sequential steps for the assistance of those who wish to work in this field on a practical level.

Given the inadequate number of reference books available in this field, this book is a considerable resource for pharmaceutical industrialists, academicians and students in this area. Although the book is mainly targeted to readers from the field of industrial pharmacy, intellectual property, pharmaceutical management, pharmaceutics and product development, we believe that readers from several allied fields such as clinical pharmacy, toxicology and even medicinal chemists will profit from its scope and content.

We are indebted to the authors for investing their time and energy to put forth this compilation in product life cycle management for pharmaceuticals. We would like to further extend our deepest gratitude to Dr L. Sesha Rao, Dr Nandkumar Bhilare, Susan Josi and Suhas Katare for their counsel on technical matters, Mahafrin Rustomjee for cover page design and to Tanvi Shah for providing valuable inputs and time towards formatting and language structuring.

Vandana B. Patravale
John I. Disouza
Maharukh T. Rustomjee

Editors

Vandana B. Patravale, PhD, is currently a professor of pharmaceutics at the Department of Pharmaceutical Sciences and Technology of the Institute of Chemical Technology, Mumbai, India. She has more than 90 refereed publications, 7 book chapters, 4 granted patents, 25 patents in the pipeline and 3 trademark registries to her credit and has handled many national and international projects. Dr Patravale has worked in close collaboration with industry and holds extensive experience of approximately 25 years in the field of pharmaceutical sciences and technology. Her areas of expertise include conventional and modified release dosage forms, formulation strategies to enhance bioavailability and targeting, medical device development (coronary stents, intrauterine devices), nanodiagnostics and novel nanocarriers with major emphasis on malaria, cancer and neurodegenerative disorders. Dr Patravale has recently published a book titled *Nanotechnology in Drug Delivery – A Perspective on Transition from Laboratory to Market* by Woodhead (now Elsevier) Publishing House.

John I. Disouza, PhD, is a professor in pharmaceutics and principal at the Tatyasaheb Kore College of Pharmacy, Kolhapur, India. He has more than 15 years of teaching and research experience. He also has an executive MBA (higher education) degree. He has written two books, *Experimental Microbiology* and *Biotechnology and Fermentation Processes*, and has published more than 50 research papers in peer-reviewed journals. Dr Disouza has worked in diverse research areas, including herbal formulations, micro/nanoparticulate, self-emulsifying, liposomal, fast and modified release drug delivery systems and so on. His research areas of interest are probiotics, novel diagnostic tools and therapies in cancer and structural modifications of natural polymers for their pharmaceutical potential. Dr Disouza is an active consultant to the pharmaceutical industry.

Maharukh T. Rustomjee is a consultant and advisor to the life sciences and healthcare industry. She is one of the founder directors of Rubicon Research Private Limited, which is an innovation-led drug delivery company in India. Being its chief operating officer for 15 years, she now continues to serve on the board of directors for the company. Spearheading Rubicon's drug delivery technology effort, Rustomjee has built a strong intellectual property knowledge base, holding numerous patents in the area of drug delivery systems, developed complex generics and worked on innovative dosage forms as LCM opportunities for multinational innovator pharma companies. She has more than 30 years of experience in dosage form development and drug delivery platform technologies in all phases of product development from product design through commercial manufacturing. Under Rustomjee's thought leadership, Rubicon developed solutions for bioavailability enhancement, gastric retention, liquid oral controlled release, taste masking and customised release profiles. Many of the products using these technologies have already been licensed out/commercialised to leading generic companies or innovator pharmaceutical companies.

Contributors

Ankit A. Agrawal
Institute of Chemical Technology
Mumbai, India

Mahalaxmi A. Andheria
Panacea Biotec Limited
Mumbai, India

Sivagami V. Bhatt
Panacea Biotec Limited
Mumbai, India

Manasi M. Chogale
Institute of Chemical Technology
Mumbai, India

Preshita P. Desai
Institute of Chemical Technology
Mumbai, India

John I. Disouza
Tatyasaheb Kore College of Pharmacy
Kolhapur, India

Sandip M. Gite
Institute of Chemical Technology
Mumbai, India

Amit G. Mirani
Institute of Chemical Technology
Mumbai, India

Abhinandan R. Patil
Tatyasaheb Kore College of Pharmacy
Kolhapur, India

Vandana B. Patravale
Institute of Chemical Technology
Mumbai, India

Priyanka S. Prabhu
Institute of Chemical Technology
Mumbai, India

Maharukh T. Rustomjee
Rubicon Research Pvt. Ltd.
Mumbai, India

Amita P. Surana
Jinstar Pharma Consultants Pvt. Ltd.
Mumbai, India

Swati S. Vyas
Institute of Chemical Technology
Mumbai, India

1 Product Life Cycle Management
Stages of New Product Development

Preshita P. Desai, John I. Disouza
and Vandana B. Patravale

CONTENTS

1.1 INTRODUCTION

The pharmaceutical industry is one of the largest, most dynamic and steadily growing industries. With a cumulative average growth rate of 5.1 per year and estimated global sales of prescription drugs to cross over a trillion dollars by 2020, it is poised for a visible rise in its growth trajectory (EvaluatePharma 2014).

The staggering success of this industry is attributed to its ability to offer the most reliable benefits to society and improve overall public health in terms of life expectancy, quality of life and affordable medical services. An increase in life expectancy of cancer patients to approximately 83% and a reduction in HIV/AIDS-associated mortality rate by 85% exemplify these facts reiteratively (PhRMA 2015).

On a macro level, the types of developmental activities performed in the pharmaceutical sector can be divided into novel developments, generic (me too) developments, and an intermediate category of improvisation of existing pharmaceutical products (either dosage form or alteration in route of administration). The same are depicted in Figure 1.1. Here, it must be noted that all these developmental activities are resource intensive because of long development timelines, stringent regulatory control and bulk investment needs. Amongst these, novel developmental activities (new therapeutic activities and innovative product development) are more time consuming (12–20 years) and resource intensive owing to a demand for extensive research and lack of a supportive scientific and developmental database (USFDA-CDER 2015). The risk intensity associated with these is well supported by the fact that only 1 in 5000–10,000 lead molecules screened in research reach the market, and amongst these, only 1/3 of the successful projects are estimated to meet the desired target performance in terms of timeline, cost input, market share and financial profit (Gibson 2009; Hein 2010).

On the other hand, generic formulations and improvised dosage form developments are considered to exhibit comparatively lower risks owing to the availability of a supportive safety and efficacy database enabling faster development

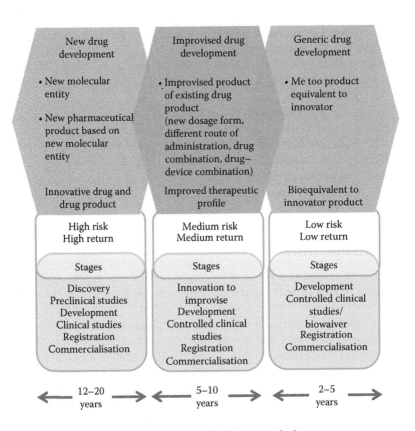

FIGURE 1.1 Types of development activities in pharmaceuticals.

and possible exemption of clinical studies. In view of this, the majority of new medications approved in recent years are variations of existing products (either generic/improved dosage forms). These products offer advantages in terms of better safety, efficacy, quality, affordability, easy access and consumer choice (USFDA-CDER 2015).

However, it must be realised that the newer therapy (new chemical entities/ new products) development is an integral part of pharmaceuticals as it brings next-generation therapeutics to fulfil unmet medical needs. The slow introduction of these new therapies (idea [lead] – candidate drug – product development – market) is attributed to low research and developmental productivity, complexity of new drug/ product design and development, high rate of attrition attributed to clinical failures and lack of pragmatic strategy, planning and execution.

As a thumb rule, each product goes through a life cycle from development to market entry to exponential phase (in terms of sales and profit) followed by a plateau stage, which, if not maintained profitable, proceeds to its decline and, eventually, its death. Thus, even for a successful pharmaceutical development, it is imperative to initially understand the life cycle of a new pharmaceutical product. Figure 1.2 describes the pharmaceutical product life cycle both from a financial aspect (a) and stepwise developmental activities (b).

1.2 NEW PRODUCT DEVELOPMENT: A NEED

The existence and the course of a business in the form of an enterprise are directly linked with the course of its products, and it only exists till the product's sales goes well. Further, it must be noted that long-term operations of an enterprise depend on *new product development* (*NPD*), as new products fuel financial and economic growth for business, and it is often referred to as the growth engine. There are many factors that make it essential for businesses to adopt NPD that includes technological advancement, change in customer's need, competition and so on. The NPD trend is now focussed more on a patient-centric approach as it presents a dual proposition to the developer that includes better market positioning (marketing proposition) and increased acceptability by the regulatory authorities (regulatory proposition) as they envision to bring in superior products over the existing pharmaceuticals to fulfil unmet medical need (EY 2014).

NPD is a complex and time-consuming process, since it holds more perils than first meets the eye. An enterprise has to maximise the innovativeness and efficiency of the NPD and minimise the cost of development and maintain the sustainability and profitability throughout the product life cycle (Figure 1.2). Thus, to maintain the effective and profitable life cycle of a new pharmaceutical product at its peak, the pharmaceutical industry is now implementing the product life cycle management (PLM) as a holistic business transformation approach to effectively manage the entire product portfolio of an enterprise right from ideation to development to commercialisation to withdrawal from market. It must be noted that PLM serves as an efficient management tool not only for NPD but also for generic and improvised dosage form development.

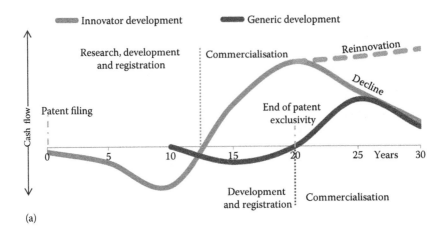

(a)

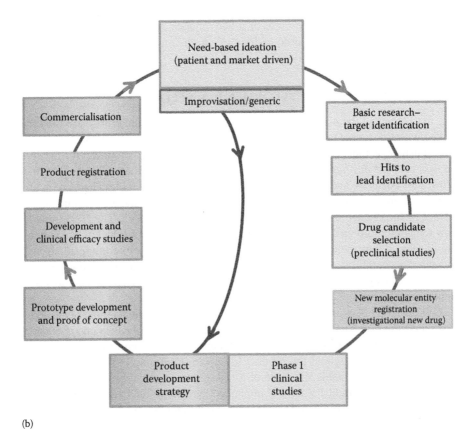

(b)

FIGURE 1.2 **(See colour insert.)** Pharmaceutical product life cycle. (a) Financial aspect. (b) Specific activities.

1.3 PRODUCT LIFE CYCLE MANAGEMENT

PLM refers to systematic synchronisation of activities that include planning, execution, effective data management and its utilisation, and cross-functional and cross-organisational control with timely improvisation throughout the product life cycle. The scope and phases of PLM are depicted in Figure 1.3.

The specific advantages offered by PLM to the pharmaceutical sector include an increase in number of new products, reduction in cost and time for development ensuring faster market entry, better regulatory compliance, avenue for product life cycle extension, efficient data management and so on (Hein 2010; Stark 2011).

For this development, a product road map is one of the most effective tools in the PLM, and if done correctly, it not only can help win and retain large customers and partners but also can guide the technical and strategic planning efforts of a company. There are different types of product road maps that can be used alone or in combination to create a comprehensive and compelling strategy (Figure 1.4).

A representative example of successful amalgamation of product road maps can be described as follows. It is a well-known fact that poorly soluble (biopharmaceutical classification system class II and IV) drugs result in poor oral bioavailability and thus require higher dose and dosing frequency to attain therapeutic efficacy. This may lead to side effects in some cases. With an understanding of the unmet need, potential market and a vision to address this issue, Elan Pharma International Ltd. developed the platform technology NanoCrystal. This technology deals with the development of a nanocrystalline drug that exhibits enhanced solubility and in turn improved bioavailability. This platform technology was then adapted by various pharmaceutical industries to develop various drug products such as Rapamune (nanocrystalline sirolimus), TriCor (nanocrystal fenofibrate) oral tablets and Emend (nanocrystal aprepitant) oral capsule. Thus, an appropriate selection of a product road map plays a very critical role in PLM (Bawa 2009).

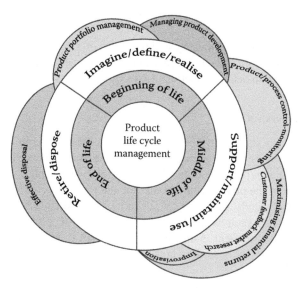

FIGURE 1.3 (See colour insert.) Overview of PLM.

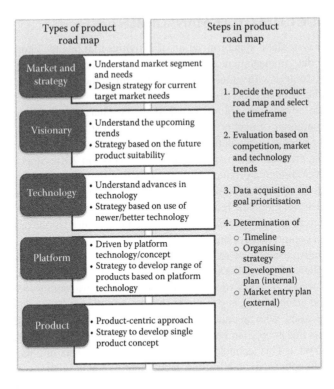

FIGURE 1.4   Types of product road maps.

1.4   PHASES OF NPD

Any NPD activity is a sequential process and the general phases are depicted in Figure 1.5.

It must be noted that the criticality and the depth of investigation for each phase may vary on the basis of product design, need, market positioning and so on. These phases are described in detail in Sections 1.4.1 through 1.4.8 (Komninos et al. 2015).

1.4.1   IDEA GENERATION

Every NPD process starts with the inception of an idea. It is a process in which innovative thinking is used to produce ideas for new products. The source for such ideas can be a market research, which is a tool to gather market information, present a picture of the pharmaceutical market and the product intended, and predict future market trends. Ideas for NPD can be inspired from both internal and external stakeholders such as market specialists, customers, partners and so on. Quality function deployment is an important concept in ideation and in further designing as it primarily considers consumer needs as a base to define product attributes.

Expos, fairs and seminars are also good sources of product ideas as the competitor's creativity, new innovative techniques or emerging technologies are showcased.

FIGURE 1.5   **(See colour insert.)** Phases of NPD.

In conclusion, customer/patient needs are gathered and are considered as the basis for NPD by transforming into new product ideas. There are various tools, techniques, or methodologies that can be used to transform the data collected into new product ideas, and these are summarised in Figure 1.6 (Chia-Chien 2007; Komninos et al. 2015).

1.4.2   IDEA SCREENING

The shortlisted ideas need further screening to assess their compliance with critical quality and efficacy parameters, manufacturing and commercial feasibility, technical compliance, regulatory and intellectual property compliance, success probability, market value, business proposition and investment need in terms of both time and money so as to select and take forward the most suitable idea.

The various tools used for this screening include risk analysis tools, qualitative research (survey based), sticking dots (voting based), SWOT analysis (identification of strengths, weaknesses, opportunities and threats), PMI analysis (identification of plus, minus and interesting aspects of an idea and their implications) and so on. Amongst these, the risk analysis tools are widely used in pharmaceutical product development and are listed in Table 1.1 along with their salient features and probable applications in the pharmaceutical sector (ICH 2005). From the table, it is clear that

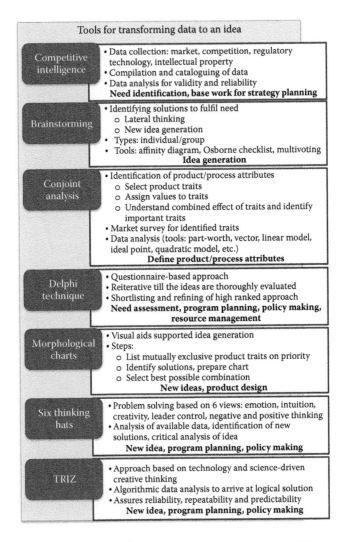

Tools for transforming data to an idea

Competitive intelligence	• Data collection: market, competition, regulatory technology, intellectual property • Compilation and cataloguing of data • Data analysis for validity and reliability **Need identification, base work for strategy planning**

Brainstorming	• Identifying solutions to fulfil need
	  o Lateral thinking
	  o New idea generation
	• Types: individual/group
	• Tools: affinity diagram, Osborne checklist, multivoting
	Idea generation

Conjoint analysis	• Identification of product/process attributes
	  o Select product traits
	  o Assign values to traits
	  o Understand combined effect of traits and identify important traits
	• Market survey for identified traits
	• Data analysis (tools: part-worth, vector, linear model, ideal point, quadratic model, etc.)
	Define product/process attributes

Delphi technique	• Questionnaire-based approach • Reiterative till the ideas are thoroughly evaluated • Shortlisting and refining of high ranked approach **Need assessment, program planning, policy making, resource management**

Morphological charts	• Visual aids supported idea generation
	• Steps:
	  o List mutually exclusive product traits on priority
	  o Identify solutions, prepare chart
	  o Select best possible combination
	New ideas, product design

Six thinking hats	• Problem solving based on 6 views: emotion, intuition, creativity, leader control, negative and positive thinking • Analysis of available data, identification of new solutions, critical analysis of idea **New idea, program planning, policy making**
TRIZ	• Approach based on technology and science-driven creative thinking • Algorithmic data analysis to arrive at logical solution • Assures reliability, repeatability and predictability **New idea, program planning, policy making**

FIGURE 1.6   Tools for idea generation.

risk analysis tools are not only used to screen the idea but also used as screening tools at every stage of product development to identify the risks in order to control them throughout the product life cycle with timely solutions. These tools have now been integrated as a mandatory requirement by all the regulatory authorities and are part of a formal regulatory guideline, ICH Q9, which is described in detail in Chapter 2. The practical examples of application of these tools are described in Chapters 4 and 9.

Recently, some Internet software-based models are also emerging as tools for idea screening and can be upgraded and used at later stages of development. An example would be Concept Screen, an Internet-based concept screening system by Decision Analyst Inc., which helps decide which idea to proceed with after it has been developed. In this, 200 to 500 consumers are requested to assess and evaluate

TABLE 1.1
Risk Analysis Tools

Risk Assessment Tool	Key Features	Application
Basic risk management facilitation method	Simple technique • Flowcharts • Check sheets • Process mapping • Cause-and-effect diagrams (Ishikawa/fish bone diagram)	• Primary risk assessment, concept development, etc.
Failure mode effects analysis (FMEA)	• Relies on prior knowledge of product, process and hazards • Qualitative analysis of modes of failures, factors causing these failures and the likely effects of these failures on desired attributes	• To prioritise risks associated with product composition, manufacturing operation, equipment and facilities
Failure mode, effects and criticality analysis (FMECA)	• Extension of FMEA (quantitative tool) to assess degree of severity probability and detectability of hazards. • Product or process specifications are prerequisite • Risk 'score' for each failure mode is given and is used to rank risk severity	• To prioritise risks associated with product composition, manufacturing operation, equipment and facilities
Fault tree analysis (FTA)	• Relies on the subject matter experts • Process understanding is prerequisite to identify causal factors • The results are represented pictorially in the form of a tree of fault modes • Single cause or combined multiple causes of failure can be pinpointed by identifying causal chains	• Investigation in case of deviation or failure
Hazard analysis and critical control points (HACCP)	Preferably used to assess and assure human safety • Determine critical control points • Set limits and establish corrective strategy in case of failure	• Physical, chemical and biological hazards management • Microbial contamination management

(Continued)

TABLE 1.1 (CONTINUED)
Risk Analysis Tools

Risk Assessment Tool	Key Features	Application
Hazard operability analysis (HAZOP)	• Based on the assumption that risk events are caused by deviations from the design or operating specification • Often uses subject matter experts to define limits • Brainstorming is a widely used tool for risk identification • Output is a list of critical operations and requires regular monitoring	• Manufacturing process, safety hazards management
Preliminary hazard analysis (PHA)	• Qualitative evaluation of the extent of damage because of hazard • Relative ranking of the hazard using a combination of severity and likelihood of occurrence • Identification of possible remedial measures	• Used at the early stages of project development when there is little known information on design details or operating procedures
Risk ranking and filtering	• Tool for comparing and ranking risks • Risk question is divided in various components and are ranked • The factors are then combined to assign a single relative score and are prioritised	• Used to prioritise manufacturing sites for inspection/audit by regulators or industry • Simultaneous assessment of quantitatively and qualitatively assessed risks within the same organisational framework (used at the managerial level)

all the ideas that are inserted in the software. The results are reviewed through a market test known as the conceptor, which evaluates the market robustness of the product idea (ConceptScreen 2015).

From the business proposition perspective, business analysis of an idea is of immense importance as financial investment and confirmed return on investment are two highly important aspects that drive the product life cycle. Briefly, business analysis looks into the cash flow and the revenue the product can generate, the investment needed, share value, and the expected market life of the product. For this, cost–benefit analysis is a widely used tool that involves comparison between the investment cost and the return on investment for the idea under consideration. Applications of cost–benefit analysis include evaluation of a new project, feasibility in raising finance and so on. This is a preferred tool for making simple business decisions and is expressed as the time taken to reach the breakeven point and profitability thereafter. For this, the cost and stages of investment for the project are determined along with a probable return timeline. This is then compared against the identified benefits, and project profitability is calculated. Based on this, the project selection is accomplished.

1.4.3   Concept Development

Once a single product idea is selected through the process described in idea generation and idea screening, a product concept has to be developed for the said idea and tested so that a complete product can emerge in later levels of the NPD road map. This can be visualised as an advanced step of pharmaceutical strategy development wherein a pharmaceutical product and probable method of manufacturing is proposed in compliance with a quality target profile. Typically, the process involves multiple steps as described in Figure 1.7. The figure also describes the relevance of each step in pharmaceutical product development.

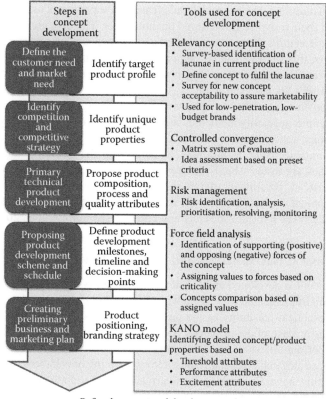

FIGURE 1.7   Concept development: tools and steps.

One important aspect of this concept development process is scheduling the activity timeline. To schedule this activity plan, tools such as Gantt charts, PERT (Program Evaluation and Review Technique) and so on are used. The Gantt chart is a form of bar chart that is used to plan the entire product development schedule, know the progress of the ongoing project and its current status, and organise the tasks associated with the project. PERT is based on critical path analysis or critical path method, which aids in scheduling and managing complex product development projects as well as in robust resource planning. The continuous assessment based on these tools also helps analyse if the progress is at par with the preset timeline (Komninos et al. 2015; Vanhoucke 2012).

1.4.4 Concept Testing

Concept testing is used to quantitatively assess the user response to a product before it is developed and launched in the market. The process typically includes the following:

Concept evaluation: The product details/features are presented to the customers and their response is measured quantitatively. The data are then evaluated to determine the market potential and acceptance. For pharmaceutical products, this survey includes opinions from physicians (key opinion leaders), scientific organisations/societies working in welfare of patients (e.g. Alzheimer's Society) and so on.

Concept positioning: The concept is assessed for the specified target market, for example, speciality medicines such as anti-infectives for region-specific epidemics, anticancer drugs and so on.

Product/concept trials: Consumers actually use the products and their responses are tested (this is generally not done for pharmaceutical products; instead, preclinical studies in animal models at an earlier stage of development or phase I/II clinical studies at the later stage of development are undertaken to establish safety and efficacy) (Komninos et al. 2015).

1.4.5 Technical Implementation and Prototype Development

Once the concept is developed and tested for its feasibility, the product and process are practically implemented to develop a product prototype. The product prototype ensures the practical feasibility of the product design and helps in identifying the risks at the early stage of development. A successful prototype design is considered as a very important milestone in NPD as important decisions with respect to project continuation, strategy changes if any, possible outsourcing or collaboration and so on are made at this stage. Particularly for pharmaceutical products, proof-of-concept, preliminary clinical trials are conducted at this stage to make *go* or *no go* decisions (for details, refer to Chapter 7). On the basis of the prototype development, further product development and optimisation is performed.

1.4.6   PRODUCT DEVELOPMENT

This is the heart of any development process and involves many cross-functional activities. Particularly for pharmaceutical products, it typically involves systematic product development at the laboratory scale using stringent quality by design tools (empirical/advanced design of experimentation approach), product and process optimisation, scale-up, commercial batch manufacture and so on (for details, refer to Chapters 2, 4, 5, 7 and 9).

It must be noted that the risk analysis and in-process testing to attain desired product attributes in terms of safety and efficacy are performed at every stage of this development process.

1.4.7   PRODUCT PERFORMANCE TESTING/ASSESSMENT

This process commences from the beginning of product development till its conclusion and includes testing at various stages of development. It helps to keep a track on the project proceedings. Major types of product development testing carried out at various stages of NPD are as described below:

The exploratory tests: Carried out in the initial development process to evaluate the primary design concepts and analyse whether it matches the consumer needs. Data collection is qualitative and based on observation, interviews and discussions. It is an important test as the further development process depends on these simulation tests. For pharmaceutical products, this survey includes opinions from physicians (key opinion leaders).

The assessment tests: It analyses the ideal solution and is performed at the latter development stages. It aims at ensuring the applicability of this solution by emphasising its utility and testing the relevance of the design options chosen for the same. It requires complex product modelling, analytical testing, simulations and working product replicas.

The validation tests: Conducted at the end stage, when the product is in its final form (i.e. ready for distribution), one evaluates whether all the targets have been achieved. It quantitatively checks the product's functionality, reliability, usability, performance, maintainability, robustness and so on. The product is compared to the predetermined standards, and the shortcomings are improved before it is released.

The comparison tests: Conducted at any stage of the NPD. Herein, the competitor/innovator product, its concepts, and execution are compared by tests such as benchmarking. It assesses the performance, transcendence and disadvantage of various designs. In the pharmaceutical sector, these tests are generally performed to prove the bioequivalence of a generic product to that of the reference product. Reference product herein refers to an innovator drug product that is approved by the regulatory authority and is referred by the generic product developer for filing and seeking an approval for the generic product.

In the general scenario, once the product is developed, scaled up and tested for quality and efficacy, it is introduced to the market (product commercialisation) on the basis of a predetermined marketing plan to capture maximum market share and thus profit. However, in pharmaceutical development, this is preceded by a regulatory approval process wherein a regional regulatory body evaluates the product development and performance report for safety, efficacy and so on and gives licence to the applicant to market the product. This is a crucial step in any pharmaceutical product development as without this approval the product cannot be introduced for use. The in-depth details of various regulatory authorities, types of applications and approval process thereof are described in Chapter 3.

1.4.8 Product Commercialisation

The successful market entry and sustainability of any product needs to be supported by a well-structured marketing portfolio. This can be defined as a road map that directs the marketing activities such as product promotion, distribution channel management, sales and so on. Figure 1.8 depicts the general input parameters considered while making the marketing plan and outcome of the same.

Here, it must be noted that the most important base input for any marketing plan is the company strategy for the said product. For example, the marketing strategy for a novel pharmaceutical product will focus more on promoting the solution for the unmet need while the generic pharmaceutical product marketing plan will focus on aspects such as product cost reduction.

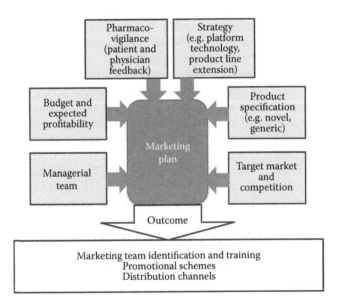

FIGURE 1.8 Overview of marketing plan.

Another important aspect of any marketing plan is distribution channel set-up. This is majorly driven by market considerations (end user considerations, patient-centric approach), producer considerations and product considerations (need of specific storage conditions, shelf life, etc.). An effective marketing plan not only is important for market entry but also plays a crucial role throughout the product life and thus needs continuous review and modification (product cost reduction, promotional offers, etc.) based on changing market and competition scenario.

In view of the complex nature of any product development process as described above and to maintain a time-bound and successful product life cycle, many cross-functional activities are undertaken. In this context, active collaboration, outsourcing, licencing and so on are gaining wide attention as these allow fast, efficient, cost-effective means to perform the task under consideration. For example, clinical studies of pharmaceutical products are a very cost-intensive activity and need subject experts, clinicians and statisticians to effectively perform the job. Outsourcing such activity to a contract research organisation (CRO) specialised in clinical studies not only resolves the burden on the product developer but also makes the process more fast and promising as the CRO holds expertise as well as necessary infrastructure ready in place.

Thus, PLM can be looked upon as a multi-organisational activity that makes maximum use of expertise and specialities of various organisations to develop and maintain the product at its best and in a most timely and cost-efficient manner.

In due course of time, after the market entry of any product, it reaches a decline phase as a result of extensive competition, market need and so on. At this stage, if not reinnovated, the product might have to be withdrawn from the market. Thus, reinnovation becomes a milestone in PLM. Product line extension (platform technology), improvisation in existing product, market exclusivity extension (patent evergreening) and so on are widely used techniques to achieve this.

Thus, to summarise, PLM is a managerial activity that plans, executes and controls the scientific, regulatory, market, personnel, business, finance and data management and analysis activities across organisations to effectively develop, deliver and maintain the product at a desired quality, efficacy, profitability and consumer acceptance and suitability.

REFERENCES

Bawa, R. 2009. Nanopharmaceuticals for drug – A review. http://www.touchophthalmology .com/sites/www.touchoncology.com/files/migrated/articles_pdfs/raj-bawa.pdf (accessed June 19, 2015).

Chia-Chien, H. 2007. The Delphi technique: Making sense of consensus. *Practical Assessment Research and Evaluation* 12(10):1–8. http://pareonline.net/pdf/v12n10.pdf (accessed June 19, 2015).

ConceptScreen. 2015. Concept Screen®. http://www.decisionanalyst.com/Services/concept .dai (accessed June 19, 2015).

EvaluatePharma. 2014. World preview 2014, outlook to 2020. http://info.evaluategroup.com /rs/evaluatepharmaltd/images/EP240614.pdf (accessed June 19, 2015).

EY. 2014. Commercial excellence in Pharma 3.0. http://www.ey.com/GL/en/Industries/Life -Sciences/EY-commercial-excellence-in-pharma-3-0 (accessed June 19, 2015).

Gibson, M. 2009. *Pharmaceutical preformulation and formulation – A practical guide from candidate drug selection to commercial dosage form*, second edition, Drugs and The Pharmaceutical Sciences, Volume 199. New York: Informa healthcare USA Inc.

Hein, T. 2010. Product lifecycle management for the pharmaceutical industry. http://www.oracle.com/us/products/applications/agile/lifecycle-mgmt-pharmaceutical-bwp-070014.pdf (accessed June 19, 2015).

ICH. 2005. International conference on harmonisation of technical requirements for registration of pharmaceuticals for human use, ICH harmonised tripartite guideline, quality risk management Q9. http://www.ich.org/fileadmin/Public_Web_Site/ICH_Products/Guidelines/Quality/Q9/Step4/Q9_Guideline.pdf (accessed June 19, 2015).

Komninos, I., Milossis, D., Komninos, N. 2015. Product life cycle management a guide to new product development. http://www.urenio.org/tools/en/Product_Life_Cycle_Management.pdf (accessed June 19, 2015).

PhRMA. 2015. 2015 profile biopharmaceutical research industry. http://www.phrma.org/sites/default/files/pdf/2014_PhRMA_PROFILE.pdf (accessed June 19, 2015).

Stark, J. 2011. *Product lifecycle management – 21st century paradigm for product realization*, second edition, London: Springer-Verlag London Limited.

USFDA-CDER. 2015. Novel new drugs 2014 summary. http://www.fda.gov/downloads/Drugs/DevelopmentApprovalProcess/DrugInnovation/UCM430299.pdf (accessed June 19, 2015).

Vanhoucke, M. 2012. *Project management with dynamic scheduling*, Springer-Verlag, Berlin Heidelberg.

2 Principal Concepts in Pharmaceutical Product Design and Development

Preshita P. Desai and Maharukh T. Rustomjee

CONTENTS

2.1 INTRODUCTION

The prime objective of any pharmaceutical product design and development is to generate an assurance that the therapeutic active in the proposed formulation or dosage form is suitable, safe and effective for intended use and holds acceptable quality. The pharmaceutical industry, since its emergence, is expected to work with this vision while converting the market opportunity into commercial reality. Since it deals with human life, safety and health, regulations covering it are very stringent as they rightfully should be. In view of this, pharmaceutical product development is a very time-consuming and resource-intensive process. It is estimated that the cost of developing a new drug is higher than 1.2 billion USD and some estimates put it as high as 2.6 billion USD (PhMRA 2015; Tufts CSDD 2014) and it takes anywhere between 10 and 15 years to do so (PhMRA 2015).

However, over the past two decades, manufacturing of pharmaceutical products has been plagued with drug shortages because of product recalls and supply issues related to manufacturing problems. This has further been compounded by the increasing presence of counterfeit medicines in the supply chain and has in part been caused by the following:

- Slower adoption of newer technologies
- Inability to predict effect of scale-ups on final product
- Absence of complete understanding of the manufacturing process and inappropriate selection of testing methods
- Inability to analyse or understand reasons for manufacturing failures, high cost due to inefficiencies (time, energy, materials) and high wastage (as high as 50% in some cases)
- Sluggish investigational drug development process

To overcome these issues, continuous *product quality assurance with timely improvement* was considered as a paradigm and the concept is evolving since then. Earlier, the term *quality* was related mainly to the finished product, but with increasing failures and recalls, controlled manufacturing process (in-process quality control, process validation) was looked upon as the framework towards reliable product quality. However, in recent years, it is thought that quality must be planned for a product and should be maintained and revised throughout the product life cycle (Juran 1992).

While this concept of designing quality over a product life cycle was emerging in the pharmaceutical arena, regulatory authorities were also upgrading the quality perspective and the chronological paradigm shift towards the same is summarised in Figure 2.1.

The steering change began with the introduction of Common Technical Document (CTD) format by the International Conference on Harmonisation (ICH) to ensure uniform dossier filing pattern under multidisciplinary guidelines (M4 guideline) in November 2000 wherein *quality* of drug and drug product was introduced as a compulsory section under module 3 (CTD-Q). The important feature under this was Section 3.2.P.2 of *pharmaceutical development* and was thought to be in consistency with designing quality into a product. In alliance with this, both the European Union and the US Food and Drug Administration (FDA) issued a territorial guideline describing what should be included in Section P.2.

By then, it was well recognised and agreed by all players that a sound pharmaceutical development program is essential for a high-quality product. It was also universally accepted that quality cannot be tested but should be 'built in' by design. Thus, understanding the importance of systematic product development and the need for a harmonised guideline for the same, ICH decided to bring in systematic and harmonised pharmaceutical quality system guidelines applicable across the life cycle of the product that begin with predetermined objectives, emphasise on product and process understanding and knowledge domain generation and utilise an integrated approach of science and quality risk management to deliver a quality product.

This put forth the concepts of *product life cycle* in greater discussion, which can be briefly described as all phases in the life of a product from the initial development through marketing until the product's discontinuation (ICH 2009). Broadly, it includes four stages, namely, pharmaceutical development, technology transfer,

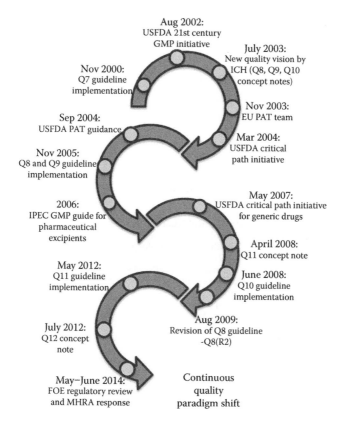

Aug 2002:
USFDA 21st century
GMP initiative

July 2003:
New quality vision by
ICH (Q8, Q9, Q10
concept notes)

Nov 2000:
Q7 guideline
implementation

Nov 2003:
EU PAT team

Sep 2004:
USFDA PAT guidance

Mar 2004:
USFDA critical
path initiative

Nov 2005:
Q8 and Q9 guideline
implementation

May 2007:
USFDA critical path initiative
for generic drugs

2006:
IPEC GMP guide for
pharmaceutical
excipients

April 2008:
Q11 concept note

May 2012:
Q11 guideline
implementation

June 2008:
Q10 guideline
implementation

July 2012:
Q12 concept
note

Aug 2009:
Revision of Q8 guideline
-Q8(R2)

May–June 2014:
FOE regulatory review
and MHRA response

Continuous
quality
paradigm shift

FIGURE 2.1   Chronological paradigm shift in quality perspective. FOE, Focus on enforcement; MHRA, the Medicines and Healthcare Products Regulatory Agency.

commercial manufacturing and product discontinuation. The specific activities undertaken under these steps are depicted in Figure 2.2.

In continuation with this agreement, it was brainstormed that a superior quality approach can be built up in the form of regulatory guidelines based upon the knowledge of product and process, quality risk management and modern pharmaceutical quality systems that resulted in the release of new ICH guidelines that are under continual improvement and are summarised in Table 2.1.

Further, these collative efforts brought in the newer concepts and principles of process analytical technology (PAT) (USFDA 2004), continuous process verification (EMEA 2014; USFDA 2011) and quality by design (QbD), risk management and life cycle quality management (ICH 2005, 2008, 2009, 2012, 2014) for delivering change. These concepts focus on using a scientific, risk-based framework for the development, manufacture, quality assurance and life cycle management of pharmaceutical drug products (ICH 2005, 2008, 2009, 2014) and drug substances (ICH 2008, 2012) to support innovation and efficiency and are discussed in Sections 2.2 through 2.4.

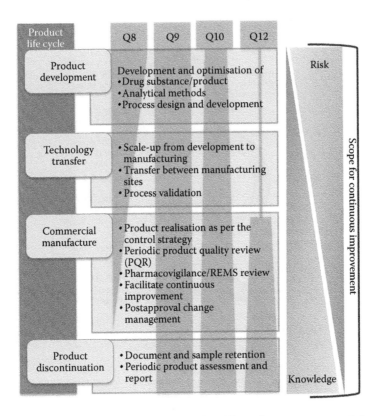

FIGURE 2.2   Key steps in pharmaceutical product life cycle management and the role of ICH quality guidelines. REMS, Risk Evaluation and Mitigation Strategy.

2.2   PHARMACEUTICAL PRODUCT DEVELOPMENT: AN UPGRADED ICH PERSPECTIVE

Implementation of a system that follows ICH guidelines (ICH 2005, 2008, 2009, 2014) plays a crucial role during the entire product life cycle to a variable extent, but it is worthwhile to note that they work in a complimentary manner to provide an ecosystem for quality planning, control and improvement in a continuous manner. Figure 2.2 describes the major activities undertaken during each stage of product life cycle management and depicts the importance and extent of application of ICH quality guidelines for pharmaceutical products (ICH 2005, 2008, 2009, 2014) in each stage. Implementation of these guidelines and extrapolating the gathered information lead to a gain in scientific knowledge along with risk minimisation.

2.2.1   ICH Q8(R2) GUIDELINE: PHARMACEUTICAL DEVELOPMENT

Pharmaceutical development is a systematic approach to design a quality product that begins with predetermined objectives and emphasises product and process understanding and determination of control strategy based on sound science and quality

TABLE 2.1

ICH Quality Guidelines for Pharmaceutical Life Cycle Management

ICH Guideline	Implementation Timeframe	Brief Scope	Application
Q8(R2): Pharmaceutical development	Q8 parent guideline, November 2005 Second revision, August 2009	Guideline for Section 3.2.P.2 (pharmaceutical development) of module 3 of CTD (CTD-Q). Introduces concept of QbD	Drug product life cycle management
Q9: Quality risk management	November 2005	Principles and tools of systematic risk analysis over product life cycle	Drug product life cycle management
Q10: Pharmaceutical quality system	June 2008	Development and implementation of an integrated quality system in the pharmaceutical industry for better quality assurance and product life cycle management	Drug substance and drug product life cycle management
Q11: Development and manufacture of drug substances	May 2012	Guideline for Sections 3.2.S.2.2 through 3.2.S.2.6 (drug substance) of module 3 of CTD (CTD-Q). Introduces concept of Q8(R2), Q9 and Q10 with reference to the drug substance	Drug substance life cycle management
Q12 concept: Technical and regulatory considerations for pharmaceutical product life cycle management	Concept note released in July 2014. Guideline under preparation	Guideline to enable swift and rationalised approach to implement postapproval changes in chemistry, manufacturing and controls	Drug product life cycle management

risk management. The implementation of this concept is predominantly seen in earlier stages of the product life cycle as described in Figure 2.2. The current guideline focuses on the concept of QbD, the objectives of which are as follows:

- To define and achieve meaningful product quality attributes that ensure clinical efficacy and safety for the patients consistently
- To achieve enhanced product and process design and understanding so as to be able to put into place suitable controls to reduce product variability and hence reduce the risk to product quality

- To achieve improved efficiencies in product development, registration and manufacturing
- To enhance capability to manage postapproval changes during the product life cycle and assist in continual improvement in the process/product

The first very obvious phase in pharmaceutical development is defining the new product quality attributes based on patients, therapy and market/business needs and regulatory, intellectual property scenario. Once the new product attributes are decided along with desired specifications, a strategy is designed to attain the same. This is followed by an actual development process that, in recent time, is undertaken using a QbD that includes

- Determination of target product profile or quality target product profile that identifies the critical quality attributes (CQAs) of the drug product
- Identification of the critical material attributes (CMAs) of the drug substance and other inactives/excipients that can have an impact on the CQAs
- Identification of critical process parameters (CPPs) that can have an impact on the CQAs
- An iterative risk assessment-based approach to achieve a thorough understanding of the product composition and process to minimise the risk to product quality and establish linkage of CMAs and CPPs to the CQAs
- Empirical or enhanced (design of experiment) approach based on the criticality and complexity of the material and process attributes to generate a design space
- A control strategy based on design space that includes specifications for drug substance, excipients and drug product as well as in-process controls during each step of the manufacturing process
- Continual improvement and life cycle management

This knowledge with respect to material and process understanding derived from pharmaceutical development studies provides a scientific understanding to establish the design space, specifications and manufacturing controls.

The very important concept of design space herein refers to the allowable quantitative combination of material (composition) and process parameters that assures quality within the specified limit. This is a very important element in any product design and development as it determines the boundary within which the product can be manufactured within the specification limits with confidence. With defined design space and updated risk analysis that ensures the risk minimisation within the acceptable limits, a control strategy is proposed. Further, any change in the control strategy requires an assessment and regulatory approval. A detailed description of these concepts, along with a few illustrative examples, is given in Chapters 4, 7 and 9 for a thorough understanding.

In addition to these very tangible product-specific goals, QbD also includes intangible goals of enhancing the knowledge management process in any organisation as it mandates the need to document knowledge gained during development, scale-up and commercial manufacturing of any product or process. This can be looked upon as an opportunity to improve certainty to meet success, to encourage reviews and self-inspection to initiate risk-based regulatory decision making, improvisation of manufacturing process within the predetermined design space and multidisciplinary application of the knowledge so gained to other product design projects that will reduce both financial and time burden in the development process.

2.2.2 ICH Q9 GUIDELINE: QUALITY RISK MANAGEMENT

Risk analysis and risk management at every stage is considered as a key component of any pharmaceutical development (Q8(R2)) and is continued throughout the product life cycle (Figure 2.2) as it not only identifies the possibility of failure but also allows the opportunity to minimise hazards. To perform this activity in a systematic and uniform manner, ICH introduced a quality risk management guideline (ICH 2005). Quality risk management is a systematic and sequential process that deals with understanding of risk and its impacts, thereby facilitating better and more scientific decision making particularly to attain, maintain and revise product quality, life cycle management and good manufacturing practices (GMP) compliance.

A comprehensive summary of quality risk management is depicted in Figure 2.3. The typical steps to be followed include the following:

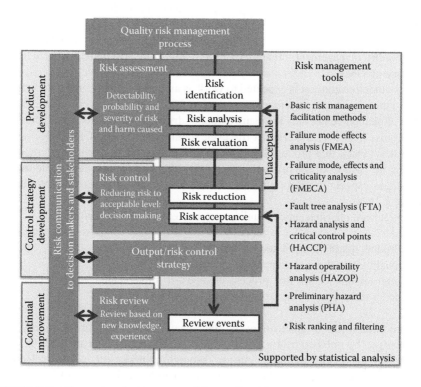

FIGURE 2.3 Comprehensive summary of the quality risk management process.

- Risk assessment (identification–analysis–evaluation), which identifies the probable hazard that may or may not be detectable, the probability of that happening and the severity of the same. In simple terms, it assesses the *criticality* of the hazard. The various systematic tools as described by ICH under the scope of the Q9 guideline are previously discussed in Chapter 1 (ICH 2005).
- Risk control is the next step that analyses if the risk is within the acceptable limit, and if not, the strategy must be planned and executed to minimise the risk to the acceptable level.
- Risk review is the process whereby risk assessment and control undergoes continual review and improvement subject to any deviation, failure, recalls or new information generation. The knowledge so generated is well documented and communicated to the developer, regulators and, in some cases, patients, which enables the rationalised, low risk-based decision policy. The knowledge domain so generated not only helps in controlling the desired quality attributes of the product throughout the product life cycle but also allows the extension of this knowledge to other product development plans.

2.2.3 ICH Q10 Guideline: Pharmaceutical Quality System

As the concept of imbibing QbD was maturing, ICH extended the quality paradigm to generate a comprehensive quality management model for the pharmaceutical industry and introduced the pharmaceutical quality system guideline – Q10. This guideline is in compliance with International Organization for Standardization (ISO) quality management guidelines (9000, 9001:2000, 9004) and earlier ICH Q7 guideline on GMP.

This guideline concept is expected to work as an interlink wherein the knowledge grasped during product development and manufacture along with risk management can be implemented not only to ensure a quality drug product but also to aid continuous improvement over the product life cycle (complimentary to ICH Q8(R2) and 9 guidelines). Thus, in a product life cycle, the pharmaceutical quality system starts building up right from the product development stage and holds great value during later stages especially in commercial manufacturing and product discontinuation (Figure 2.2).

The relevant pharmaceutical quality system elements are as follows:

- Product quality control and assurance
- Quality model maintenance and upgradation
- Internal verification that process changes, if implemented, are successful under the suggested control strategy

In addition to the product life cycle, the concept of pharmaceutical quality system needs to be imbibed as a general strategy throughout the organisation at each level including quality policy, planning, resource management, vendor selection, outsourcing, purchase, ownership changes, acquisitions and so on and is expected to be controlled by an organisation managerial board. The managerial board herein is held responsible for quality and is liable to maintain the standards throughout the product life cycle.

2.3 DRUG SUBSTANCE DEVELOPMENT: AN UPGRADED ICH PERSPECTIVE

The need for change in quality paradigm and harmonisation in dossier filing and review in pharmaceutical products developed triggered the development of modern life cycle management concepts as described in Section 2.2. While these changes were implemented, it was obvious to the regulators that similar guidelines are mandatory for the life cycle management of a drug substance and should go in accordance with the drug substance quality section (3.2.S.2.2 through 3.2.S.2.6) of module 3 – Quality of the CTD. With this perspective, the Q11 guideline on Development and Manufacture of Drug Substances (Chemical Entities and Biotechnological/ Biological Entities) was implemented in May 2012. To summarise, this guideline describes the same principal concepts as mentioned in the Q8(R2), Q9 and Q10 guidelines for pharmaceutical product in relation to the development and manufacture of a drug substance. Thus, the concepts described in Section 2.2 need consideration and implementation even when designing and developing a drug substance.

2.4 PHARMACEUTICAL PRODUCT DESIGN AND DEVELOPMENT: EMERGING CONCEPTS

2.4.1 ICH Q12 Concept Guideline: Technical and Regulatory Considerations for Pharmaceutical Product Life Cycle Management

In very recent times, it was realised that the guidelines till now provide a framework for life cycle management in a systematic manner and more so till the commercial manufacturing stage of a drug product. These guidelines define the quality paradigms and controls well within the design space but do not cover the systematic aspect to be implemented in case of variation or change (i.e. working within/ outside the defined design space). Thus, it was envisioned to propose a guideline that will enable a swift and rationalised approach to implement postapproval changes in chemistry, manufacturing and controls.

In view of this, the concept of the Q12 guideline entitled 'Technical and Regulatory Considerations for Pharmaceutical Product Lifecycle Management' was proposed and approved by ICH in July 2014 and the guideline is under construction. In brief, it aims to introduce and implement a concept of *postapproval change management plans and protocols* that will streamline a strategy to identify and investigate the change in QbD approach and successful implementation of the changes with quality assurance.

Once in force, the guideline is expected to bring rationalised and swift implementation of postapproval changes and can be looked upon as a probable replacement to the current scale-up and postapproval changes guideline. Further, it will also promote innovation and continuous improvement as a systematic and specific framework will be available to do so (refer to Chapter 8 for current postapproval change process).

2.4.2 PAT AND CONTINUOUS MANUFACTURING

Currently, the majority of pharmaceutical manufacturing is performed as a batch process wherein the product quality is controlled via sample analysis for both in-process and final product quality control. This assures the product quality within the specified limit with an assumption that sampling represents the entire batch and the batch is uniform. With the emergence of newer product development and control strategies, it is proposed that the quality can be better controlled if in-line, on-line or at-line continuous and real-time analysis is performed over the entire batch. With this, the new concept of PAT is gaining attention and a regulatory guidance for the same is also available (USFDA 2004).

PAT refers to a set of tools for designing, assessing and controlling the quality and performance of raw materials, in-process materials and processes via timely analysis during the process that in turn assures the finished product quality. The major advantages of PAT include flexible process opportunity, reduction in processing time because of in-process real-time analysis, increased automation that reduces manual errors, better understanding and control of the process and integrated analysis of physical, chemical and microbiological critical attributes as well as reduced regulatory intervention owing to more scientific and realistic data generation and analysis. The primary PAT tools and their applications are summarised in Figure 2.4.

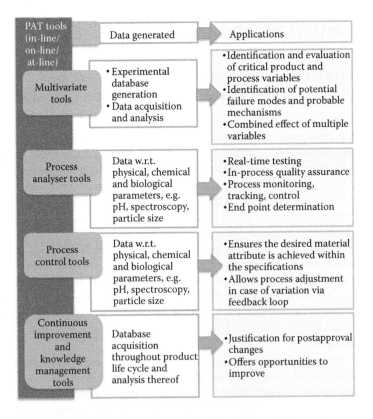

FIGURE 2.4 Primary PAT tools and their applications. W.r.t., with respect to.

For instance, consider a hypothetical example of a mixing unit operation performed for mixture 'A' using an octagonal blender wherein the desired quality target is a uniform blend, which can be analysed using near-infrared (NIR) spectroscopy. Here, the CPPs were identified to be loading volume, mixer speed and time of mixing based on initial studies. The optimised process revealed that uniform mixing can be achieved at a blender speed of 10 rpm for 10 min and the end point can be determined using NIR spectroscopy. Application of PAT in this scenario will ensure uniform mixing at the end of the process. This is considered as an application of PAT for process analysis. Now, consider a situation wherein the raw material density or particle size has varied and thus at the end of specified process parameters, nonuniform blend results. In such a situation, instead of discarding the batch, the mixing operation can be continued till the NIR spectrum assures uniform blending, and this is known as using PAT application as a control tool. Additionally, this can be used as an easy tool to implement changes in case of postapproval variation (such as, in this case, change in raw material properties because of vendor change, etc.). Thus, PAT can be visualised as a versatile tool to ensure product quality.

Owing to these advantages, PAT-assisted process control will be the near future of all the pharmaceutical manufacturing processes enabling more efficient in-process quality measurements and control. With efficient implementation of PAT, the concept of *continuous manufacturing* is getting established (USFDA 2012). Continuous manufacturing implies a sequential integration of individual unit operations in a closed circuit. This allows simultaneous charging and discharging of the input materials and output product. Thus, in a continuous manufacturing process, the production scale can be defined in terms of rate of production in contrast to the absolute batch size as in the case of a batch process.

This is possible in the pharmaceutical sector only if it is integrated with PAT as it will allow in-process quality control, track and trace mechanism to identify any unacceptable changes in desired quality and thus rejection of under-quality products online. This will in turn ensure 100% correction of the end product with assured quality. In addition to PAT-enabled advantages, continuous manufacturing will additionally minimise the issues of contamination attributed to manual handling and will reduce the transfer time between each unit operation. Moreover, it can be also viewed as an opportunity to skip the scale-up step in the pharmaceutical product life cycle. Briefly, it can be explained as follows: Pharmaceutical development and optimisation is generally performed at the laboratory scale and scale-up is the next crucial step. The multi-fold increase in batch size, changes in equipment parameters and control at the production site (size, geometry and operational variation between laboratory scale and production scale machinery) offer many challenges during technology transfer to achieve the desired quality product. With the introduction of continuous manufacturing, the production can be conducted using the same machinery used during development, which in turn will avoid scale-up and related issues. This will turn out as an asset to the pharmaceutical industry in terms of reducing both manufacturing cost and time, which will result in increased productivity. This concept is currently being implemented in a few pharmaceutical industries and will soon be adapted throughout the pharmaceutical arena as regulatory authorities (such as the USFDA) have extended a helping hand to support implementation of continuous manufacturing units.

PAT and continuous manufacturing concepts need faster adaptation and execution in the pharmaceutical sector as novel control strategy approaches because they are consistent with current ICH quality guidelines and are best representatives of inclusion of innovative strategies in pharmaceutical product development in the current setting.

Along with implementation of foolproof concepts in pharmaceutical product design and development, effective management is a must towards successful development and market reality of pharmaceuticals.

2.5 PHARMACEUTICAL PROJECT MANAGEMENT

Project management is the group effort that involves meticulous planning, decision making, execution, review and resource utilisation to attain desired goals. Since the past decade, with the profound understanding of pharmaceutical product life cycle management and in view of the stringent regulatory requirements, it has been realised that the traditional case-to-case project management approach needs revision with a systematic and uniform project management policy. This is further expected to enable swift mergers and acquisitions that are witnessed more frequently in recent time by pharmaceutical industries. Thus, it is envisaged that the effective project management techniques must be capitalised in every pharmaceutical industry.

In addition to the basic project management concepts covered in Chapter 1, the pharmaceutical project management concept needs to be looked upon as a specialised field. This is required as pharmaceutical products directly relate to health and safety, the design–development–commercialisation timelines are lengthy (anywhere between 10 and 15 years for a new pharmaceutical product) and anticipate high-cost investment as a result of developmental and clinical studies involved and the development process requires many partnerships and outsourcing. This demands a meticulous selection of a project management team composed of people from managerial, scientific and regulatory backgrounds.

When applying general project management techniques in the pharmaceutical sector, it must be remembered that the basic steps in project management such as project kick-off, planning, execution, monitoring, tracking, controlling and project closure in an unfavourable scenario remain the same but the approach should be focussed on quality management and productivity in terms of clinical testing and development as well as on a stringent timeline and effective resource management. Figure 2.5 depicts the key deliverables of an effective pharmaceutical management system (Bateman 2012; Brown and Grundy 2011; Khatri 2012).

With financial investment being a high-demand and high-risk aspect of the pharmaceutical product life cycle, budgeting and cost monitoring become a very important part of this management process. Amongst the various models, the cost–benefit analysis and variance-based analysis between actual and budgeted costs are the key cost control policies used in the pharmaceutical industry. When managing the financial aspects of a pharmaceutical product, some special considerations are necessary and the managerial team should be well equipped to understand those parameters. For example, from a new product development perspective, the major cost to be

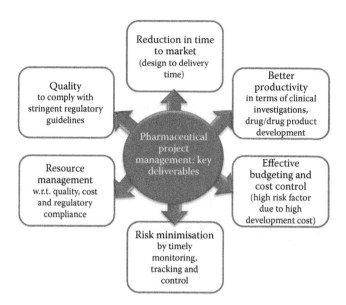

FIGURE 2.5   Key deliverables of effective pharmaceutical project management system.

considered is clinical studies cost. This demands a stringent monitoring and controlling management policy to take go or no-go decisions during pharmaceutical development in a realistic manner on the basis of the preliminary results, which, if not considered, may lead to unnecessary financial burden.

Another important aspect of pharmaceutical management is meticulous vendor selection. This involves identification of vendors that meet the predefined specifications for the material under consideration. After identification, vendors are scrutinised for consistency in material quality, GMP facility used for material generation, capability of vendor to supply material to meet the need, material cost, and so on, and based on suitability, generally two vendors are selected. This is very important in pharmaceutical product development as alteration in vendors is considered as a postapproval change and demands a battery of studies and assurance of quality maintenance and reapproval from the regulatory bodies.

It is thus clear that the pharmaceutical project management concept needs special consideration while planning for successful product development and is slowly getting integrated with the pharmaceutical industries worldwide.

Thus, to summarise, the pharmaceutical product design and development process is witnessing the implementation of many newer concepts not only in terms of development but also in terms of data generation analysis and management. This will help the developers to better understand the product scientifically and to manufacture it with assured quality and will encourage them to bring in more innovations in product development. On the other hand, it will make the job of regulatory authorities easier in evaluating dossiers and product performance, and this will speed up the drug approval process, creating a win–win situation for both industry and regulatory authorities.

REFERENCES

Bateman, L. 2012. The benefits of applying project management in the pharmaceutical industry. http://projectmgmt.brandeis.edu/downloads/BRU_MSMPP_WP_Feb2012_Pharma.pdf (accessed June 19, 2015).

Brown, L., and Grundy, T. 2011. Project management for the pharmaceutical industry. https://www.gowerpublishing.com/pdf/SamplePages/Project-Management-for-the-Pharmaceutical-Industry-Pre.pdf (accessed June 19, 2015).

EMEA. 2014. Guideline on process validation for finished products – Information and data to be provided in regulatory submissions. http://www.ema.europa.eu/docs/en_GB/document_library/Scientific_guideline/2014/02/WC500162136.pdf (accessed June 19, 2015).

ICH. 2005. International conference on harmonisation of technical requirements for registration of pharmaceuticals for human use, ICH harmonised tripartite guideline, quality risk management Q9. http://www.ich.org/fileadmin/Public_Web_Site/ICH_Products/Guidelines/Quality/Q9/Step4/Q9_Guideline.pdf (accessed June 19, 2015).

ICH. 2008. International conference on harmonisation of technical requirements for registration of pharmaceuticals for human use, ICH harmonised tripartite guideline, pharmaceutical quality system Q10. http://www.ich.org/fileadmin/Public_Web_Site/ICH_Products/Guidelines/Quality/Q10/Step4/Q10_Guideline.pdf (accessed June 19, 2015).

ICH. 2009. International conference on harmonisation of technical requirements for registration of pharmaceuticals for human use, ICH harmonised tripartite guideline, pharmaceutical development Q8(R2). http://www.ich.org/fileadmin/Public_Web_Site/ICH_Products/Guidelines/Quality/Q8_R1/Step4/Q8_R2_Guideline.pdf (accessed June 19, 2015).

ICH. 2012. International conference on harmonisation of technical requirements for registration of pharmaceuticals for human use, ICH harmonised tripartite guideline, development and manufacture of drug substances (chemical entities and biotechnological/biological entities) Q11. http://www.ich.org/fileadmin/Public_Web_Site/ICH_Products/Guidelines/Quality/Q11/Q11_Step_4.pdf (accessed June 19, 2015).

ICH. 2014. International conference on harmonisation, final concept paper Q12: Technical and regulatory considerations for pharmaceutical product lifecycle management. http://www.ich.org/fileadmin/Public_Web_Site/ICH_Products/Guidelines/Quality/Q12/Q12_Final_Concept_Paper_July_2014.pdf (accessed June 19, 2015).

Juran, J.M. 1992. *Juran on quality by design: The new steps for planning quality into goods and services*, revised edition. New York: Free Press.

Khatri, J. 2012. Project management in pharmaceutical generic industry basics and standards. http://www.slideshare.net/JayeshKhatri1/project-management-in-pharmaceutical-generic-industry-basics-and-standard (accessed June 19, 2015).

PhMRA. 2015. Profile biopharmaceutical research industry. http://www.phrma.org/sites/default/files/pdf/2014_PhRMA_PROFILE.pdf (accessed June 19, 2015).

Tufts CSDD. 2014. Tufts center for the study-briefing -cost of developing a new drug. http://csdd.tufts.edu/files/uploads/Tufts_CSDD_briefing_on_RD_cost_study_-_Nov_18,_2014.pdf (accessed June 19, 2015).

USFDA. 2004. FDA guidance for industry, PAT – A framework for innovative pharmaceutical development, manufacturing, and quality assurance. http://www.fda.gov/downloads/Drugs/Guidances/ucm070305.pdf (accessed June 19, 2015).

USFDA. 2011. FDA guidance for industry, process validation: General principles and practices. http://www.fda.gov/downloads/RegulatoryInformation/Guidances/ucm125125.pdf (accessed June 19, 2015).

USFDA. 2012. FDA perspective on continuous manufacturing. http://www.fda.gov/downloads/AboutFDA/CentersOffices/OfficeofMedicalProductsandTobacco/CDER/UCM341197.pdf (accessed June 19, 2015).

3 Regulatory and Intellectual Property Aspects during Pharmaceutical Product Development

Preshita P. Desai, Sivagami V. Bhatt and Mahalaxmi A. Andheria

CONTENTS

3.1   INTRODUCTION

Regulatory affairs and *intellectual property (IP) strategies* are two vital aspects of any pharmaceutical product development that interlink idea or market opportunity to the end stage of product development, that is, commercialisation of the product.

A return on investment only occurs after the product has been approved by the regulatory authorities for marketing and on sale of the product, which in turn is directly or indirectly based on the patent protection and strategies towards elimination of such barriers for an early market entry.

3.2   REGULATORY AFFAIRS

Public health being the prime concern, it is necessary that the drug/drug product available for human/veterinary use and medical devices must not only be effective but also be safe for the intended use. To ensure this, various territorial regulatory bodies came into existence.

Major regulatory agencies include World Health Organization (WHO), United States Food and Drug Administration (USFDA, United States), European Medicines Agency (EMA, European Union), Medicines and Healthcare Products Regulatory Agency (MHRA, UK), Therapeutic Goods Administration (TGA, Australia), Health Canada (Canada), Pharmaceuticals and Medical Devices Agency (PMDA, Japan) and Central Drugs Standard Control Organization (CDSCO, India).

It was observed that regulatory guidelines differ with respect to territorial requirements; this demanded the need for *universal harmonisation*. Thus, there was the emergence of The International Conference on Harmonization of Technical Requirements for Registration of Pharmaceuticals for Human Use (ICH) in 1990 by united efforts of the United States, Europe and Japan. Since its inception, the ICH has evolved gradually with a mission to attain better harmonisation towards development and registration of medicines with a higher degree of safety, efficacy and quality worldwide (ICH 2015a). The guidelines laid down by

ICH are broadly classified into four categories that are briefly depicted in Figure 3.1 (ICH 2015b).

Although ICH has harmonised the drug regulatory aspects worldwide, the regional regulatory bodies continue to play a pivotal role in drug approvals across the territory. The drug development and approval protocols for some of the major territorial regulatory authorities are described herewith.

Q (quality guidelines)		E (efficacy guidelines)	
Q1A	Stability	E1	Clinical safety for drugs used in long-term treatment
Q1F			
Q2	Analytical validation	E2A	Pharmacovigilance
		E2F	
Q3A	Impurities	E3	Clinical study reports
Q3D		E4	Dose–response studies
Q4	Pharmacopoeias	E5	Ethnic factors
Q4B		E6	Good clinical practice
Q5A	Quality of biotechnological	E7	Clinical trials in geriatric population
Q5E	products		
Q6A	Specifications	E8	General considerations for clinical trials
Q6B			
Q7	Good manufacturing practice	E9	Statistical principles for clinical trials
Q8	Pharmaceutical development	E10	Choice of control group in clinical trials
Q9	Quality risk management	E11	Clinical trials in pediatric population
Q10	Pharmaceutical quality system		
Q11	Development and manufacture of drug substances	E12	Clinical evaluation by therapeutic category
		E14	Clinical evaluation
Q12	Life cycle management	E15	Definitions in pharmacogenetics/pharmacogenomics
S (safety guidelines)		E16	Qualification of genomic biomarkers
S1A	Carcinogenicity studies	E17	Multiregional clinical trials
S1C		E18	Genomic sampling methodologies
S2	Genotoxicity studies		
S3A	Toxicokinetics and pharmacokinetics	**M (multidisciplinary guidelines)**	
S3B		M1	MedDRA terminology
S4	Toxicity testing	M2	Electronic standards
S5	Reproductive toxicology	M3	Nonclinical safety studies
S6	Biotechnological products	M4	Common technical document
S7A	Pharmacology studies		
S7B		M5	Data elements and standards for drug dictionaries
S8	Immunotoxicology studies		
S9	Nonclinical evaluation for anticancer pharmaceuticals	M6	Gene therapy
		M7	Genotoxic impurities
S10	Photosafety evaluation	M8	Electronic common technical document (eCTD)
S11	Nonclinical safety testing		

FIGURE 3.1 Types of ICH guidelines.

TABLE 3.1

Types of Drug Approvals under the USFDA

Drug Application	Section	Application Criteria	Review Period	Market Exclusivity
Investigational New Drug (IND)	NA	Drugs under investigation	30 days	NA
New Drug Application (NDA)	505(b)(1)	Drug for which there is no previous NDA filing	As per newer guidelines, 90% of the applications must be reviewed and acted upon within 10 months from date of application	5 years new chemical entity (NCE)
	505(b)(2)	Drug with some alteration in previously approved NDA		3 years NDA or CI. Also includes OD/PE exclusivity
Abbreviated New Drug Application (ANDA)	505(j)	Generic product	Approximately 27–30 months, will be reduced in the future	6-month exclusivity to first to file para IV-ANDA applicant

Note: CI, clinical investigation; OD, orphan drug; PE, pediatric.

3.2.1 UNITED STATES FOOD AND DRUG ADMINISTRATION

The USFDA regulates pharmaceutical drug/drug products in the United States. These fall under the regulatory supervision of the Center for Drug Evaluation and Research (CDER), a division of the USFDA that monitors the approval process in compliance with the Food, Drug and Cosmetic Act (FD&C Act) and ICH guidelines. In alliance with the scope of this discussion, this chapter shall focus on the drug approvals under various sections of the USFDA as described in Table 3.1 (ANDA 2015; IND 2015; NDA 2015).

3.2.1.1 Investigational New Drug Application

An investigational new drug (IND) application is meant for molecules that are in the investigation stage and desire the safety and efficacy testing in human volunteers. An IND approval gives a legal permission to the developer to transport and distribute drug across the territory for testing purpose only and to test the diagnostic or therapeutic potential of investigational molecules in humans.

It should be strictly noted that drugs with IND approval are not allowed to be marketed for clinical/commercial use. This is mandatory as the federal law requires that a drug be the subject of an approved marketing application before it is transported or distributed across state lines.

Considering the early development stage, practically, it is advised that the developer should consult FDA before IND proposal submission. The schematic representation of the IND approval process is described in Figure 3.2 (IND 2015). After approval of IND, the developer can conduct clinical studies in human volunteers in successive stages as described in Table 3.2.

Data generated at each phase will support requirements for the marketing application. After a successful Phase 3, a new drug application (NDA) is submitted.

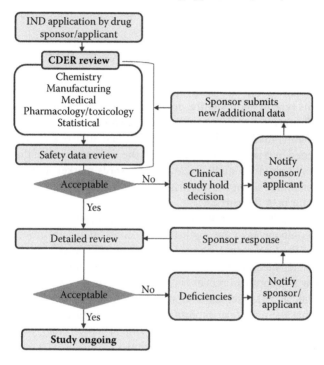

FIGURE 3.2 IND by the USFDA – review process.

TABLE 3.2

Clinical Study Design

Clinical Trial	Type of Human Volunteers	Number of Volunteers	Evaluation Parameter
Phase 1	Healthy volunteers	20–80	Safety (understanding the side effects, metabolic pathways, etc.)
Phase 2[a]	Diseased/affected volunteers	12–300	Safety and efficacy in a small population
Phase 3[a]	Diseased/affected volunteers	Hundreds to 3000	Safety and efficacy in a large population

[a] The clinical phase trial is conducted subject to (1) satisfactory results of previous clinical phase trials and (2) discussion and approval from FDA to commence the next trial.

3.2.1.2 New Drug Application

After a successful investigation of an IND-approved drug towards safety and efficacy, the next step is to obtain a marketing licence for the same. For this, the applicant submits the NDA to FDA authorities under Section 505(b). Under this, there are two categories: 505(b)(1) and 505(b)(2) as discussed in Sections 3.2.1.2.1 and 3.2.1.2.2.

3.2.1.2.1 505(b)(1)

The NDA under Section 505(b)(1) of the FD&C Act [21 U.S.C. § 355(b)(1)] is a comprehensive application submitted by a brand-name or innovator company, for the active ingredient of the new drug that has not been previously approved by the USFDA. For this type of approval, the developer conducts and investigates the complete battery of safety and efficacy studies and submits the elaborate data on chemistry, pharmacology, medicine, biopharmaceutics and statistics for review. This is generally preceded by a pre-NDA meeting with the USFDA wherein the views of the USFDA on data of previous studies, feasibility towards market approval, additional data requirements and so on are discussed, enabling the smooth and faster NDA review process for both the parties.

Although the depth and quantity of data submitted under NDA vary from case to case, the basic components of the application are constant and are covered under various modules of the Common Technical Document (CTD), which is filed in paper/electronically is depicted in Figure 3.3, and an interactive flow chart presentation of the NDA review and approval process is illustrated in Figure 3.4.

Post approval, as FDA serves as a surveillance agency ensuring the safety, it is mandatory that the developer reports the postmarketing clinical data (Phase 4), adverse drug reaction information and NDA annual report every year to the FDA authority to ensure safety, efficacy and optimal use.

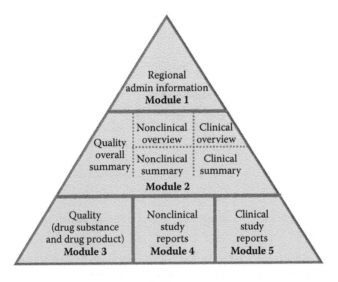

FIGURE 3.3 Schematic representation of CTD.

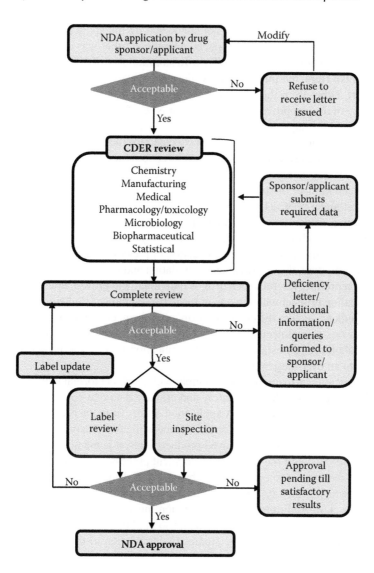

FIGURE 3.4 NDA by the USFDA – review process.

Upon approval, the NDA under 505(b)(1) gets nonpatent market exclusivity for 5 years and indication and orphan drug exclusivity of 3 years and 7 years, respectively, as applicable.

3.2.1.2.2 505(b)(2)

The NDA under Section 505(b)(2) of the FDCA [21 U.S.C. § 355(b)(2)] is a type of submission containing a complete safety and efficacy profile along with the drug substance and product quality and the nonclinical and clinical data of an intended NDA, but it differs from the standard NDA application under 505(b)(1), as a part of the data comes from the studies not conducted by the applicant himself for the said NDA. Thus, it is viewed as a special amendment that was enforced under the Drug Price Competition and Patent Term Restoration Act of 1984 (Hatch-Waxman Amendments). Here, the intended NDA filer under 505(b)(2) has a prior knowledge and support, either from the published literature or the agency's findings on safety and efficacy of drug approved under innovator NDA [505(b)(1)]. This was done with an aim to avoid the duplication of work and to save resources that are wasted in conducting studies that have previously been executed for the drug under consideration. Such an application presents the benefits as it reduces the risk attributed to available safety and efficacy data, reduces the cost and speeds up the approval process owing to the reduced number of studies. The typical difference in drug development timeline and cost under 505(b)(1) and 505(b)(2) are depicted in Table 3.3.

Thus, it is considered as a fast-track approval process for a product that has limited alterations/modifications from the previously approved drug (NDA), which broadly includes the following: new indication of previously approved drug; change in dosage form, dose, dosage regimen, formulation and route of administration; new combination product wherein each drug is individually approved; alteration in existing combination product wherein one of the active ingredient is altered; alteration in active ingredient, such as different salt form, ester form, enantiomer, chelate, derivative and so on; new molecular entity wherein the drug is either a pro-drug or an active metabolite of an established drug; application to alter the prescription (Rx) indication to an over-the-counter indication.

This application also continues to comply with the CTD format (Figure 3.3). The differentiating points include the safety and efficacy data that come from the previously approved NDA. As this NDA application differs from the previously approved NDAs in one or more aspects, the applicant needs to conduct studies to supplement the changes in the previously approved product. Thus, it is mandatory for the applicant to provide in detail the product development process, chemistry manufacturing and control data with respect to drug substance and drug product, stability studies and manufacturing facility data ensuring identity, quality, strength and purity of drug (Figure 3.3). Additionally, the application must contain the satisfactory label data to indicate use of the drug, regimen and so on. Further, the applicant may conduct bioavailability/bioequivalence studies in comparison to the listed drug as per the case-specific needs by FDA.

Upon approval, under normal scenario, the NDA under 505(b)(2) gets a market exclusivity for 3 years, which may be extended up to 5–7 years depending upon the

TABLE 3.3
505(b)(1) and 505(b)(2): Timeline and Cost Comparison

	Discovery (Years)	Preclinical Studies (Years)	Clinical Studies (Years)	Cost Estimate (Amount in USD)
505(b)(1)	2–5	1–3	8–15	≤3 billion (majority of cases)
505(b)(2)	<1–3	<1–2	<2–5	<10 million to >100 million

clinical data and the extent of alteration from the existing NDA. This pathway is now being envisaged as a low-cost and low time-to-market avenue by most of the pharmaceutical companies as it not only allows evading the competition by generic market but also provides better market prospects via product differentiation and branding (Camargo 2015; NDA 2015).

3.2.1.3   Abbreviated New Drug Application

Under Section 505(j) of the FD&C Act [21 U.S.C. § 355(j)], abbreviated new drug applications (ANDAs) are filed to obtain approval for generic versions of innovator drugs and generally do not include new clinical data. Instead, an ANDA relies on the FDA's finding of safety and efficacy for a previously approved drug, which is known as the *reference listed drug* (RLD). In other words, the point of an ANDA is not to demonstrate safety or effectiveness but to establish that the generic product is equivalent to an RLD already known to be safe and effective. Thus, the coined word is *abbreviated* for a generic application. The ANDA review and approval is based mainly on bioequivalence, chemistry, manufacturing and control information, drug label information and plant inspection. The USFDA maintains a list of drug products that are eligible for submission as ANDAs. The list, Approved Drug Products with Therapeutic Equivalence Evaluations, referred to as the *Orange Book*, contains a detailed description of USFDA-approved drug products. The Orange Book also designates an RLD to which all the generics must be compared and on which an ANDA applicant relies in seeking approval of its application. A generic drug product must be identical to an RLD in parameters such as active ingredient, dosage form, strength, dose, route of administration, quality, performance characteristics and intended use (i.e. having pharmaceutical and therapeutic equivalence). Such an ANDA is filed at CDER's Office of Generic Drugs (OGD) under Section 505(j) for review, which upon approval gives a licence to the applicant to manufacture and market the generic drug product to provide a safe, effective, low-cost alternative to the patients.

An important aspect of ANDA submissions is *patent certifications*. The Orange Book provides patent and exclusivity information concerning the listed drug products. The innovator drug companies are required to submit patent information on their drug product to the USFDA, which is listed in the Orange Book. The patents listed are patents claiming the active ingredient(s), drug product patents including formulation/composition patents and use patents for an approved indication. An ANDA applicant, seeking to market a therapeutic equivalent to the innovator drug product, is required to certify against each of the Orange Book listed patents on one of the following para as described in Table 3.4.

It was the Hatch–Waxman Act of 1984 and the Medicare Modernization Act of 2003 that established the above-described ANDA approval processes. Amongst these, a paragraph IV certification is the most interesting one, which questions the validity or enforceability of the listed patent or alleges noninfringement of the listed patent by the generic product. The ANDA applicant is required to notify the patent owner and the innovator company (NDA holder) about the patent challenge within 20 days of receiving the acceptance letter from the FDA. The innovator has 45 days from receipt of notice to initiate a patent infringement lawsuit against the ANDA applicant and the USFDA may not give final approval to the ANDA

TABLE 3.4

Type of Certifications and Timeline for Generic Product Market Entry

Type	Criteria	Market Entry Date
Para I	No patent listed in the Orange Book	Immediate market entry on approval
Para II	Listed patent has already expired	Immediate market entry on approval
Para III	Listed patent will expire on a stated date	Generic market entry after patent expiry on approval
Para IV	Listed patent is invalid or will not be infringed	Generally legal intervention and verdict thereof or settlements govern the market entry

Note: An ANDA applicant may also submit a *Section VIII* statement certifying that the ANDA does not seek approval for any indications claimed by relevant patents listed in the Orange Book indicating that the generic product shall be marketed for the said indication only after the expiry of the patent. Such an indication is carved out from the label during the ANDA filing.

for at least 30 months from the date of the notice. If successfully challenged or proven noninfringing, the first para IV generic applicant is awarded an incentive of 180 days of market exclusivity on approval (ANDA 2015; Hatch-Waxman 2015; Orange Book 2015).

An interactive flow chart presentation of the ANDA review process is illustrated in Figure 3.5. In order to balance between incentives for early generic entry and to innovate, apart from patent and patent term extensions, the Act provides additional nonpatent market exclusivity to innovator companies. Such exclusivities are listed in Table 3.5.

A summary of key features of ANDA versus 505(b)(2) applications is provided in Table 3.6.

3.2.2 EUROPEAN MEDICINES AGENCY

EMA was established by the European Union (EU) and pharmaceutical industries in 1995. EMA is responsible for granting manufacturing and marketing licence for drug and drug products in the European territory. Unlike the USFDA, the EU allows the decentralised regulatory approval process applicable for each member state. To attain harmonisation in the European member states, the EC Guideline Volume III was adapted under the Committee on Proprietary Medicinal Products. In the EU, the Hybrid application [an analog of NDA under 505(b)(2) of the USFDA] is filed

- For products that do not meet the definition of a *generic medicinal product*
- When bioequivalence is not established with respect to the reference/innovator product
- When there is alteration in the active ingredient, indication, strength, dosage form and route of administration with respect to the reference/innovator product

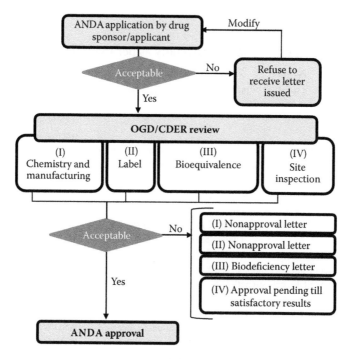

FIGURE 3.5   ANDA by the USFDA – review process.

On the other hand, the Generic Medicinal Product (an analog of ANDA under the USFDA) application is filed for products exhibiting pharmaceutical and therapeutic equivalence to that of a previously approved innovator product.

3.2.2.1   Drug Product Approval in the EU

Broadly, European regulatory authorities accept and access two kinds of applications, namely, clinical trial application (similar to IND under the USFDA) and marketing authorisation application (MAA) (includes application equivalent to NDA and ANDA under the USFDA). At EMA, the MAA is evaluated by the Committee for Medicinal Products for Human Use (CHMP). Considering the large number of member states under EMA, the sponsor can submit the drug approval application through multiple pathways that include Centralised Procedure, National Authorisation Procedure, Decentralised Procedure and Mutual Recognition Procedure, and the details are depicted in Table 3.7 for thorough understanding (EMA 2015).

The MAA under the Centralised Procedure is submitted to EMA in compliance with the CTD format (Figure 3.3), and the general approval process is depicted in Figure 3.6.

Here, it must be noted that, under EU and USFDA guidelines, although the overall development process, safety–efficacy studies, CTD modules and so on go hand in hand, some differences are observed in the application criteria, test and specification requirements and acceptance criteria. Table 3.8 gives a brief comparison of various regulatory applications filed in the United States and the EU.

TABLE 3.5
Types of Exclusivities, Period and Eligibility as per the USFDA

Exclusivity	Period	Eligibility	Bars
New chemical exclusivity (NCE)	5 years	First time approval of new chemical entity	Submission of ANDA or 505(b)(2) for the same drug (except for para IV filings where it is 4 years)

Note: In October 2014, FDA finalised a new policy that allows new fixed-dose combination drugs consisting of at least one new drug product to be eligible for 5 years of so-called new chemical entity (NCE) exclusivity.

New clinical investigation exclusivity (NCI)	3 years	NDAs that contain reports of new clinical investigations that are essential to the approval and are conducted or sponsored by the applicant	Approval of ANDA or 505(b)(2) application for same conditions of approval of the drug

Includes new ester or salt or enantiomer (NE), new product (NP), new combination, new indication (I), new dosage form (NDF), new strength (NS), new dosing schedule (D), new route of administration (NR), and new patient population (NPP)

Orphan drug exclusivity (ODE)	7 years	First NDA approved for a rare disease or a condition, i.e. one that affects less than 200,000 persons in the United States	Approval of another NDA, ANDA, 505(b)(2) for the same drug for the same disease or condition
Pediatric exclusivity (PE)	6 months to end of all exclusivities	Timely submission of studies in pediatric patients upon written request of FDA for drugs that are approved (studies need not be successful)	Submission of ANDA or 505(b)(2) in case of NCE and approval of application in case of NCI and OD

Similar to the USFDA, the EU has also implemented provisions to balance between incentives for early generic entry and the incentives to innovate, by providing nonpatent market exclusivities to innovator companies. Such exclusivities are listed in Table 3.9.

3.2.3 CENTRAL DRUGS STANDARD CONTROL ORGANISATION

The Drugs and Cosmetics Act of 1940 and the Drugs and Cosmetics Rules of 1945 were passed by the Indian Parliament to regulate the import, manufacture, distribution and sale of drugs and cosmetics, and the CDSCO and the office of its leader, the Drugs Controller General (India) (DCGI), were established. Thus, the national regulatory body for Indian pharmaceuticals and medical devices is CDSCO. In 1988, the Indian government added Schedule Y to the Drugs and Cosmetics Rules of 1945. Schedule Y provides the guidelines and requirements for clinical trials, which was further revised in 2005 to bring it at par with the internationally accepted procedure.

TABLE 3.6
Comparison of ANDA with 505(b)(2)

Parameter	ANDA	505(b)(2)
Basic differentiating criteria (data requirement)	Approval for duplicates of approved products with only BE data, safety and effectiveness can be established through bioavailability	Any applicant submitting a change to an approved drug will not be able to submit an ANDA if its application requires FDA to review new clinical data; an applicant in this position should seek approval under 505(b)(2)
Full reports of efficacy/ safety studies	No	Maybe in some cases
PK/bioavailability/ bioequivalence studies with RLD	Yes	Yes
Timeline of filing in case of NCE date	No ANDA submission until 5 years after NCE (4 years if para IV certification); can be extended to 7.5 years by para IV certification	No 505(b)(2) submission until 5 years after NCE (4 years if para IV certification); can be extended to 7.5 years by para IV certification
Effect of PDE	Final ANDA approval delayed for 6 months	Final 505(b)(2) approval delayed for 6 months
Effect of ODE	Final ANDA approval delayed for 7 years	Final 505(b)(2) approval delayed for 7 years
Patent certifications	To be submitted	To be submitted
Para IV challenge, notice letter and litigation	Applicable	Applicable
30-Month stay	Applicable	Applicable
Exclusivity	180-day marketing exclusivity	180-day exclusivity not applicable; since a 505(b)(2) approval is not subject to competition, there is no reason to grant this separate exclusivity period A 505(b)(2) application may obtain 3 years of Hatch-Waxman exclusivity if one or more of the clinical investigations, other than BA/BE studies, was essential to approval of the application and was conducted or sponsored by the applicant

(Continued)

TABLE 3.6 (CONTINUED)
Comparison of ANDA with 505(b)(2)

Parameter	ANDA	505(b)(2)
FDA records of filing	Para IV filing record maintained for ANDA filings	None
Eligibility for ODE/ PDE	Not eligible	A 505(b)(2) application may be eligible for orphan drug exclusivity or pediatric exclusivity
Patent submission by ANDA/505(b)(2) applicant	Not eligible as ANDA is a copy	505(b)(2) applicant may submit its own patent information that can be listed in the Orange Book, thereby potentially delaying market entry by competitors

Note: BA, Bioavailability; BE, Bioequivalence; PDE, Pediatric drug exclusivity; PK, pharmacokinetic.

The filing process slightly varies in the Indian scenario wherein it is not mandatory to prepare the dossier in CTD format in contrast to that of the United States and the EU. However, the drug approval process is more or less synonymous to standard guidelines but varies in some aspects. To understand these differences, a comprehensive comparison of regulatory requirements of the United States, EU and Indian agencies is summarised in Table 3.10 (CDSCO 2015; EMA 2015; USFDA 2015).

3.2.4 Case Studies

Sample case studies are depicted in this section to help understand the regulatory scenario better and apply it for drug product development to ascertain timely approvals.

3.2.4.1 Case Study 1

RLD Choice Policy – Selection of the Most Pharmaceutically Equivalent Listed Drug as the RLD (Venlafaxine Case): On October 20, 1997, Wyeth obtained approval for Effexor XR (venlafaxine HCl) as multiple-strength extended-release capsules for the treatment of major depressive disorder. The drug was later approved for additional indications of generalised anxiety disorder, social anxiety disorder and panic disorder.

In April 2003, Sun Pharma submitted a suitability petition requesting the USFDA to grant permission for filing of ANDA on venlafaxine HCl extended release as a different dosage form, namely, tablets instead of RLD capsules. This petition was granted by the USFDA in March 2005.

While Suns ANDA was pending or rather before its approval by the USFDA, on May 20, 2008, Osmotica's 505(b)(2) application for multiple strengths of venlafaxine HCl extended-release tablets was approved for treatment of major depressive

TABLE 3.7

Types of Application for Regulatory Authorisation of Product in the EU

Application Procedure	Application Process	Product Specification	Application Assessment Time
Centralised	Single application submitted to EMA Upon approval, licence to manufacture and market the drug in all EU member states plus Iceland, Liechtenstein and Norway	Mandatory for some of the applications that include • All biologic agents/ products prepared using high-technology procedures • Products indicated for HIV/ AIDS, cancer, diabetes, neurodegenerative diseases, autoimmune and other immune dysfunctions and viral diseases	EMA opinion issued within 6–7 weeks followed by submission to the European Commission for final approval
National	Individual application submitted to the competent authority of each member state, e.g. MHRA in the United Kingdom	For products outside the mandatory scope of Centralised Procedure	Approximately 6–7 weeks
Decentralised	Simultaneous application and authorisation in more than one EU state (of products that have not yet been authorised in any EU country)	For products outside the mandatory scope of Centralised Procedure that have not yet been authorised in any EU state	Approximately 6–7 weeks
Mutual	Approval in EU state subject to mutual agreement on the basis of prior product approval in other EU states under national procedure	For products outside the mandatory scope of Centralised Procedure that are authorised in one or more EU states Preferred for generic products	Approximately 12–13 weeks

disorder and social anxiety disorder. The approval of Osmotica's NDA was based on the Agency's finding of safety and effectiveness for Effexor XR extended-release capsules and supported by comparative bioavailability data.

Pursuant to its NDA approval, in early 2008, Osmotica filed a citizen petition with the USFDA stating that a suitability petition may not be used as the basis for ANDA submission after an NDA is approved for the same drug product described in the approved suitability petition.

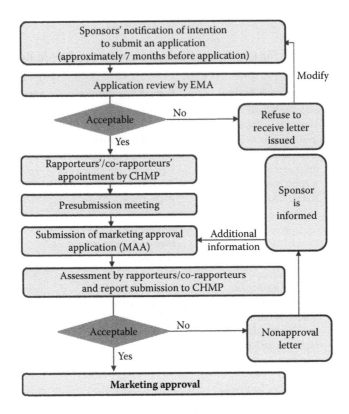

FIGURE 3.6   Marketing approval by EMA – review process.

TABLE 3.8
Comparison of US and EU Applications

Application Type in the United States			Application Type in EU	
Application submitted to USFDA	Investigational new drug	⟷	Clinical trial application	Application submitted to EU state member authority
	New drug application [Section 505(b)(1)]	⟷	Marketing authorisation application	Application submitted to EU state member authority or EMA based on drug category
	New drug application [Section 505(b)(2)]	⟷	Marketing authorisation application (hybrid application)	
	Abbreviated new drug application [Section 505(j)]	⟷	Marketing authorisation application (generic medicinal product application)	

TABLE 3.9

Types of Regulatory Exclusivities, Period and Eligibility as per EMA

Exclusivity	Period	Eligibility
Data exclusivity (DE)	11 years [8 + 2 + 1]; i.e. 8 years of data exclusivity for an NCE, additional 2 years of marketing exclusivity, and 1 additional year of exclusivity for new indication	In 2005, this DE directive was brought into force for NCE. A generic applicant can file a request for a marketing authorisation after the 8-year period, but it can market only after 10 years from date of approval of NCE. NCE holder can obtain an additional 1-year extension of marketing exclusivity if he obtains an additional indication authorisation during the initial 8 years of exclusivity.
Orphan drug exclusivity (ODE)	10 years from date of approval of NCE	In 2000, Orphan Drug Regulations became effective. This is applicable for disease that affects no more than 5 people out of 10,000 in the EU. During this 10-year period, the EU or a member state will not accept another application for a marketing authorisation, for the same therapeutic indication with respect to a similar medicinal product.
Pediatric exclusivity (PE)	6 months extension to the products' supplementary patent protection (SPP)	It is an incentive to innovator companies to carry out tests on pediatric population. PE will apply not only to the product's pediatric indications but also to all indications of the product having the same active ingredient and it includes additional 6 months extension to existing patent term extensions (SPPs).
Supplementary patent protection (SPP)	Maximum up to 5 years	SPP extends the patent expiry by up to 5 years in order to compensate them for the lengthy period required to obtain regulatory approval of their products. A generic product can be marketed only after the SPP expiry.

In September 2008, the USFDA granted Osmotica's citizen petition and stated that Sun would be required to submit to the USFDA a new ANDA citing Osmotica's approved venlafaxine HCl extended-release tablets drug product as the RLD, and that the USFDA requires any such ANDA applicant to conduct new bioequivalence studies comparing its proposed drug product to Osmotica's approved drug product. In this case, FDA determined that 'an ANDA for venlafaxine HCl extended-release tablets submitted based upon an approved suitability petition that was pending at the time of approval of Osmotica's NDA and that seeks approval for a pharmaceutically equivalent drug product must be withdrawn, as the ANDA is required to reference the corresponding approved strengths of Osmotica's venlafaxine HCl

TABLE 3.10

Comparative Regulatory Requirements for Drug Product Application Submission in the United States, EU and India

Registration Requirement	United States	EU	India
Approval agency	Only USFDA	Multiple agencies • EMA • CHMP • National Health Agencies	Only DCGI
Registration process	Single process	Multiple processes • Centralised • National • Decentralised • Mutual	Single process
Plant GMP approval	Required	Required	Required
Stability requirement	$25°C \pm 2°C/60\% \pm 5\%$ RH (long term) $30°C \pm 2°C/65\% \pm 5\%$ RH (intermediate/long term) $45°C \pm 2°C/75\% \pm 5\%$ RH (accelerated)		$30°C \pm 2°C/70\% \pm 5\%$ RH (long term) $45°C \pm 2°C/75\% \pm 5\%$ RH (accelerated)
No. of submission batches	3 primary batches, out of which min 2 are pilot scale; one batch can be smaller but not less than 25% of pilot scale	3 primary batches, out of which min 2 are pilot scale	2 pilot scale/production scale (if API is stable) 3 primary batches (if API is unstable)
Minimum stability data during submission	6 months	6 months	6 months
BE requirement and acceptance criteria[a]	Against US innovator 90% confidence interval on log transformed data C_{max}: 80%–125% AUC_{0-t}: 80%–125% $AUC_{0-\infty}$: 80%–25%	Against EU innovator 90% confidence interval on log transformed data C_{max}: 80%–125% AUC_{0-t}: 80%–125% $AUC_{0-\infty}$: not applicable	Against US/EU/Australian innovator 90% confidence interval on log transformed data C_{max}: 80%–125% AUC_{0-t}: 80%–125% $AUC_{0-\infty}$: not applicable
Dossier format	e-CTD/paper	e-CTD	Paper

(Continued)

TABLE 3.10 (CONTINUED)

Comparative Regulatory Requirements for Drug Product Application Submission in the United States, EU and India

Registration Requirement	United States	EU	India
Multimedia dissolution	Not compulsory. Depends upon product property. OGD media dissolution data for RLD and test product on 12 units required	Required	Not compulsory
Drug substance DMF	2.3.S and 3.2.S part of the eCTD is known as DMF	2.3.S and 3.2.S part of the eCTD is known as ASMF	Yes, required
CEP requirement	CEP certificate is not applicable	CEP certificate from EDQM is required	CEP certificate from EDQM is required
IID and elemental iron	Needs to be followed	Not applicable	Not applicable
Release specification for assay limit	90%–110% of label claim	95%–105% of label claim	90%–110% of label claim
Colour identification	Not required	Required	Required
Water content	Required	Not required	Required
Label code	Braille code is not a must on the label	Braille code is a must on the label	Braille code is not a must on the label

Note: API, active pharmaceutical ingredient; CEP, certificate of suitability; EDQM, European Directorate for the Quality of Medicine; IID, inactive ingredient database; RH, relative humidity.

ª Acceptance limit differs for highly variable and narrow therapeutic window drugs.

extended-release tablets as its RLD, and section 505(j)(2)(D)(i) of the Act precludes such a change from being submitted as an amendment to an ANDA. If Sun Pharma or any other applicant wishes to pursue approval of an ANDA for venlafaxine HCl extended-release tablets, it must submit a new ANDA containing data and information required by section 505(j) of the Act for approval (including, but not limited to, a demonstration of bioequivalence to the RLD, Osmotica's venlafaxine HCl extended-release tablets)' (USFDA-Blog 2015).

3.2.4.2   Case Study 2

Nonavailability of RLD in the Market – How to File a Generic: Clindamycin Phosphate and Benzoyl Peroxide Gel, 1.2%/5% RLD, DUAC, designated to Stiefel Laboratories, Inc. is not available in the marketplace. Although the drug has not been officially discontinued by Stiefel Laboratories, Inc., the product is not available for commercial distribution from Stiefel Laboratories, Inc. with the wholesale distributors.

There is another product, Clindamycin Phosphate and Benzoyl Peroxide Gel, 1.2%/5% marketed by Prasco Laboratories with the same NDA number as that of RLD but not listed as RLD. This is an authorised generic product licensed by Stiefel.

In a suitability petition, company Z requested FDA to determine that Prasco's product be deemed as suitable to use as an alternate RLD as the current Orange Book RLD. Company Z also quoted the FDA's prior decision wherein it was clarified that 'An ANDA applicant may use the authorised generic version of the current RLD as the reference standard in the *in vivo* bioequivalence study with proper documentation, because the RLD is still available as an authorised generic'.

The USFDA granted company Z's petition under the following grounds 'Based on our records, it does not appear that the brand name drug, DUAC, is currently commercially available. Our records indicate further that the authorised generic version of DUAC entered the market on June 27, 2012, and is currently available for commercial distribution. As we explained in our response to a previous citizen petition, an ANDA applicant may, with proper documentation, use the authorised generic version of the current RLD as the reference standard for *in vivo* BE testing. Accordingly, you may use the authorised generic version of DUAC as the reference standard for BE testing. We recommend that you provide complete lot information on the authorised generic product to FDA for confirmation of its acceptability for use as the reference standard in BE studies before conducting the studies' (USFDA-Notice 2014).

3.2.4.3   Case Study 3

Patent Challenge Cases – Decided in Favour of Innovator (Takeda v. Alphapharm Pioglitazone Case): In cases involving new chemical compounds, it remains necessary to identify some reason that would have led a chemist to modify a known compound in a particular manner to establish prima facie obviousness of a new claimed compound.

In the *Takeda Chemical Industries, Ltd. v. Alphapharm Pty. Ltd.* case, Alphapharm, a generic drug manufacturer, had filed an ANDA to make a generic version of Takeda's pioglitazone.

Alphapharm argued that Takeda's patent on pioglitazone (US4687777) was invalid because a prior art compound, which differed only slightly from pioglitazone (identified as compound b), rendered pioglitazone obvious. Both compounds include a ring of five carbons and one nitrogen, a pyridyl ring. Compound b has a methyl group (a group containing only one carbon) at position 6 on the pyridyl ring. Pioglitazone has an ethyl group at position 5 on its pyridyl ring (one carbon over from position 6). Alphapharm argued that these changes were structurally obvious because they were

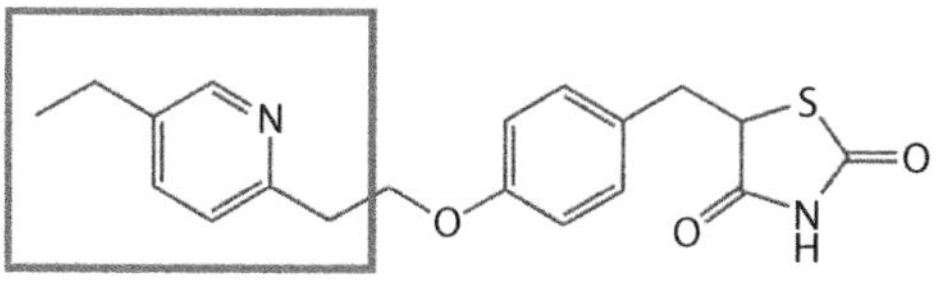

Pioglitazone

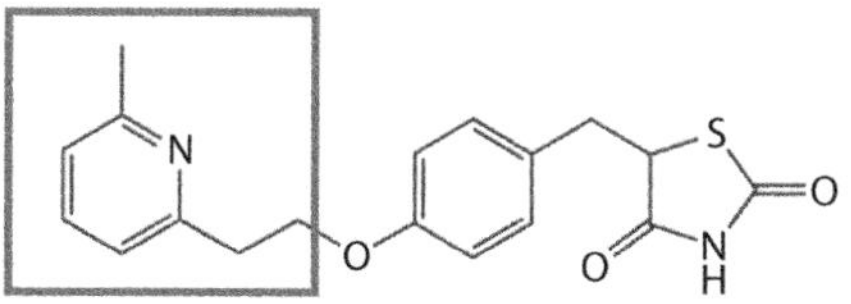

Prior art – compound B

FIGURE 3.7   Chemical structures of pioglitazone and prior art – compound b.

examples of routine experimental steps, namely, replacing one group with a similar group (methyl to ethyl), or homologisation, and *ring-walking* the substituent group from position 6 to position 5 on the pyridyl ring (Figure 3.7). In this case, prior art identified three specific compounds that were deemed most favourable in terms of toxicity and activity. The compound structurally closest to pioglitazone, compound b, was identified as one of the three most favourable compounds but was singled out as causing considerable increases in body weight and brown fat weight. The District Court found that compound b was not an obvious choice to modify to make an anti-diabetic because one of its side effects is weight gain. Weight gain, while generally undesirable, is even less so in diabetic patients. Additionally, compound b is toxic and therefore less suitable for treatment of chronic diseases, such as diabetes. This finding was affirmed by the Federal Circuit.

The toxicity and other side effects of compound b, the closest prior art, taught away from the claimed invention and so the prior art did not suggest modifying compound b either by homologisation or ring-walking (USCAF 2015).

3.2.4.4   Case Study 4

Decided in Favour of Generics (Entecavir Case): As part of its ANDA application to market a generic version of BMS's anti-hepatitis B drug, entecavir (marketed as Baraclude), Teva certified a paragraph IV to US5206244 (*the 244 patent*) stating that it was invalid.

In a subsequent case, the US District Court of Delaware, after hearing all the evidence presented including expert testimony from both sides, concluded that the 244 patent was invalid under 35 U.S.C. §103 as obvious in view of the prior art. More specifically, the court concluded that a person of ordinary skill would have been motivated by the prior art to select a compound known as 2′-CDG as a *lead compound* and to modify it in such a way as to arrive at entecavir and that the evidence

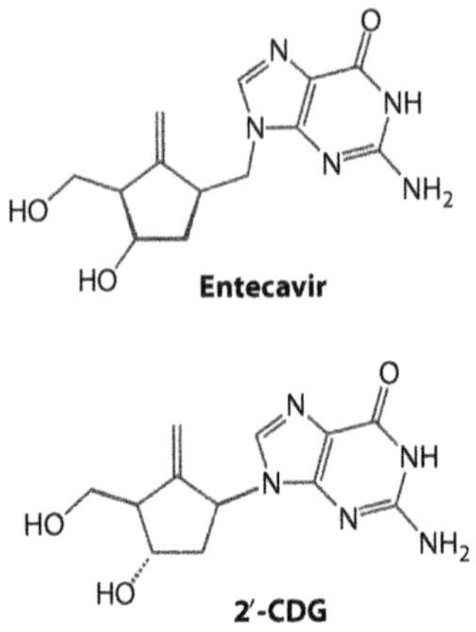

FIGURE 3.8   Chemical structures of entecavir and 2′-CDG.

of secondary considerations, including evidence that entecavir possessed unexpected properties over 2′-CDG, did not overcome the prima facie case that claim 8 was obvious in view of the prior art (Figure 3.8).

The Federal Circuit found that there was ample evidence to support the District Court's finding that 2′-CDG was a suitable lead compound, in no small part because Bristol-Myers's own expert admitted that 'medicinal chemists during the relevant time frame were actually treating and using 2′-CDG as a lead compound' to search for new antivirals at the time, and three prior art references demonstrated as much.

The Federal Circuit also found that the trial record likewise supported the motivation to modify 2′-CDG to arrive at entecavir and an expectation of success in making that change. Importantly, a prior art reference demonstrated that making the change to a similar structure to that of the lead compound resulted in increased potency, and here, again, Bristol-Myers's expert made critical concessions that the prior art reference 'would not dissuade' the skilled chemist from making the change and that the prior art 'could have led' the skilled chemist to seek new antivirals. The fact that the prior art suggested increased potency served as the basis for the finding of the expectation of success (US Court 2013).

3.3   INTELLECTUAL PROPERTY

Authors' Note – Building of an intellectual property (IP) portfolio along with a robust business model is an efficient strategy for a successful business irrespective of the size of the enterprise. Recognising the IP and devising an effective strategy for protecting, managing and enforcing the IP are critical to making the most out of the IP. It is the first and not the last step in an entrepreneur's journey.

IP and patent protection/information play an important role in new product development and its life cycle management. It encourages research and innovation and is fundamental to the success of enterprises and institutes/universities, by way of providing return on investment on the drug development expenditure by giving the company monopoly over the invention for a defined period.

In a pharmaceutical product development process, the role of IP is integral at every stage in the process, starting from idea or opportunity analysis to the development of the product, dossier filing and ultimately the commercialisation of the product.

In case of new drug discovery, although thousands of experimental drugs undergo clinical testing, only a handful of new drugs/dosage forms qualify for regulatory approval, that too after decades of clinical testing and expenditure of billions (recent report indicates 1–5 billion USD as expenditure for one new molecule discovery). At the grassroots level of such experimental drugs are patents and data protections that majorly determine the path for development of the drug.

In case of generic development, early market entry and strategies around it are the main criteria, determining the path of development. Thus, there are two major aspects of IP during any drug development that includes creating new IP and designing around the existing IP.

3.3.1   CREATING NEW IP

One incentive for a company spending billions of dollars on new drug development is the return on investment by way of IP, which can help curtail competition and also allows revenue generation by licensing activities. Drafting and filing patents are an essential aspect of new product development.

3.3.2   DESIGNING AROUND THE EXISTING IP

Once a product is selected for development, the cardinal exercise that is carried out to determine the desired route is patent study or entire patent landscaping on the selected product. The results of the patent study indicate the key path to be adopted for pharmaceutical development. The selected path should be feasible for scale-up, should have minimum concerns related to patent infringement aspects, should lead to a desired industrial yield and should be economically feasible. Managing IP during pharmaceutical drug development is hence a multifactorial and multifunctional task that calls for alignment of multiple faculties including business development, R&D, regulatory, clinical, production and marketing.

3.3.3   FREEDOM-TO-OPERATE OPINIONS (PATENTS)

Freedom-to-operate (FTO) opinions generated by skilled personnel at the early stage of product development (immediately after conceptualisation and selection process) provide the basis and hence the freedom to proceed with the research, development, production and commercialisation of a selected process. FTO opinions help in ascertaining the complexity of process development and the risks involved in development via various plausible routes. FTO is very important to reduce or, in many cases, eliminate the risk of infringing the IP rights of third parties. A skilled person meticulously performs an FTO by relying on search databases and by analysing the hits obtained vis-à-vis the selected product. The product essentially is split into its fundamental components such as the drug per se, process for making the drug, formulations and process for making the formulations, method of use and so on, and then scrutinised against crucial IP rights. The early reliance on an FTO analysis is critical because it lays the road map for what happens later on during the product development and commercialisation. An FTO may be variously designated as patent landscaping, patent/product clearance reports and so on.

As a result of entire patent landscaping and analysis against a selected process, there are a set of patent(s) that would then need a separate noninfringement or invalidation detailing. Hence, an immediate outcome of an FTO is identification of such relevant patents for independent treatment.

3.3.3.1 Noninfringement Opinion

Claim construction is the central aspect of noninfringement opinions and often effectively determines whether there is infringement. The claims of a patent delineate the scope of the property right, and it is the patent claims that are infringed, held invalid or unenforceable.

Key decisions on whether to continue R&D on a selected product, whether to invest money on the project, whether the existing process needs to be altered in any manner and so on emerge from such patent noninfringement opinions. A claim-to-claim analysis and a country-based analysis are key to any noninfringement opinion. Claims of a same patent may be different in scope in different countries or it is even possible that a key patent is not filed in some countries of interest; hence, incorporating these controls is essential while drafting a patent(s) noninfringement opinion.

Typically, a noninfringement opinion will include an analysis of whether there is infringement of the patent claims either literally or under the doctrine of equivalents. For determination of literal infringement, the proposed product/formulation will be compared with a logically construed claim to determine if every element of the claim is present in the product/formulation. The study would determine whether there are differences between the product/formulation and what is claimed in the patent and whether the extent of these differences would lead one to conclude that the product/formulation does not include all of the claimed elements.

A logical and correct claim construction would mean a proper understanding of what the claim embodies. In construing claims, the claim terms are generally given their ordinary and customary meaning at the time of the invention. However, where there is no ordinary and customary meaning that can be given to the claim terms, reference to the claim terms in the claims themselves, the patents specification and the patents prosecution history (all three sources commonly referred to as *intrinsic evidence*) are reviewed. Claim construction requires consideration of whether the inventor/patentee defined a claim term in the patent specification in a specific way or if the inventor limited the scope of the invention during patent prosecution. In either situation, the claim terms will be given a meaning consistent with the inventor's intended use of the terms and how the inventor understood what was being claimed.

A determination of infringement under doctrine of equivalents requires establishing equivalency between the elements of the claimed invention and the elements comprising the proposed product/formulation. In the reverse determination of noninfringement under design of experiment, all the elements of the patent claim must be considered, and it must be shown that there are substantial differences between each element of the proposed product/formulation and each element of the patent claim or that the proposed product/formulation does not perform substantially the same function, in substantially the same way, to achieve substantially the same result as the claimed invention.

A key tool at play during the drafting of patent noninfringement opinions is a thorough analysis and understanding of the file wrappers to determine if the patentee gave up certain subject matter or intended to cover only a specific subject matter that may not be immediately apparent after plain reading of the claim. The prosecution history of a patent is the complete record of proceedings before the patent office including express representations made by an applicant regarding the scope of various claims and special definitions attributed to particular terms or phrases. Arguments and amendments made during the prosecution of a patent limit the interpretation of claim terms so as to exclude any interpretation that was disclaimed during prosecution.

Prosecution history estoppel requires that the claims of a patent be interpreted in light of the proceedings in the Patent and Trademark Office during the patent application process. This estoppel typically acts as a disclaimer of certain subject matter and applies when an applicant amends or cancels claims rejected by an examiner as unpatentable in light of the prior art to obtain allowance of the claims. Such claim amendments can inform the general public of limits on the scope of the protection that has been agreed to by the patent applicant to which they cannot claim ownership at a later date. There are two forms of prosecution history estoppel: argument based and amendment based.

Further prosecution history estoppel, however, is not limited to the applicant's own explicit words but may embrace the applicant's implied responses to the examiner's actions as well. If the patentee does not rebut an examiner's comment or acquiesces to an examiner's request, the patentee's unambiguous acts or omissions can create an estoppel.

3.3.3.2 Invalidation Opinion

In many cases, it may be determined that the process chosen for development of a drug formulation or the drug formulation itself does not avoid one or more claims of a patent after interpretation and analysis of the terms of those claims. This calls for an invalidity analysis to be performed separately. Invalidity is analysed on a claim-by-claim basis. Each claim of a patent must be separately assessed for validity. For example, an independent claim might be invalid because of the fact that there was a prior art product with the claimed limitations, but a dependent claim, or another independent claim, might not be invalid because it could contain limitations that are not in the prior art product.

In many jurisdictions, patents are presumed valid, and the party asserting invalidity has the burden of proof by clear and convincing evidence. An invalidity analysis seeks to determine whether the patent is invalid under one or more of the following: not an invention, not novel or anticipated from prior art, obviousness, double patenting, indefiniteness, lack of enablement, wrong inventorship, inequitable conduct, unclean hands and prosecution laches.

3.4 RELATIONSHIP BETWEEN PHARMACEUTICAL PRODUCT DEVELOPMENT AND INTELLECTUAL PROPERTY RIGHTS

In the pharmaceutical industry, product development is monitored and managed by a sophisticated project management system. IP is an integral part of such a project management system. The decision on product selection for development is generally a collaborative process involving R&D, the IP department and business managers. Patent strategies are chartered in detail by the IP department. On the basis of its value, strength, competitive advantage, enforcement, right fit in product life cycle and commercial benefits, patents are filed at appropriate stages of product development and a number of patents are filed on a single molecule covering its various properties and applications to optimise a product patent life. Apart from offering protection and market monopoly, a patent also provides a good source of revenue generation by way of licensing and can be used as a good negotiation tool during settlement discussion with other parties.

Let us look at the various stages of product development in a drug discovery R&D and formulation development R&D [505(b)(2) or generic product development] and the role and function of the IP department and types of patents filed at each stage of development.

3.4.1 DRUG DISCOVERY

The key steps in new drug discovery program are discussed in Sections 3.4.1.1 through 3.4.1.4.

3.4.1.1 Lead Finding

At this stage, detailed patent landscaping and mining is done using the various databases to conduct prior art search and patentability studies and a patent application covering the drug substance as a Markush product patent or specific product patent is filed at the patent office. These patents confer the broadest and highest degree of protection to a product. Generic entry in the market is prevented until expiry of the product patent (unless challenged and successful). For example, everolimus (brand name Zortress) of Novartis is covered under US5665772 expiring on March 9, 2020. Imatinib mesylate (brand name Gleevec) of Novartis is claimed in US5521184 expiring on July 4, 2015 (Orange Book 2015).

3.4.1.2 Preclinical Trials

At this stage, the complete patent application of National Phase entry of the already filed provisional application of the product is filed or a product patent may be filed at this stage too if not filed earlier.

3.4.1.3   Clinical Trials

At this stage, generally the method of use patents are filed claiming the primary medical use (if not claimed in the product patent). These are equally strong and broad patents blocking entry of a generic version of the molecule until expiry of the patent (unless challenged and successful). For example, everolimus (brand name Affinitor) of Novartis is protected by a method of treatment patent US8436010 claiming method of inhibiting growth of solid tumors of the breast and by US8778962 claiming method of inhibiting growth of nonmalignant solid tumors of the brain. Sitagliptin (brand name Januvia) of Merck is protected by US6890898 claiming method of treating type II diabetes using DPPP-IV inhibitor expiring on February 2, 2019 (Orange Book 2015).

3.4.1.4   Registration and Commercialisation

At this stage, the dossier is compiled and filed with the regulatory authorities, and on approval, the new product is launched in the respective market.

Secondary patents to improve the life cycle of the drug are filed at various stages of development; these are explained in detail in Section 3.4.2. Table 3.11 provides correlation of drug discovery steps and patent filing stages and the types of patents filed as an illustrative example.

3.4.2   Formulation Development

The key steps in formulation research and development are discussed in Sections 3.4.2.1 through 3.4.2.4.

3.4.2.1   Idea or Opportunity Analysis

The type of product to be developed, whether generic (ANDA filing) or new product (505(b)(2)) filing, is discussed and decided at this stage. Every company has its

TABLE 3.11
Correlation of Drug Discovery Steps and Patent Filings

Drug Discovery Steps	Patent Steps	Type of Patent Filed
Ideas shortlisted	Patent landscape and mining	Product Markush
Target identification and validation	Detailed patent search and patentability study	Product-specific MOU
Hit identification	Provisional patent filing	
Lead identification	Updated patent search	
Lead optimisation	Complete patent filing	
Preclinical candidate	National phase entry	
Drug development	Develop secondary patent strategy	Salt, polymorphic forms, hydrate,
Preclinical testing	File secondary patents	crystal aspect, particle size, purity,
Clinical studies	File secondary patents	composition, combination, API
Registration	Patent grant and listing and enforcement	process, devices, kits, etc.

unique criteria and methodology for product identification and selection on the basis of various permutations and combinations of factors associated with feasibility of development, IP and business areas. These are explained briefly in Table 3.12 and examples are provided below.

A document that includes product information is generally generated, which contains information on type of dosage form to be developed, strength, market scenario, patent status and strategy, R&D feasibility, sales forecast, facility availability and other relevant information that helps in selection of a project. A proper evaluation of the above information helps in the selection of products for development.

TABLE 3.12
Factors to Be Considered for Product Selection

Factors	Points to Consider
General	• Selection of therapeutic class for selecting the drug candidates
	• Selection of candidates based on market demand
	• Selection of candidates from special categories such as ODE, opioids, etc.
	• Selecting candidates with a threshold revenue generation of say more than 10 million USD for innovator
	• Selection of candidates based on complexity of drug development
	• Selection of candidates based on anticipated competition in the targeted market
	• Selection of candidates based on market of interest
	• Selection of candidates based on net present value and projected forecasts
	• Selection of candidates based on complexity of patent protection around the candidate
	• Selection of candidates based on complexity of potential patent licensing deals
Product related	• Sales, dose form and strength, patents, patent challenges, labelling, chemistry, synthesis, data exclusivity, market exclusivity, approval process, API sources, manufacturing capabilities, bioequivalence criteria, stability, pharmacodynamics, toxicity, country regulations, pack pricing
	• Reimbursement
IP	• Number of patents in the Orange Book
	• Type of patent in the Orange Book (product, MOU, composition, polymorph) and its expiry date
	• IP strategy (noninfringing or invalidation)
	• Patent certification (para III or para IV)
	• Type of exclusivities on the product
	• First-to-File opportunity or second/third generic
	• 180-day exclusivity possibility and whether shared exclusivity
	• 30 months stay possibility
	• Litigation possibility
	• Patent opinion and litigation cost
	• Market entry date
	• Number of other generics
	• Scope of settlement
	• Licensing and partnering possibility

At this stage, the IP department provides preliminary patent information on the molecules under evaluation, including information on product, method of use and other relevant patents and a brief strategy to determine the IP constraint date or market entry date for the product. For a US-marketed product, the Orange Book patents are majorly evaluated at this stage and equivalent patents in other markets of interest are studied to provide market entry date in the respective countries of interest.

Each company designs its own unique product selection criteria. In Table 3.12, we have tried to list most of the important factors to be considered during product selection, especially for a US ANDA product development.

3.4.2.2   API Sourcing

Sourcing of active pharmaceutical ingredient (API) that caters to various regulatory requirements such as HPLC (high-performance liquid chromatography) purity, isomer ratio, stability and so on in addition to being noninfringing to patents is a critical step in the drug development process. An important decision would be to determine whether the API can be developed in-house or should be outsourced from a third-party supplier. This is a technical as well as a strategic decision that is taken after considering the number of drug master file (DMF) filers, time and cost of API development in-house vis-à-vis outsourcing, patent noninfringement aspects, compliance to regulatory aspects, the impact the specific form or state of API may have on formulation and availability of manufacturing facility and its capacity.

At this stage, a bulk drug structure centric search and patent landscaping is carried out to understand which processes are free of IP issues. Many aspects on designing the API, including specific isomeric form, impurity content, polymorphic form, moisture content, structure–aspect ratio, reaction conditions, use of specific intermediates, purity of intermediates and final API, associated chirality and so on, are completely addressed as a result of such patent landscaping. The searches and results are periodically updated to ensure that the process under development is always free of IP concerns. Once the API is developed in-house and finalised for scale-up, a final API IP clearance is issued after reviewing and updating all patents around the chosen molecule, its starting materials and intermediates. In case of an outsourced API, a list of all relevant patents is exchanged with the supplier seeking its views on noninfringement of their API against the listed patents and their in-house (or preferably attorney) invalidation opinion on specific patents if broadly claimed along with a patent certification of clearance of their API in said markets of interest. Based on a satisfactory response from suppliers, their API is cleared by the IP department for sourcing and use in product development. For ICH markets, a US DMF API is preferable.

For an in-house API, after a thorough evaluation of patentability, if found novel, patent applications are filed on novel process of manufacturing, polymorphic forms, isomers, purity, hydrates, solvates, particle size or new salts. Such patents may be avoidable by other generics but at times better and stronger protection can be availed on unique processes of high efficiency, purity and stability. For example, cabazitaxel (brand name Jevtana) of Sanofi has a patent claiming its acetone solvate form in US7241907 expiring in December 2025 (Orange Book 2015). Omeprazole was a blockbuster molecule protected by several patents before its basic patent expiry in 2002. Astrazeneca introduced the single (s) enantiomer of omeprazole, esomeprazole, which was covered by a separate patent US5714504 extending the patent protection until February 2014. Fexofenadine is claimed as a substantially pure form in US5578610 expiring in November 2013 by Sanofi Aventis (Orange Book 2015). When acetaminophen was introduced as an injection solution, the new product was protected by a product by process patent US6992218 expiring only in June 2021 (Orange Book 2015). Adefovir dipivoxil (brand name Hepsera) is claimed broadly as a crystalline drug by Gilead in US6451340 expiring in July 2018, claiming a composition comprising crystalline adefovir dipivoxil (Orange Book 2015).

3.4.2.3 Preformulation (Prototype)/Pilot Bioequivalence/ Proof-of-Concept Stages

Various formulations are developed at this stage at R&D to establish the proof of concept. A pilot biostudy is conducted to finalise the formulation to be taken up for further development.

At this stage, a preliminary patent report is first generated by using a combination of patent searches such as SciFinder structure search and key word searches in subscribed and free databases, listing all patents and applications relevant to the product under development. The patents are segregated into product, memorandum of understanding (MOU), composition, process, API related and others, and each is analysed against the prototype formulations. The patents are discussed with R&D and a strategy is laid down to develop a noninfringing composition and process. Once the formulation is finalised, the FTO opinion is generated to give IP clearance to proceed with the proof-of-concept biostudies. Patents with broad blocking claims are studied by the IP department for prior arts and both factual and legal basis invalidation opinions are generated, which may be validated through attorneys as required. FTO opinions and invalidation opinions are two cardinal IP documents generated during any product development, clearing the path to progress from idea to commercialisation.

At this stage, a new formulation/composition, process of manufacturing, dissolution parameters, stability, devices, new excipients used, new dosage form, route of administration, combinations and so on are patented as secondary patents to ward off competition. Such claims may confer substantial protection on the basis of the width and breadth of claims obtained on the composition.

New formulations with improved patient compliance, through reduced dosing or favourable side effects or better clinical profile such as an extended-release dosage form, are developed for most new molecules and are patent protected leading to evergreening of a molecule. For example, metformin XR (Glucophage XR), fluoxetine SR (Prozac once weekly), verapamil HCl ER (Covera-HS), oxybutynin chloride ER (Ditropan XL), metoprolol succinate ER (Toprol-XL), memantine XR (Namenda XR) and many such molecules are covered by once daily or once weekly composition patents. Methyl phenidate suspension (brand name Quillivant XR) of NextWave Pharma is covered as composition claims in US8062667 expiring in March 2029, claiming the modified release formulation containing drug–ion exchange resin complexes. After introducing sitagliptin tablets (Januvia), Merck also got approval for a combination product, sitagliptin phosphate + metformin HCl (Janumet), and an extended-release composition, sitagliptin phosphate + metformin HCl ER (Janumet XR), and claimed the combination of sitagliptin + metformin through a series of patents including US6890898 expiring in February 2019. When sumatriptan (Imitrex), useful for migraine treatment, was losing its patent protection on the compound in 2006, GSK developed and introduced a new route of administration of this known drug as an intranasal formulation and protected it by filing a number of patents, thus extending the life cycle of the molecule (Gupta et al. 2010).

3.4.2.4   Scale-Up and Optimisation/Technology Transfer/ Exhibit Batch/Pivotal Biostudies or Clinical Trial Stages

Once a proof of concept is established, the product is manufactured in large scale through technology transfer and exhibit batches are taken to conduct pivotal biostudy for submission in dossiers to obtain regulatory approvals. At this stage, an updated FTO opinion is generated and a final IP clearance is given to take the exhibit batches. If any new use or new combination is identified at this stage, then patents are filed on such new findings.

Bupropion (brand name Wellbutrin XL) was initially indicated for treatment of depression. Later, GSK realised its potential as an aid to smoking cessation treatment and patented this use and introduced this use as an extended-release tablet (Zyban) (Gupta et al. 2010).

3.4.2.5   Dossier Compilation/Filing/Launch of Product

After clearing pivotal bio and stability, the generated data are compiled and reviewed by the Regulatory Department and filed as a dossier to the regulatory authorities, and on obtaining approval, the product is launched in the market.

At the stage of dossier compilation, the label or the package insert is sent to the IP department for clearance to check whether the indication and dosage administration mentioned in the label is covered by a patent or is free to be included. If found to be protected by a patent, then the secondary indications are carved out from the final label until the expiry of the patent. Additionally, in the case of a US dossier, a patent certification is provided, which provides information on type of certification (whether, para I, II, III, or IV or section VIII) for all the patents listed in the Orange Book. A para IV filing triggers a series of activities before approval of the dossier and launch of product. This is explained as a flow chart in Figure 3.9.

Table 3.13 provides correlation of formulation development steps from idea to launch, the types of patent reports generated and types of patents filed at various stages, as an illustrative example, the details of which has been explained above.

3.5   LEGAL PROCEEDINGS IN PHARMACEUTICAL PRODUCT DEVELOPMENT

Sections 3.2 through 3.4 dealt with regulatory and IP considerations related to drug product development and approval, especially the generic (ANDA) and 505(b)(2)

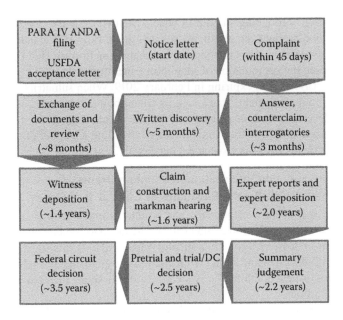

FIGURE 3.9 Legal activities post-para IV ANDA filing (USFDA) – litigation and timelines.

products. In the United States, product approval is linked to patents and to the legal proceedings associated with it.

As mentioned above, para IV filing triggers a series of activities before approval of the dossier and launch of the product. On receiving the Acceptance Letter from the USFDA (which presently takes approximately 6–8 months from the filing of the dossier), the generic filer sends a Notice Letter to the patentee and innovator, within 20 days; such a Notice Letter provides factual and legal bases stating that the patent is noninfringed or is invalid or is unenforceable. Upon receiving the Notice Letter, the innovator or NDA holder has 45 days to bring in suit, which automatically brings in a 30-month stay in the dossier approval by the USFDA. After the 30-month period or if the patents involved are deemed invalid or not infringed (whichever occurs sooner), the FDA may approve the ANDA. As an incentive to the generic filer, the Hatch-Waxman provides a 180-day period of exclusivity for the first para IV filer on successful challenge of the patent and approval. Briefly, the steps in litigation and their timelines are represented in Figure 3.9.

Patent protection and patent portfolio created on a developed product/technology generates greater economic value to an organisation by establishing monopoly in business. It determines the innovation capabilities and intellectual assets of the organisation and forms the foundation of its business. Patent portfolio is built by the process of patent filing, maintenance and enforcement of such patents. Commercialisation of the product is the key essential aspect, to maximise the potential of patent protection, leading to exclusive marketing rights, higher profit margins and FTO. In today's product development scenario, patent protection and patent strategies provide high value in licensing, acquisitions and collaborations, which is a successful emerging business model of the pharmaceutical industry.

TABLE 3.13

Patent Landscape during Formulation Development

Formulation Development Stages	Types of Patent Reports Generated	Types of Patents Filed
Idea/opportunity analysis	Preliminary patent information included in the product information form	New composition
API sourcing	API IP clearance FTO report	Novel process of manufacturing, polymorphic forms, isomers, purity, hydrates, solvates, particle size or new salts
Preformulation (prototype)/ pilot BE/proof of concept	Preliminary patent report and formulation, IP clearance and FTO reports	New formulation/composition, process of manufacturing, dissolution parameters, stability, devices, new excipients used, new dosage form, route of administration, combinations, etc.
Scale-up and optimisation/ technology transfer/ exhibit batch/pivotal biostudies or clinical trial	Updated final formulation, IP clearance, FTO, opinions and invalidation opinions	New use or new combination patents
Dossier compilation/filing/ launch of product	Label clearance Patent certifications	–

REFERENCES

ANDA. 2015. Abbreviated New Drug Application. http://www.fda.gov/Drugs/Development ApprovalProcess/HowDrugsareDevelopedandApproved/ApprovalApplications /AbbreviatedNewDrugApplicationANDAGenerics/ (accessed June 19, 2015).

Camargo. 2015. http://www.camargopharma.com/ (accessed June 19, 2015).

CDSCO. 2015. Central Drugs Standard Control Organization. http://cdsco.nic.in/forms /Default.aspx (accessed June 19, 2015).

EMA. 2015. European Medical Agency. http://www.ema.europa.eu/ema/ (accessed June 19, 2015).

Gupta, H., Kumar, H., Roy, S. K., Gaud, R. S. 2010. Patent protection strategies. *J Pharm Bioallied Sci* 2(1):2–7.

Hatch-Waxman. 2015. Drug Price Competition and Patent Term Restoration Act of 1984 (Hatch-Waxman amendments). http://www.fda.gov/newsevents/testimony/ucm115033 .htm (accessed June 19, 2015).

ICH. 2015a. http://www.ich.org/home.html (accessed June 19, 2015).

ICH. 2015b. http://www.ich.org/products/guidelines.html (accessed June 19, 2015).

IND. 2015. Investigational New Drug Application. http://www.fda.gov/drugs/development approvalprocess/howdrugsaredevelopedandapproved/approvalapplications/investiga tionalnewdrugindapplication/default.htm (accessed June 19, 2015).

NDA. 2015. New Drug Application. http://www.fda.gov/Drugs/DevelopmentApprovalProcess
   /HowDrugsareDevelopedandApproved/ApprovalApplications/NewDrugApplication
   NDA/ (accessed June 19, 2015).

Orange Book. 2015. http://www.accessdata.fda.gov/scripts/cder/ob/ (accessed June 19, 2015).

USCAF. 2015. United States Court of Appeals for the Federal Circuit. http://www.ub.edu
   /centredepatents/pdf/doc_dilluns_CP/Takeda-vs-Alphapharm_USFedCir-June28-2007
   _refers%20to%20KSR.pdf (accessed June 19, 2015).

USFDA. 2015. U.S. Food and Drug Administration. http://www.fda.gov/ (accessed June 19,
   2015).

USFDA-Blog. 2015. FDA grants Osmotica citizen petition; Decision affirms RLD change
   and choice policies and implements MMA RLD change prohibition provisions. http://
   www.fdalawblog.net/fda_law_blog_hyman_phelps/2008/11/fda-grants-osmotica
   -citizen-petition-decision-affirms-rld-change-and-choice-policies-and-implements
   -.html (accessed June 19, 2015).

USFDA-Notice. 2014. FDA Docket No.: FDA-2014-P-1065. http://www.noticeandcomment
   .com/FDA-2014-P-1065-fdt-42415.aspx (accessed June 19, 2015).

US Court. 2013. United States District Court for the District of Delaware; BMS V. Teva;
   Civil Action No. 10-805-CJB. http://www.ded.uscourts.gov/sites/default/files/opinions
   /cjb/2013/february/10-805.pdf (accessed June 19, 2015).

4 Strategies in Pharmaceutical Product Development

Maharukh T. Rustomjee

CONTENTS

4.1 INTRODUCTION

Research and development is a critical activity for all pharmaceutical companies. During research and development, once a lead drug candidate with required therapeutic action has been selected for further development, it is optimised as a part of the early development process. These studies include pharmacological and toxicological studies that help prove the safety and efficacy of the lead candidate. The chemical characteristics and structural specificity of drug–target interactions determine the potency and efficacy of a drug (Smith and O'Donnell 2006; PhMRA 2014; Tufts CSDD 2014). Dosage form design is crucial for maintaining usefulness of the drug candidate as its role is to ensure that the drug is able to withstand the hostile environment of the body, gets released and gets delivered at the site of absorption or action (in case of locally acting dosage forms) in the required amount and form. Further, it should be able to do this consistently for every lot and within each lot, for the shelf life of the product. If the development scientists are unable to design a dosage form that will ensure the above requirements, it could mean a premature death for the drug candidate, which could be ascribed to a variety of reasons such as extensive degradation in the body, suboptimal absorption and blood concentrations, extensive hepatic metabolism and so on. Drug dosage form design and delivery, therefore, is very important for ensuring optimum utilisation of the drug candidate by the human body.

The quality of the strategy/design phase process cannot be emphasised any more. It seems very obvious that before any serious development activities are undertaken, the product to be developed should be well defined. In the excitement and hurry to start development and get to the market quickly, the value of the design phase is very often underestimated (Gibson 2009).

4.2 DOSAGE FORM DESIGN

4.2.1 Preformulation Considerations

Strategising and planning for the design and development of a suitable dosage form for any drug is one of the most important tasks that determine the success of the development program. The effort and energy put into this activity never go to waste and, if done well, invariably lead to commercial success.

The dosage form development team needs to consider the information and know-how available in the areas given in the list below.

- Preformulation, screening and preclinical data of the drug substance that provide relevant physicochemical and pharmacokinetic/dynamic parameters of the drug substance
- Impact of preformulation on the ability of the dosage form to meet therapeutic requirements of the patient, the route of administration and the type of dosage forms
- Identification of specific biopharmaceutical targets and challenges and difficulties in meeting them
- Identification and definition of suitable test methods and designs to correlate the biopharmaceutical targets to experimental data

- Inactive ingredients/excipients – an understanding of their functionality, properties, quality and availability
- In-depth understanding of the unit operations used in the manufacture of drug products, their inherent limitations and performance essential from the raw materials and other ingredients proposed to be used
- Understanding and knowledge of how all of the above work together to be able to manufacture a consistent drug product that delivers the drug to the patient as per the design criteria and is stable over the product shelf life

There may be different options possible, and it is imperative that the right one be selected for the commercial success of the product in treating ailments. Drug substances are most of the time administered as part of a formulation manufactured in combination with one or more inactive agents that serve various functions in the formulation. There are various types of formulations or dosage forms. These inactive ingredients or excipients are selected on the basis of the type of dosage form being developed. They function as solubilisers, disintegrants, fillers, lubricants, antiadhesives, suspending agents, thickeners, emulsifiers, preservatives and many such roles and transform the drug substances into safe, efficacious and appealing dosage forms for the patients. They offer choice of form and delivery system for doctors to prescribe. The drug substances and inactive ingredients must be compatible with each other, and these materials should have the desired physicochemical properties to make them amenable to the unit operations of the manufacturing process (Gibson 2009; Hassan 2012). More details on the preformulation studies and drug–excipient compatibility studies are given in Chapter 6.

4.2.2 BIOPHARMACEUTICAL AND FORMULATION CONSIDERATIONS

The physical, chemical, biological and toxicological characteristics of the drug substances and inactive ingredients need to be considered while designing the dosage form. Biopharmaceutics, including pharmacokinetics (PK) and pharmacodynamics (PD), plays a very important role in selection of the type of dosage form for a drug candidate. The PK profile of a new molecule describes the factors affecting the relationship of dose to active-site concentration and PD describes the relationship between active-site concentration and pharmacological effect. Pharmacokinetic studies measure parameters such as the blood level concentrations over time, area under the curve (AUC), the concentration at maximum and minimum (C_{max} and C_{min}) and other parameters such as clearance, half-life and volume of distribution, which are calculated from the observed data. These parameters reflect the absorption (A), distribution (D), metabolism (M) and excretion (E) of a drug from the body. The overall systemic exposure of the body to a drug and its metabolites after drug administration is controlled by its absorption–distribution–metabolism–excretion–toxicity (ADMET) profile. This systemic exposure, reflected in plasma concentrations of the drug and its metabolites, is generally used to correlate the dose to its efficacy and side effects. The release of drug molecules from the dosage form has a very significant effect on its ADMET profile.

Dosage forms are needed to ensure safe and convenient delivery of accurate dosage to the human body. Drug doses can range from as low as 0.1 µg to 1000 mg and above, and therefore a formulation that is stable and can deliver the accurate dose, which is then released into the body to give the desired efficacy, is essential for any drug candidate. Depending on the biopharmaceutical, PK and PD characteristics of the drug candidate, there are various routes of administration and dosage form designs utilised to achieve the same. The various dosage forms available could be classified as depicted in Figure 4.1.

Since oral dosage forms (viz., tablets and capsules) make up more than 70% of the marketed products, the discussions further on are exemplified using this dosage form. However, the same principles could apply to all types of dosage forms.

The Biopharmaceutical Classification System (BCS) proposed by Gordon L. Amidon in 1995 (Amidon et al. 1995) classifies drugs into four categories

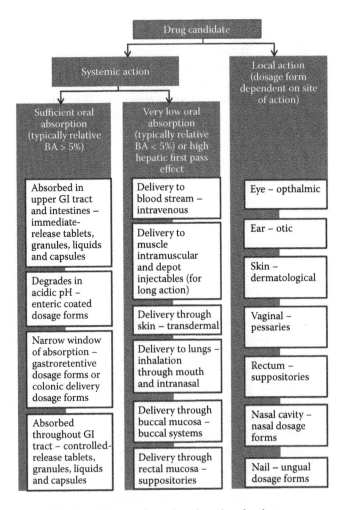

FIGURE 4.1 Classification of dosage forms based on site of action.

according to their aqueous solubility and permeability: Class I: High Solubility–High Permeability, Class II: Low Solubility–High Permeability, Class III: High Solubility–Low Permeability, and Class IV: Low Solubility–Low Permeability.

The BCS proposes a scientific rationale and justification for qualifying oral drug products for a waiver of in vivo bioequivalence studies. This rationale is based on in vitro dissolution methodology, with qualifications related to pH and dissolution, centrally embracing the permeability and solubility characteristics of the drug. This in turn helps drug companies to cut down on cost and time for in vivo studies and also cuts down on unnecessary exposure of human subjects in clinical studies. The objective of the BCS is to predict in vivo performance of drug products from in vitro measurements of permeability and solubility. The BCS is a valuable tool in the arsenal of formulation scientists. The application of the principles of the BCS helps in the design of the biopharmaceutical performance from oral dosage forms.

Many a time the dosage form development strategies have to be designed specifically to overcome the biopharmaceutical problems to enhance realisation of the therapy needs for the patients. Examples of such dosage forms are given in Table 4.1.

Apart from the biopharmaceutical considerations, there are other aspects that also need to be considered when designing a product development strategy as given in the list below.

- The nature of the illness, the manner in which it is treated (locally or through systemic route), e.g. eye drops for local eye infection, but for delivery to the vitreous humor and retina of the eye, intraocular injectables are used

TABLE 4.1
Examples of Biopharmaceutical Problems and Solutions for the Same

Biopharmaceutical Problem	Design Elements to Overcome the Problem
Low bioavailability or high variability in absorption	Increase drug release using surfactants, increase permeability by changing drug properties, ensure drug release at site of absorption, protect from degradation in GI tract
Low half-life needing frequent administration (3–4 times a day)	Prolonged effect using controlled/sustained/extended-release strategy
Narrow therapeutic window with increased chances of side effect owing to C_{max} peak in blood plasma level	Develop sustained/controlled release system to taper the peak values to lower side effect profile
Narrow window of absorption/required at specific site for local action in GIT	Gastroretentive, colonic delivery, enteric delivery strategies
Extensive hepatic bypass effect leading to loss of drug from circulation	Strategies that help bypass the liver, for example, buccal, nasal, lymphatic delivery

- The age of the patient, for example, for children below 5 years and old people with dysphagia, palatable liquid forms or capsules whose contents can be sprinkled over food or drinks are preferred
- Therapeutic situation, for example, if patient is comatose and unable to take oral forms, then the injectable form is required. For nausea and vomiting or motion sickness, oral forms that can be taken without water or transdermal forms or suppositories are preferred
- Old people on multiple medications find it better if the shape, size and colour are distinctive so that identification is easier
- Duration of therapy, that is, chronic or acute; for example, injectables may be acceptable for acute therapy, but for chronic therapy, self-administered dosage forms such as tablets/capsules are preferred
- Dosage form specific requirements:
  - Multidose aqueous liquid dosage forms such as oral syrups or multidose eye drops and injectables will need suitable preservatives
  - Oral syrups, suspensions, chewable tablets and so on will need suitable sweeteners and flavouring agents as well as colorants to enhance palatability of the drug product so that patient acceptance is ensured
  - Moisture or oxidation labile drug substances will need inclusion of appropriate stabilisers in order to ensure stability of the product; for example, aspirin needs excipients with low amount of free moisture or excipients that can act as internal desiccant; simvastatin tablets need antioxidants such as ascorbic acid and butylated hydroxyanisole to stabilise the tablets
  - Products for local site applications such as topical, eye drops/ear drops and nasal drops need to be nonirritant and suitably designed for patient comfort and acceptance

Over the last two decades, the regulators and pharmaceutical industry have come together to radically change the approach towards development of drug substances and drug products using the quality by design (QbD) approach. Most of these changes (as explained in Chapter 2) are targeted to the use of scientific, risk-based framework for the development, manufacture, quality assurance and life cycle management of pharmaceutical drug substances and drug products to support innovation and efficiency.

4.3 QUALITY BY DESIGN

4.3.1 INTRODUCTION TO QbD

The pharmaceutical QbD approach to product development requires identification of the product characteristics that are critical for assuring efficacy and safety for the patient. These patient-centric characteristics are translated into the critical quality attributes (CQAs) of the drug product, and the product development process establishes the relationship between the formulation/manufacturing process variables and the CQAs. QbD is a concept that was first developed by quality pioneer Dr. Joseph Juran. Juran was one of the first to write about the cost of poor quality (Juran 1992).

This was illustrated by his 'Juran trilogy', an approach to cross-functional management, which is composed of three managerial processes:

- *Quality planning*: Quality should be designed into the process to ensure delivery of the product that consistently meets the needs of your customer
- *Quality control*: A system implemented and maintained to ensure that the process used to make the product remains in a state of statistical control
- *Quality improvement*: A culture should be established that rewards the continual improvement of the process/product

The root cause of all quality problems with any product or process is most often the variations in inputs causing a variable output. This variation drives risk, quality, costs and eventually customer satisfaction. Traditionally, pharmaceutical products are manufactured with a fixed process, and therefore, when the inputs are variable, it is evident that output will also be variable. QbD can be considered as a process for identifying and controlling variability that is holistic and systems based. Figure 4.2 depicts this paradigm shift in the approach towards product development.

Chapter 2 explains the goals/advantages of integrating the QbD concept into the development program for pharmaceutical products. It also discusses the key elements of this process and explains how the Quality Risk management (ICH Q9) and Pharmaceutical Quality System (ICH Q10) principles work in tandem with QbD throughout the product life cycle to provide an ecosystem for continuous control and improvement in product quality (ICH 2005, 2008).

Usage of the science-based and risk-based approach results in a deeper understanding of a product and its manufacturing process, and this helps companies develop a robust and more efficient manufacturing process. Further, this enhanced understanding should also create opportunities for scientific rationale for supporting flexible regulatory options and other business gains. Companies may wish to apply elements of this approach selectively on the basis of their business strategy for a particular product (ICH 2005, 2008, 2009; ISPE 2010, 2011a,b; Yu 2008; Van Buskirk et al. 2014; Yu et al. 2014).

Figure 4.3 depicts a conceptual application of QbD through a product's life cycle. In practice, product and process development and continual improvement are not

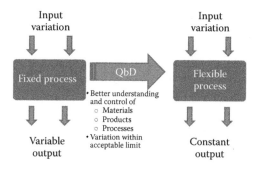

FIGURE 4.2 Relationship of QbD with variability.

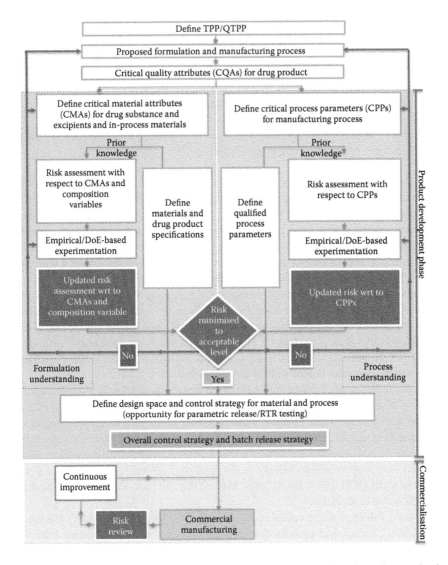

FIGURE 4.3 **(See colour insert.)** The iterative nature of QbD applied through a product's life cycle.

linear processes. Development and continual improvement processes usually consist of several parallel activities and are typically iterative, and this iterative nature is represented in this figure as cyclic arrows.

4.3.2 Target Product Profile and Quality Target Product Profile

The role of a Target Product Profile (TPP) is to serve as a tool for 'quality planning' for the drug product with 'the end in mind', that is, a summary of the drug development program described in the context of prescribing information goals. When a dosage form for a new drug candidate is being designed, initially a TPP is drawn up from the inputs on the desired therapeutic profile from the clinical research team. The TPP enumerates the product attributes that need to be designed into the product so that it can attain the target profile. At this stage, it is often difficult to nail down as to which attributes are exactly linked to the quality of the product. As development progresses and more experimental data are generated, the relationship between the product attributes, target profile and efficacy and safety in patients is better understood and hence the TPP evolves into the Quality Target Product Profile (QTPP). QTPP is a natural extension of TPP for product quality. A QTPP relates to the quality of a drug substance or drug product that is necessary to achieve a desired therapeutic effect. QTPP is a predetermined prospective summary of the characteristics of the drug product that will ideally be essential to ensure the desired quality with respect to safety and efficacy of the product. The QTPP for a new product at the start of its development is likely to be qualitative or semiquantitative and be reflective of the patient needs. These predetermined QTPPs evolve during the life cycle – during drug development and commercial manufacture – to incorporate new knowledge, as is warranted by ongoing clinical studies such as dose effect and toxicology data, identification of new patient needs and so on.

QTPP forms the basis of design for the drug product and could include the intended use in clinical setting, route of administration, dosage form and delivery system, dosage strengths, container closure system, therapeutic moiety release or delivery and attributes affecting pharmacokinetic characteristics (e.g. dissolution and aerodynamic performance) appropriate to the drug product dosage form being developed, drug product quality criteria (e.g. purity, sterility, stability and drug release) appropriate for the intended marketed product and requirements of the healthcare professionals, regulatory authorities and internal customers.

Table 4.2 gives an example of how the QTPP of an immediate-release tablet can be correlated to its safety and efficacy requirements (ICH 2009; ISPE 2010, 2011a,b).

Table 4.3 gives an example of how a TPP for a hypothetical extended-release dosage form for an antidiabetic product that is currently marketed as an immediate-release product could be arrived at.

This TPP will further evolve during development, and with a better understanding of the product substantiated with scientific data, it will evolve into the QTPP for the product.

A product TPP/QTPP provides guidance for correlating the product manufacturing process, input materials and formulation to the therapeutic outcome in patients.

4.3.3 Critical Quality Attributes

Identification of the CQAs of the drug product is the next step in product development. CQAs are those attributes of the drug product or drug substance that assess the quality in alignment with the predetermined QTPP. They link the product quality with its desired therapeutic performance. A CQA is a physical, chemical, biological or microbiological property or characteristic of the finished drug product that should be within an appropriate limit, range or distribution to ensure the desired product quality. The quality attributes of a drug product may include identity, assay, content uniformity, degradation products, residual solvents, drug release or dissolution,

TABLE 4.2

QTPP Correlation to Safety and Efficacy Requirements for a Tablet

TPP Element	Target Based on Patient-Centric Approach	Corresponding CQA
Dose	30 mg	Identity, assay, uniformity of content
Physical properties	Colour, shape, size, no off taste or odour	Appearance, size and shape, hardness
Patient efficacy	PSD of drug substance that ensures bioperformance and robust manufacturing	Acceptable PSD of API, in vitro drug release
Patient safety	Chemical purity – degradant and impurities below ICH limits or qualified	Related substances control in drug substance, control of manufacturing process and packing to ensure control at release
Chemical and drug product stability	2 years shelf-life (worldwide = 30°C), no change in bioperformance and degradation within qualified limits up to expiry date	Appropriate packing to control exposure of the product to ensure that degradation and dissolution are within limits up to the expiry date

moisture content, microbial limits and physical attributes such as colour, shape, size, odour, score configuration and friability.

Although the CQAs are deemed product quality or process relevant, it is important to understand how they can potentially affect patient therapeutic outcomes when establishing 'acceptance criteria'. In this way, the QTPP guides the selection of appropriate CQAs and their specifications that are necessary to ensure the desired therapeutic effect in patients. More recently, the emphasis has been to progress towards performance-based quality specifications that can give confidence about the performance of the product in actual clinical conditions. The Food and Drug Administration (FDA) guidance on 'Size of beads in drug products labeled for sprinkling' and that on 'Size, Shape, and Other Physical Attributes of Generic Tablets and Capsules' are examples of critical attributes that have been identified as essential for clinical performance of the drug product.

The iterative nature of product and process development mandates that due attention is given as early as possible in the development to distill down the QTPP into potential product CQA and to move towards identifying those critical for performance and establish acceptance criteria for the same (ICH 2008; ISPE 2010, 2011a,b).

4.3.4 Risk Assessment

Once all the components of the QTPP have been laid down and CQAs have been defined, it is necessary to perform a preliminary risk assessment before starting any

TABLE 4.3

TPP for Extended-Release Tablets for an Antidiabetic Product

TPP Element	Target	Product Attribute
	Physical Parameters	
Dosage form	Tablet/capsule	Appearance, shape and size, hardness
	Chemical and Microbiological	
Dosage strengths	250 and 500 mg (similar to existing immediate-release dosage form)	Identity, assay and uniformity of dose
Microbial quality	No contamination of the product with pathogenic organisms	Microbial enumeration test and absence of pathogens
	Biological	
Dosage administration	Once a day	–
Indications and usage	For the treatment of diabetes similar to existing immediate-release product but once a day dosing	–
PK target – efficacy and safety	• The C_{max} should be above the minimum effective concentration in the blood and below the minimum toxic concentration of the drug • The T_{max} should be extended significantly as compared to the immediate-release tablets • The T_{mec} should not be significantly different from the immediate-release tablet • The area under the drug plasma concentration curve should not be significantly different from the immediate-release tablet	In vitro drug release test showing a suitable extended-release profile
Adverse reaction profile	Improved or similar to current marketed product	–
Patient safety	Product should be stable in an appropriate selected pack	Degradation products during release and shelf-life
	Packing and Storage Related	
Container closure system and storage condition	Appropriate multidose and unit dose pack based on product characteristics	Compatibility and stability test supporting shelf-life

(Continued)

TABLE 4.3 (CONTINUED)
TPP for Extended-Release Tablets for an Antidiabetic Product

TPP Element	Target	Product Attribute
Administration and concurrence with labeling	Similar to current marketed product	–
Use in specific population	Similar to current marketed product	–
Overdose	Similar to current marketed product	–
Nonclinical toxicity	Similar to current marketed product	–

development work. In any given product, the input variables relating to materials and processes that can have an impact on product quality are numerous, and it is not possible to study and understand the impact of all the variables. Risk assessment is a tool used to prioritise the input variables that can cause the highest risk to product quality so that development studies can focus on the high-risk variables. A suitable risk management tool, such as preliminary hazard analysis (PHA), is used to identify potential CQAs and classify them based on a severity rating as varying levels of 'criticality'. On the basis of the risk ranking, experimental studies are then prioritised by allocating resources given to understanding the impact of material attributes and process parameters on a particular CQA. It is very important to understand that risk assessment needs to be carried out at different stages in product development as further knowledge is gained and needs to be continued till all the risks are minimised (ICH 2005; ISPE 2010, 2011a,b).

4.3.5 Product Design and Understanding

The objective of product design and development is to develop a robust product that can meet the therapeutic objectives (QTPP), can be manufactured consistently on a commercial basis and meet all quality and regulatory requirements and maintain its quality over the shelf life.

To design and develop a robust product, the development team must give serious consideration to the following:

- Physical, chemical and biological properties of the drug substance such as particle size distribution and morphology, melting point, polymorphic form, hygroscopicity, pK_a, chemical stability in solid state and solution, membrane permeability, bioavailability and so on.
- Excipients to be used and their functional role in the drug product, safety limits, compatibility and their physical and chemical properties. Many

excipients are also available from different vendors and as different grades from each vendor. Selection of the excipients should be carried out using a scientific approach to select the best possible option for the drug product.

Formulation optimisation studies are essential in developing a formulation that is robust and can be manufactured commercially such that it will always conform to the quality specifications that have been laid down without frequent failures. These studies provide important information and understanding on the following:

- Identification of critical material attributes (CMAs) of the drug substance, excipients and in-process materials
- Correlating the CMAs with the attainment of the CQAs
- Developing control strategies for the product revolving around control on the attributes of the input materials and in-process materials

The number of optimisation studies conducted is not as important as the relevance of the studies and the knowledge gained through these to allow for developing a robust and quality product. Design of experiments is a statistical experimental tool that is often used to carry out multivariate experiments. It helps in understanding the impact of variables both singly and in combination on the product CQAs and helps define the design and control space (for details, see Chapters 7 and 9).

Drug substances, excipients and in-process materials can have many CMAs. A CMA is a physical, chemical, biological or microbiological attribute of an input material that should be within an appropriate limit in order to achieve the desired quality of that input material. Further, these CMAs of the input materials are those which through the development studies are shown to be necessary for achieving the desired CQAs of the drug product.

Since there are many attributes of the drug substance and excipients that could potentially affect the CQAs of the intermediates or the finished drug product, it is not practical to study the impact of all the attributes of all the input materials, and therefore a risk assessment is a very valuable tool in prioritising those material attributes that are expected to have a significant effect on the product and hence warrant further study and optimisation. A material attribute is critical when a realistic change in that attribute can lead to a significant impact on product quality. Understanding the linkage of CMAs of input materials to the CQAs is one of the achievements of a good product design and development (ICH 2005, 2008, 2009; ISPE 2010, 2011a,b).

4.3.6 Process Design and Understanding

A manufacturing process is made up of unit operations (in batch mode or in continuous processing mode) that lead to the manufacture of the finished drug product. Typically for a solid dosage form, the unit operations include mixing, milling, granulation, drying, compression and coating. A process is supposed to be well understood when (a) all critical process variables are identified and explained and (b) the process controls enable management of this variability so that product quality attributes can be accurately and reliably achieved.

Process robustness is the ability of a process to deliver a product meeting all desired CQAs overcoming the variability in input materials and process. The effects of variation in process parameters and material attributes (especially in-process materials) are investigated in process evaluation studies. The analysis of these experiments helps in identifying the critical process parameters (CPPs) that could affect the drug product CQA and establishes limits for these CPPs (and CMAs of in-process materials) within which the quality of the product is assured (for details, see Chapters 8 and 9).

Process understanding/evaluation studies (USFDA 2011) include the following steps:

- Identification of all possible known process parameters that could affect the performance of the process and identification of potentially high-risk parameters using risk assessment and scientific knowledge (including prior knowledge)
- Establishing levels/range of these identified high-risk process parameters using empirical or statistical experimentation
- Analysing the data to determine if the process parameter is critical and, if possible, linking the CMA and CPP to the CQA
- Developing a control strategy and setting suitable acceptance ranges for critical parameters; for noncritical parameters, using the studied range as the acceptable range

In a nutshell, the ultimate goal of QbD is to reduce product variability and defects, thereby enhancing patient efficacy and safety (ICH 2005, 2008, 2009; ISPE 2010, 2011a,b).

4.4 PROCESS FOR BUILDING AN EFFECTIVE PRODUCT DEVELOPMENT STRATEGY

The need for developing a robust process to drive the development of the product development strategy cannot be more emphasised and is the key to achieve sustained growth for all companies. There are several challenges in developing and executing a product strategy, the major ones being the following:

- The product development strategy is very important in ensuring the success of the product. If the strategy is not well thought out and well planned, then the execution of the actual product development also suffers as detours and reworks often result out of a flawed strategy.
- The market environment is complex and affected by external variables such as competition, reimbursements from insurance companies, regulations and so on. The strategy has to be developed and executed within this environment.
- A realistic product strategy can only be developed if there is contribution from all stakeholders from within the company such as quality, scale-up and manufacturing, engineering, marketing, packaging and so on. The challenge is to get buy-in and agreement from all of them before finalising the strategy although they may all have conflicting goals and requirements.

Every company needs a well-defined process for developing the product strategy, which is repeatable, is measurable, has definable checkpoints and involves all the stakeholders in the process. The strategy is not static but dynamic and must have a mechanism to be updated and adjusted in response to the knowledge and understanding gained during the development and the external and internal factors that could change during the execution phase.

The product development strategy may be very simple or very complex depending on the product involved, the internal resources available, competition, intellectual property landscape, time factor and so on; therefore, it is very case dependent, and the 'one shoe fits all' approach cannot work.

Figure 4.4 depicts the process and the major steps that are usually followed in most companies to arrive at a suitable strategy for developing a pharmaceutical product. It is also normally expected that a suitable team of subject matter experts (SMEs) will be assembled for discussion and brainstorming sessions to arrive at the strategy. Typically, the team includes product development experts along with SMEs from clinical, safety and drug metabolism, regulatory, quality, manufacturing, process engineering and sourcing functions. Sometimes, if the product requires some specific technology or expertise, then SMEs with that background are also invited on the team, for example, platform technologies such as lyophilised oral solids or soft gelatin capsules or special devices.

The objective of this exercise is to arrive at a document that systematically evaluates the available prior knowledge and understanding of the product requirements from the patient's perspective. Based on this, it defines the QTPP and CQAs of the product and a suitable formulation and process for manufacture of the product and, using risk assessment, identifies the material attributes and process parameters that can be expected to have an effect on the CQAs. The strategy document is usually approved by quality and regulatory experts before initiation of the development work (ICH 2005, 2008, 2009; ISPE 2010, 2011a,b).

4.4.1 Prior Knowledge

The first step is the documentation of prior knowledge available for a product to be developed. Prior knowledge is the knowledge that stems from previous experience. It can come from literature, company experience, an individual's experience, previous work on the product or similar products (e.g. earlier experience with the same drug substance used for development of another dosage form). Prior knowledge also includes experience in some platform technologies, for example, multiparticulate coatings or dry granulations, and the team can use learning from earlier extensive knowledge of previous products and processes for predicting potential CPPs from these unit operations. Prior knowledge along with risk assessment is used to arrive at a list of potential CQAs, material attributes and process parameters to be studied in depth during development.

The team usually tries to gather the most relevant knowledge and information available on the drug substance and its physicochemical, biopharmaceutical and other important properties as part of this exercise. The more the information is available, the greater are the chances of evolving a realistic and achievable formulation

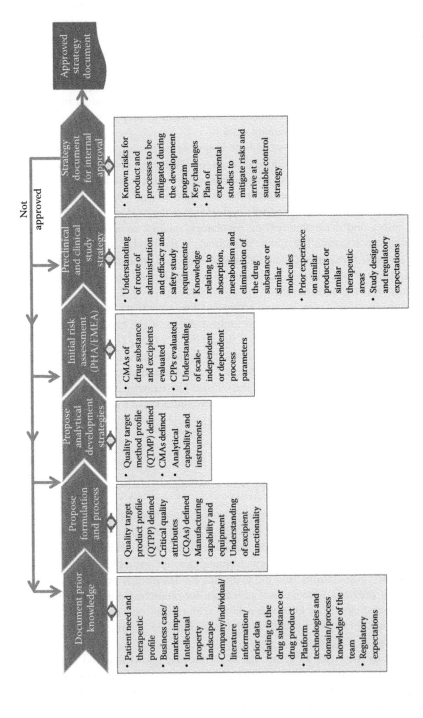

FIGURE 4.4 Process for building the product development strategy.

development strategy. For mature products, wherein the drug substance is already approved for human use, this information (drug substance and drug product) is easily available from literature and from the technical file and studies undertaken by the drug substance manufacturer. For investigational drug substances, this knowledge is gained as a part of the early stage research and Phase 1 studies, and this can be used to help in arriving at a strategy for the dosage form development although the final dosage forms and doses may not be known at that time, and it evolves as more studies are conducted.

As an example, Table 4.4 illustrates the information that would normally be required and compiled for drawing up a strategy for the product using ibuprofen as an example.

Other valuable information that could help in formulation design and linking the CMAs and CPPs to CQAs and clinical performance could be permeability of the drug substance and possibility of efflux occurring, solubility in simulated gastric and intestinal fluids (e.g. fasted state simulated intestinal fluid and fed state simulated intestinal fluid), effect of surfactant on solubility, stability of drug substance in solid state and aqueous solution, solid state form and polymorphs and surface properties, for example, wettability, hydrophilicity, porosity, surface area and so on.

Ibuprofen is a BCS Class II drug, and based on literature knowledge, it is well understood that, for such drugs, there exists a possibility of the formulation and process having a significant impact on bioavailability. It is also known that the particle size of the drug substance plays a pivotal role in bioavailability, and as particle size decreases, bioavailability increases. This sort of understanding and insight helps build the risk profile for this formulation and process and drives the experimentation during development.

During this phase, an in-depth study of the intellectual property landscape is also required to understand the boundaries within which freedom to operate exists for the company. In consultation with the business and marketing team, strategies are also worked out for the protection of intellectual property expected to be generated during the development to ensure market access and if possible exclusivity.

4.4.2 Proposed Formulation and Process

4.4.2.1 Defining the QTPP

Establishing a QTPP for the product is the first step towards proposing a formulation and process that can achieve this target. General concepts and principles governing definition of TPP/QTPP are explained in Section 4.3.2. Table 4.5 elaborates on when a QTPP can be initially defined, and this depends on the available basic pharmaceutical information (documented as prior knowledge) available to draw some realistic targets for the development.

A further iteration and fine tuning of the QTPP results from the development study and usually after a proof of concept study, it is finalised prior to development of the proposed commercial dosage form.

Table 4.6 depicts a typical QTPP for immediate-release, film-coated tablets of ibuprofen, which is drawn up on the basis of prior knowledge (Table 4.4) available on

TABLE 4.4
Physicochemical and Biopharmaceutical Properties of Ibuprofen

Attribute	Description
	Physicochemical Properties
Structural formula	
Chemical name	(RS)-2-(4-Isobutylphenyl) propionic acid
Molecular formula	$C_{13}H_{18}O_2$
Molecular weight	206.3
Description	A white to almost white, crystalline powder or colourless crystals, slightly waxy with a strong and characteristic taste. The drug also produces a burning sensation or 'kick' in the back of the throat when swallowed.
Melting point	74°C–77°C
Solubility	Practically insoluble in water, freely soluble in acetone, in methanol and in methylene chloride. It dissolves in dilute solutions of alkali hydroxides and carbonates.
pH solubility profile	Ibuprofen has low solubility in acidic pH below 4 and it increases proportionally in alkaline pH from 4 to 7.4.
pK_a	Acidic compound, pK_a 4.5–4.6
Hygroscopicity	Nonhygroscopic
Particle size distribution (PSD)	Available in various PSD grades from the manufacturer with average particle size ranging from 20 to 200 microns.
Flow properties	Dependent on the PSD, but most have Hausner's ratio in the range of 1.3 to 1.45 depicting poor flow properties.
Bulk density	Apparent bulk density ranges from 0.3 to 0.5 g/ml and tapped density between 0.45 to 0.70 g/ml depending on PSD grade.
Log P	3.68
BCS class	Class II

(Continued)

TABLE 4.4 (CONTINUED)
Physicochemical and Biopharmaceutical Properties of Ibuprofen

Attribute	Description
Therapeutic category	Nonsteroidal anti-inflammatory drug, analgesic
Details of known impurities as per Ph. Eur. 5.0	Specified impurities: A, B, C, D, E Other detectable impurities: F, G, H, I, J, K, L, M, N, O, P, Q, R

Biopharmaceutical Properties

Mechanism of action	Acts by inhibiting the activity of cyclooxygenase enzyme, which is involved in the formation of prostaglandins that cause inflammation, swelling, pain and fever.
Absorption	Ibuprofen is rapidly absorbed after administration and is rapidly distributed throughout the whole body.
T_{max}	Maximum plasma concentration is reached 45 min after ingestion if taken on an empty stomach. When taken with food, peak levels are observed after 1–2 h.
Food effect	When ibuprofen is administered immediately after a meal, there is a reduction in the rate of absorption but no appreciable decrease in the extent of absorption. The bioavailability of the drug is minimally altered by the presence of food.
Volume of distribution	Apparent volume of distribution is 0.14 L/kg.
Protein binding	99% of ibuprofen is protein bound. The high protein binding of the drug should be borne in mind when prescribing ibuprofen together with other protein-bound drugs that bind to the same site on human serum albumin.
Metabolism	About 90% of ibuprofen is metabolised to two major metabolites (A and B), and these are as follows: Metabolite A: (+) 2-4'-(2 hydroxy-2-methylpropylphenyl) propionic acid Metabolite B: (+) 2-4'-(2-carboxypropylphenyl) propionic acid Both metabolites are dextrorotatory and are devoid of anti-inflammatory and analgesic activity.
Excretion	Kidney is the major route of excretion. 95% of the drug was excreted in the urine within 24 h of a single dose of 500 mg.
Half-life	Plasma half-life of ibuprofen is in the range 1.9 to 2.2 h.

(Continued)

TABLE 4.4 (CONTINUED)

Physicochemical and Biopharmaceutical Properties of Ibuprofen

Attribute	Description
	Marketed Product Information
Brand name	Boots Ibuprofen Caplet 200 and 400 mg
Manufacturer	The Boots Company PLC
Label claim	Each film-coated tablet contains ibuprofen 200/400 mg
Dosage form	Film-coated tablet
Strength	200/400 mg
Pharmacokinetics	Maximum plasma concentrations are reached 45 min after ingestion if taken on an empty stomach. When taken with food, peak levels are observed after 1 to 2 h. These times may vary with different dosage forms. The half-life of ibuprofen is approximately 2 h.
Therapeutic indications	Nonsteroidal anti-inflammatory drug with analgesic, antipyretic and anti-inflammatory action
Inactive excipients	Microcrystalline cellulose, croscarmellose sodium, lactose, colloidal silicon dioxide, sodium lauryl sulfate, magnesium stearate, hypromellose, french chalk, titanium dioxide (E171)
Storage	The drug should be stored out of reach of children, protected from light, at a temperature not exceeding 30°C.
Pack and shelf-life	36 months – Amber glass bottle 36 months – Aluminium blister (cold form aluminium) 36 months – Aluminium blister (PVC/PVDC and aluminium foil) 24 months – HDPE bottle

the drug substance and drug product already approved for human consumption and available in the market (ICH 2005, 2008, 2009; ISPE 2010, 2011a,b).

4.4.2.2 Proposing a Formulation and Process

In this case, as it is a mature product already marketed for human consumption, the QTPP specifies that a tablet is required and so alternative dosage forms are not considered for development. A film-coated tablet is needed as it helps in swallowing and also covers up the acidic and bitter taste of ibuprofen, which is reported to cause a burning sensation at the back of the mouth. However, in case of a new chemical entity based on the biopharmaceutical considerations elaborated in Section 4.2, various dosage forms (multiple) could be proposed and studied as part of the exploratory

TABLE 4.5

Development Stage When QTPP Can Be Initially Defined

Type of Drug Product to Be Developed	Development Phase at Which the QTPP Can Be Initially Defined
New chemical or biological entity	Recommended to be defined after the animal and Phase 1 clinical (first in human) studies data are available and before commencement of Phase 2 clinical studies
New dosage form of a drug substance already approved for human consumption Dosage form and drug substance already approved for human consumption	Recommended to be defined at the project conceptualisation phase as information of the drug substance properties and dosage form requirements are usually known at this stage

studies, which could later be narrowed down on the basis of the experimental data generated and risk assessment at that stage.

Film-coated tablets are easier to swallow, and the shape, size, imprints and colour can be designed such that it helps the patient for identification and improves compliance. Ibuprofen is a low-melting, hydrophobic substance with poor flow characteristics because of its micronised particle size, which is required to ensure drug release. Further, the doses are 200 and 400 mg, and there is a market need for a compact dosage form with minimum tablet weight resulting in >50% w/w drug load in the tablets. Considering these properties, a wet granulation process is considered low risk and is best suited for this type of product, and the company can draw upon its knowledge base of similar products and processes to propose a formulation and process for this development. The availability of equipment suitable for a wet granulation process further confirms this choice for the product.

The excipient selection is based partly on the information on similar marketed products and known functionality of the excipients gained by using these excipients in other products. Microcrystalline cellulose (MCC) is proposed as a filler as it is known to have good compressibility; sodium starch glycolate is proposed as the disintegrant as fast disintegration is the key to achieving the desired drug release from the tablets. Copovidone is selected as the binder as it is a strong binder and ibuprofen is known to have sticking problems and low compressibility, and with a drug load of >50%, an efficient binder is essential for good compressibility without hindering the drug release from the tablets. Colloidal silicon dioxide and magnesium stearate are proposed to be added as glidants, antiadherents and lubricants for enabling a smooth compression process (Gibson 2009; EMEA 2014).

For each excipient, there are various grades offered by many suppliers, and on the basis of the anticipated need of the product, prior experience and advice from suppliers, the grades are selected. For example:

- MCC is available as different particle size grades based on particle size, porosity, flow, water content, bulk density and compressibility from FMC Biopolymer (FMC Biopolymer 2015); the most commonly used are Avicel

TABLE 4.6

QTPP for Film-Coated Tablets of Ibuprofen

QTPP Element	Target/Proposed Specification		Justification	Reference
	Physical Parameters			
Dosage form	Immediate-release film-coated tablet		Pharmaceutical equivalence to marketed product	Request from business team based on marketed product
Route of administration	Oral			
Indication	Analgesic			
Dosage form strength	400, 200 mg			
Appearance	White, oval/oblong, compact tablet		Manufacturability, patient acceptability for ease of swallowing and marketing need	US CDER guidance: size, shape and other physical attributes of generic tablets and capsules
Tablet weight	400 mg strength	200 mg strength	Manufacturability	
	Not more than (NMT) 750 mg	NMT 375 mg		
Length	NMT 18 mm	NMT 15 mm		
Breadth	NMT 8 mm	NMT 5 mm		
Hardness	100–200 N	50–150 N		
Thickness	6.2 ± 0.4 mm	4.2 ± 0.4 mm		
Disintegration time	NMT 15 min		Meets the pharmacopoeial limit	Ph. Eur.

(Continued)

TABLE 4.6 (CONTINUED)
QTPP for Film-Coated Tablets of Ibuprofen

QTPP Element	Target/Proposed Specification	Justification	Reference
Friability	NMT 1% w/w	Meets the pharmacopoeial limit	Ph. Eur.
Stability	At least 24 months shelf-life at room temperature	Stability condition as per ICH Q1(A)	ICH
Photo stability study	Should be photo stable	Stability condition as per ICH Q1(B)	ICH
Chemical and Microbiological			
Identification	Positive for ibuprofen	Needed for patient safety and clinical effectiveness	ICH Q6A and 3AQ11a requirements
Assay of ibuprofen	95%–105% of label claim	Needed for clinical effectiveness and for pharmaceutical equivalence	Ph. Eur.
Uniformity of dosage units	Should meet the requirements as per Ph. Eur.	Targeted for consistent clinical effectiveness	Ph. Eur. General Chapter 2.9.40. Uniformity of Dosage Units

(Continued)

TABLE 4.6 (CONTINUED)
QTPP for Film-Coated Tablets of Ibuprofen

QTPP Element	Target/Proposed Specification	Justification	Reference
Related substances	Impurity J: NMT 0.15% Impurity L: NMT 0.15% Impurity M: NMT 0.15% Any individual unspecified impurity NMT: 0.10% Total impurities: NMT 0.5%	To ensure patient safety and needed for clinical effectiveness Individual known impurity limits will be finalised on the basis of stability data of drug product	ICH Q3B(R2) guideline on impurities in new drug products Nomenclature of impurities as per ibuprofen monograph in Ph. Eur. 5.0
Drug release QC media	NLT 80% (Q) in 60 min in pH 7.2 phosphate buffer, paddle, 50 RPM	For making the test and reference product bioequivalent	USP monograph for ibuprofen tablets
Water content	NMT 1.0% above the equilibrium moisture content calculated for the composition	Excess water can cause chemical degradation	—
Residual solvents	Based upon general chapter for residual solvents Ph. Eur. ⟨5.4⟩ Section 4	Needed for patient safety	Ph. Eur. ⟨5.4⟩
Microbial enumeration test i. Total aerobic microbial count ii. Total yeast and moulds count iii. Test for specified pathogenic microorganisms	NMT 1000 cfu/g NMT 100 cfu/g Should be absent	As per Ph. Eur. general chapter ⟨5.1.4⟩	Ph. Eur. 2.6.12 and 2.6.13

(Continued)

TABLE 4.6 (CONTINUED)
QTPP for Film-Coated Tablets of Ibuprofen

QTPP Element	Target/Proposed Specification	Justification	Reference
	Biological		
Pharmacokinetics	C_{max} T_{max} AUC 90% CI for % T/R of C_{max} and AUC	Should match marketed product Should match marketed product Should match marketed product 80%–125%	Bioequivalence requirement to show safety and efficacy of proposed product equivalent to individual reference products. Waiver request for in vivo study of 200 mg strength can be made based on (i) acceptable bioequivalence studies on the 400-mg strength, (ii) proportional similarity of the formulations across both strengths and (iii) acceptable in vitro dissolution testing of both strengths.	EMEA Guideline on the Investigation of Bioequivalence
	Packaging and Storage Related		
Container closure system	PVC/Alu (200 μ/25 μ) Blister OR PVDC coated PVC/Alu 40 gsm or 60 gsm/200 μ/ 25 μ blister based on the actual stability data obtained for the prototype.	For the stability of the product	Pack of marketed product
Storage condition	Store at a temperature of 15°C–25°C.	Should be similar or more stable than the marketed product and for the stability of the product.	

(Continued)

TABLE 4.6 (CONTINUED)
QTPP for Film-Coated Tablets of Ibuprofen

QTPP Element	Target/Proposed Specification	Justification	Reference
Administration/concurrence with labeling	Do not exceed 2400 mg total daily dose for ibuprofen. If gastrointestinal complaints occur, administer ibuprofen tablets with meals or milk.	Should be similar to marketed product.	Patient information leaflet of marketed products
Alternative methods of administration	None	Should be similar to the marketed product.	NA

Note: NA, not applicable; Ph. Eur., European pharmacopoeia.

PH101 for wet granulation process and Avicel PH102 and Avicel HFE102 for direct compression process.

- Povidones from BASF are available as normal Kollidon K30 and a special low peroxide and low formaldehyde content Kollidon K30LP, which is specifically used for drug substances prone to oxidative degradation such as simvastatin or rizatriptan (Signet 2015).

The proposed composition is elaborated in Table 4.7.

The selection of the primary pack can be quite an extensive exercise for a complex product. In this case, the proposed packing material for this product is a PVC-based blister pack as per the market requirement. The type of blister forming film, that is, plain PVC or moisture-protective PVDC-coated PVC, will be decided on the basis of stability studies of the product. Other more protective films (see Chapter 10) could also be evaluated if a need for additional moisture protection is felt necessary for the product.

The equipment and process proposed for this product are as depicted in Figure 4.5.

The potential CQAs for the product can now be easily defined on the basis of the proposed formulation and process selected.

TABLE 4.7
Proposed Quantitative Composition of Ibuprofen Tablets

Ingredient	Function	Quantity in mg/Tablet	
		200 mg Strength	400 mg Strength
Ibuprofen	Active	200	400
Microcrystalline cellulose (Avicel PH 101)	Filler/diluent	60–100	120–200
Sodium starch glycolate	Disintegrant	30–50	60–100
Copovidone (Kollidon VA64)	Binder	5–10	10–20
Colloidal silicon dioxide	Glidant	4–6	8–12
Magnesium stearate	Antiadherent, lubricant	1–4	2–8
Core tablet		300–370	600–740
HPMC-based readymade coating mixture (Opadry) coating agent		7–10	14–20
Coated tablets		307–380	614–760

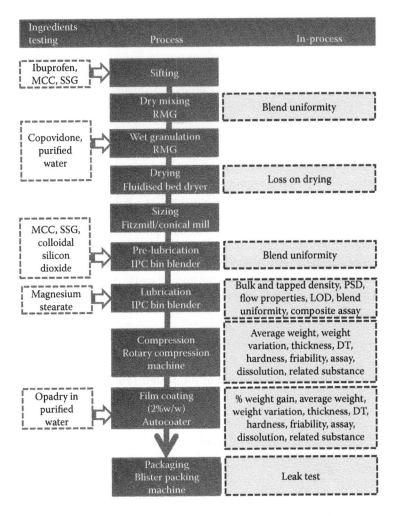

FIGURE 4.5 Proposed process flow chart with the equipment to be used.

4.4.2.3 Defining the CQA

On the basis of the enlisted targets in the QTPP and the proposed formulation and process, a list of potential CQAs is derived. An explanation of the concepts and principles used for defining CQAs is given in Section 4.3.3. As per ICH Q8 (R2), a CQA is defined as a physical, chemical or microbiological property or characteristic that should be within an appropriate limit, range or distribution to ensure the desired product quality. It further states that CQAs are generally associated with drug substance, excipient, intermediates (in-process materials) and drug product. Every attribute and parameter is critical as a product cannot be made without controlling each and every one of these. Therefore, there is a need to rank the criticality of the attributes based on the severity of the risk involved (ICH 2009).

Criticality of an attribute is primarily based on the *severity* of the risk of harm to the patient in the event that the product falls outside the acceptable range for that

TABLE 4.8

Scale of Severity Used to Assess Criticality of Potential CQAs

Scale	Situation
High severity	Adverse event or lack of efficacy or of impact on patient safety or efficacy is unknown/uncertain
Medium severity	Other CQAs that could cause batch failure even if no harm to patient, for example, customer complaints (not able to remove from pack, soft tablets, etc.)
Low severity	No measurable effect on patient

attribute. However, severity of impact is unlikely to change regardless of increased understanding. For this reason, severity and, in some cases, uncertainty are most important for assessing the criticality of the product quality attribute. Uncertainty is a risk factor to consider when there is no clear relationship between a potential CQA and harm to patient. Relatively, detectability or controllability does not influence the criticality of an attribute.

A list of potential CQAs for ibuprofen tablets is given in Figure 4.6, and these are further assessed with respect to their *criticality* based on their assessed risk to safety, efficacy and quality based on *severity*. The rating of severity is based on a qualitative scale where impact on patient in case of noncompliance of the CQA is assessed, as given in Table 4.8.

A justification based on prior knowledge for assessing each attribute is also provided for each CQA in Figure 4.6.

4.4.3 Proposed Analytical Strategies and Methods

The process for the development of analytical methods also benefits greatly from application of QbD concepts such that the output is of acceptable quality. Similarly, the life cycle validation and management concept for continual improvement in manufacture can also be applied to analytical methods. For analytical methods, the strategy can be applied in three stages: method design by defining the analytical target profile or quality target method profile that identifies the critical method attributes, method qualification by validating and confirming that the method is capable of meeting its design intent and continued method verification by gaining ongoing data and review to confirm that state of control is maintained. See Chapter 7 for more details.

4.4.4 Initial Risk Assessment

Once the formulation and process for the product to be developed have been proposed, it is essential to carry out a risk assessment of the proposed process so as to understand which formulation composition and unit processes bear the maximum risk to CQAs. This further guides the emphasis to be placed on the different

Drug product quality attributes	Target	Criticality	Justification
	Dried granules		
Appearance	White to off-white dried granules	No (Low)	Appearance of dried granules is not directly linked to the safety and efficacy. Therefore, it has low criticality.
LOD/water by KF	NMT 1.0% above the equilibrium moisture content calculated for the composition	Yes (High)	LOD/water content of the dried granules may affect the stability of the finished product. Hence, it is highly critical.
Lubricated blend			
Appearance	White to off-white granules with desired flow properties	No (Low)	Appearance of lubricated granules is not directly linked to the safety and efficacy. Therefore, its criticality is low.
LOD/water by KF	NMT 1.0% above the equilibrium moisture content calculated for the composition	Yes (High)	LOD/water content of the lubricated granules may affect the stability of the finished product. Hence, it is highly critical.
Bulk density and tapped density	0.3–0.6 g/ml	Yes (Medium)	Bulk density/tapped density will affect the compressibility of the blend. Hence, its criticality is medium.
Particle size distribution	To be decided based on the developmental data	Yes (Medium)	Particle size distribution of the lubricated blend will affect the flow characteristics and compressibility of the blend. Therefore, its criticality is medium.
Blend uniformity for Ibuprofen	(1) 90.0% to 110.0% of the label claim (2) RSD NMT 5.0%	Yes (Medium)	Dose uniformity is essential for efficacy but as drug content is >50% of blend, the uncertainty is not high. Hence, criticality is medium.
Blend assay for Ibuprofen	95.0–105.0% of the label claim	Yes (High)	Dose accuracy will determine the efficacy and hence, this is highly critical.

FIGURE 4.6   Potential CQAs and their ranking for ibuprofen tablets.   (*Continued*)

variables when planning the experimental study design and plan. The study designs need to be carefully drawn up to ensure optimum use of resources (time, equipment and materials) and to be assured that the results will be of great value in developing a relevant control strategy. In many formulations, there is an interaction between different variables such as material attributes, composition variables and CPPs, and by using empirical/univariate experiments, it is not possible to confirm these interactions. In the case of very complex products or processes, the univariate approach may mislead the team into fixing operating ranges such that it will make manufacture

Drug product quality attributes	Target		Criticality	Justification
Core tablets				
Description	White to off-white, round/oval/oblong tablets		No (Low)	Appearance of core tablet is not directly linked to the safety and efficacy. Therefore, its criticality is low.
Average weight	400 mg	200 mg	Yes (High)	Average weight will have an impact on assay and uniformity of dosage units. Hence, it is highly critical.
	650–750 mg	325–425 mg		
Size Length	17.0–18.0 mm	14.0–15.0 mm	No (Low)	Tablet size is not directly linked to the safety and efficacy. Hence, its criticality is low.
Breadth	7.0–8.0 mm	4.0–5.0 mm		
Thickness	6.2 ± 0.4 mm	4.2 ± 0.4 mm		
Uniformity of dosage units for Ibuprofen	Should meet the requirements as per Ph. Eur. 2.9.40, Acceptance value ≤15		Yes (Medium)	Dose uniformity is essential for efficacy but as drug content is >50% of blend, the uncertainty is not high. Hence, criticality is medium.
Weight variation	Average weight ±5%. Should meet the requirements as per Ph. Eur. 2.9.5.		Yes (Medium)	Weight variation will have an impact on assay and uniformity of dosage units. But it is controllable, hence, criticality is medium.
Disintegration time	NMT 15 minutes		Yes (Medium)	Disintegration time will have an impact on drug release. Hence, criticality is medium.
Dissolution of Ibuprofen	NLT 80% (Q) in 60 minutes in pH 7.2 Phosphate Buffer, Paddle, 50 RPM		Yes (High)	Ibuprofen is a low soluble drug which may affect its release characteristics. Failure to meet the dissolution specification may impact bioavailability. Both formulation and process variables affect the dissolution profile. This CQA will be investigated throughout formulation and process development. Hence, it is highly critical.

FIGURE 4.6 (CONTINUED)   Potential CQAs and their ranking for ibuprofen tablets.

(Continued)

of the product difficult and even then still lead to batch failures. However, in the early screening studies and in simple formulations where the interactions are known to be absent, empirical/univariate approaches can still be used.

During development, experiments are usually conducted as a series of experiments, and each of these is an attempt to reduce risk and the output is evaluated by the risk assessment team whether the risk can be accepted. If the risk is accepted,

Drug product quality attributes		Target	Criticality	Justification
Coated tablets				
	Appearance	White, Round/oval/oblong, compact, un-scored tablet	No (Low)	Appearance is not directly linked to the safety and efficacy. Therefore, its criticality is low. The target is set to ensure patient acceptability.
Physical attributes	Size – 400 mg strength	Length: 17.0– 8.0 mm Breadth: 7.0–8.0 mm	No (Low)	The dose is 400 and 200 mg. The shape and size should be amenable for ease of swallowing and patient acceptance. Does not impact efficacy or safety and hence, criticality is low.
	Size – 200 mg strength	Length: 14.0– 15.0 mm Breadth: 4.0–5.0 mm		
	Shape	Round/oval/oblong	No (Low)	
	Odour	No unpleasant odour	No (Low)	In general, a noticeable odour is not directly linked to safety and efficacy, but odour can affect patient acceptability. For this product, neither the drug substances nor the excipients have an unpleasant odour. Its criticality is low.
	Friability	NMT 1% w/w	No (Low)	Friability target as per pharmacopoeia assures a low customer complaint and has no impact on patient safety and efficacy. Hence, criticality is low.
Identification		Positive for Ibuprofen	No (Low)	Though identification is critical for safety and efficacy, this CQA can be effectively controlled by the quality management system and will be monitored at drug product release. Formulation and process variables do not impact identity. Therefore, this CQA will not be discussed during formulation and process development studies.
Assay for Ibuprofen		95 to 105% of the label claim	Yes (High)	Assay variability will affect safety and efficacy. Process variables may affect the assay. of the drug product. Thus assay will be evaluated throughout the product and process development. Criticality is high.

FIGURE 4.6 (CONTINUED) Potential CQAs and their ranking for ibuprofen tablets.
(*Continued*)

Drug product quality attributes	Target	Criticality	Justification
Dissolution of Ibuprofen	NLT 80% (Q) in 60 minutes in pH 7.2 Phosphate Buffer, Paddle, 50 RPM	Yes (High)	Ibuprofen is a low soluble drug, which may affect its release characteristics. Failure to meet the dissolution specification may impact bioavailability. Both formulation and process variables affect the dissolution profile. This CQA will be investigated throughout formulation and process development. Hence, it is highly critical.
Related substances	Impurity J: NMT 0.15% Impurity L: NMT 0.15% Impurity M: NMT 0.15% Any individual unspecified impurity NMT: 0.10% Total Impurities: NMT 0.5%	Yes (High)	Impurities have direct impact on safety for patient. They will be monitored based on ICH identification and qualification threshold. This CQA will be investigated throughout development and stability studies. Hence, it is highly critical.
Residual solvents	Based upon general chapter for Residual solvents Ph. Eur. <5.4> section 4. Limits of residual solvents; 4.2 for solvents to be limited; class 2 solvents and 4.3 solvents with low toxic potential; class 3 solvents	No (Low)	Residual solvents can impact patient safety. However, no organic solvents are proposed to be used in the drug product manufacturing process. The residual solvents in the drug substance and excipients are well within ICH limits. Therefore the formulation and process variables are unlikely to impact this CQA and so criticality is low.
Water	NMT 1.0% above the equilibrium moisture content calculated for the composition	Yes (High)	Water content may affect the stability of the finished product. Hence, it is highly critical.
Microbial enumeration test	Complies as per Ph. Eur. 2.6.12 and 2.6.13	No (Low)	Non-compliance with this test will impact patient safety. In this case, materials proposed to be used have low microbial load and the water content of the product is controlled which ensure low risk of microbial growth. Therefore this CQA is considered as low in criticality and will not be discussed in detail during the product development.

FIGURE 4.6 (CONTINUED) Potential CQAs and their ranking for ibuprofen tablets.

then a robust control strategy is further built around it and this is validated later on in the development. If the risk is not accepted, then more studies (as further iteration) are done to develop greater understanding and possible risk mitigation measures. In this way, the risk assessment tool is used throughout the development and life cycle of the product.

Risk assessment is usually carried out by a team of multidisciplinary experts, and deciding the constitution of the risk assessment team is very important as explained earlier in Section 4.4. Inputs from seasoned and experienced experts are very valuable in anticipating and judging the risks to the product from the formulation and process variables. In most companies, this is a formal process with team members being designated for each product and reports being finally signed off by regulatory and quality experts.

As per ICH Q9, risk is defined as the combination of the probability of occurrence of harm and the severity of that harm. The team identifies various hazards and brainstorms to analyse and assess their criticality by asking some simple questions such as

- What can go wrong?
- What are the chances (probability) that it can happen?
- If it happens, what are the consequences (severity)?

Risk assessment is a systematic process of organising information to support a risk decision to be made within a risk management process. It consists of the identification of hazards and the analysis and evaluation of risks associated with exposure to those hazards. Risk assessment is done using various tools as described in ICH Q9 and also in Chapter 1, and different teams use different tools for risk assessment (ICH 2005). In this chapter, we will discuss two of the most commonly used tools, that is, PHA and Failure Mode and Effects Analysis (FMEA), for risk assessment of the example of ibuprofen tablets for which the QTPP, CQAs and proposed formulation and process have been discussed in Section 4.4.2. In this exercise, PHA is used for risk assessment of the CMAs of the drug substance ibuprofen and the composition variables on the CQAs. FMEA is used for risk assessment of the unit processes involved in the manufacturing.

4.4.4.1 Preliminary Hazard Analysis

The first step in the risk assessment process involves an analysis of the potential hazards from the product so as to identify the different types of hazards. The probability of the occurrence of a potential risk situation/condition is termed a *hazard*. The risks attributed to the hazard are classified considering a worst-case scenario and the severity of the end condition. Potential risk/failure scenarios could be multiple and are a combination of probability and severity. A hazard analysis identifies the preliminary risk level, which is then assessed in depth with a precise prediction to arrive at the category of the risk. Some identified risks could be within the acceptable level. The main goal of this risk identification and assessment is to arrive at the best possible means of controlling or eliminating the risk.

The PHA method uses the Uncertainty (based on probability) and Severity scores to evaluate the level of risk in the composition. Different companies have different approaches to defining this scale and Figure 4.7 is an example of one such scale that could be used. Assignment of Uncertainty (U) and Severity (S) scores is done on the basis of the below-mentioned understanding of formulation development and general knowledge of science (ICH 2005).

In order to perform a risk evaluation, the risks need to be ranked in context of the overall risk as a combination of both factors – Uncertainty/Probability and Severity. One way to do this is to use a risk ranking based on the chart shown in Figure 4.8 to classify the risk as low, medium or high.

	Uncertainty/Probability level	
1	Full knowledge. The parameter set-point/range/limit has already been properly evaluated on this particular product.	Negligible. Less than 1 in 200 chance of failure.
2	Prior experience. The parameter set-point/range/limit has been properly evaluated on sufficiently similar product (same or related API, dosage form, formulation, process), or the evaluation is based on proven mechanistic or statistical models.	Low. Less than 1 in 50 chance of failure.
3	Published literature. The influence of the parameter can be inferred from literature (articles, patents, general knowledge of pharmaceutical technology.	High. Less than 1 in 20 chance of failure.
4	Insufficient knowledge. The effect of this parameter is not properly understood.	Very high. More than 1 in 20 chance of failure.
	Severity levels	
1	The parameter has no impact on patient.	No impact
2	Some impact on patient, but reversible.	Minor
3	Impact on patient, but not life threatening.	Major
4	High impact on patient which is irreversible and potentially life threatening.	Critical

FIGURE 4.7   Scale for Uncertainty/Probability and Severity.

		Uncertainty or probability			
		1	2	3	4
Severity	1	Low	Low	Med	High
	2	Low	Low	Med	High
	3	Low	Med	High	High
	4	Med	High	High	High

FIGURE 4.8   Risk ranking chart.

Going further, the risks identified as medium and high are focussed upon in the experimental study plan.

For each factor/parameter/attribute, the allocation of a risk rank as high, medium and low is arrived at using this chart after discussions and debate within the risk assessment team.

- If the resulting criticality is *low*, no follow-up action is mandatory. However, the ALARP (as low as reasonably possible) principle still applies.
- If the resulting criticality is *medium*, follow-up action is recommended, applying the ALARP principle.
- If the resulting criticality is *high*, follow-up action is mandatory. The risk must be reduced.

Initial Risk Assessment of Drug Substance (Ibuprofen) Material Attributes: Using this PHA tool kit in the current example for ibuprofen tablets, the impact of the material attributes of the drug substance ibuprofen on the CQAs of the tablets was evaluated and the assigned risk ranking is given in Figure 4.9.

The justification/explanation for the risk ranking is explained in Table 4.9.

Ibuprofen has low solubility in pH below its pK_a value and high solubility in pH above its pK_a value, which indicates pH-dependent solubility and thus would have an impact on the dissolution profile. Therefore, the particle size distribution of the drug substance may affect the dissolution of the tablet. Based on the above initial risk assessment, the particle size, solubility, assay and related substances of ibuprofen have been identified as high-risk material attributes that may affect product quality. Hence, these CMAs will need to be critically monitored during the development.

Initial Risk Assessment of Formulation Variables: Initial risk assessment of formulation variables was also done on the basis of PHA. Based on scoring and references from a pharmaceutical understanding, a risk assessment matrix (Figure 4.10) was arrived at for the composition of this product.

The justification and rationale for arriving at this risk ranking is explained in Table 4.10.

All of the identified formulation variables have a potential risk on product CQAs, especially on the dissolution. As ibuprofen has a pH-dependent solubility, selection of binder and its level would be expected to play a crucial role for uniform granule formation. Level of disintegrant and lubricant would have an impact on the disintegration pattern and further the dissolution profile. Level of surfactant would need to be studied to improve the wettability of the drug, which could enhance the drug release profile. The experimental plan will therefore incorporate studies to understand the impact of these variables alone and in combination with the process parameters to gain an understanding and finally suggest a control strategy.

4.4.4.2 Failure Mode and Effects Analysis

As explained in ICH Q9, FMEA provides for an evaluation of potential failure modes for processes and their likely effect on outcomes and product performance. Once failure modes are established, risk reduction can be used to eliminate, contain, reduce or control the potential failures. FMEA relies on product and process

Material Attributes of Drug Substance Ibuprofen

CQA of drug product	Particle size				Hygroscopicity/ photo stability/ oxidation				Assay				Related substance				Residual solvents				Flow properties				Solid state form				Solubility			
	U	S	Risk	Ref.*	U	S	Risk	Ref.*	U	S	Risk	Ref.*	U	S	Risk	Ref.*	U	S	Risk	Ref.*	U	S	Risk	Ref.*	U	S	Risk	Ref.*	U	S	Risk	Ref.*
Physical attributes	2	1	Low	1	2	1	Low	2	2	1	Low	3	2	1	Low	4	2	1	Low	5	2	1	Low	6	2	1	Low	7	2	1	Low	8
Assay	2	3	Med	1	2	1	Low	2	3	4	High	3	2	1	Low	4	2	1	Low	5	2	3	Med	6	2	1	Low	7	2	1	Low	8
Content uniformity	2	3	Med	1	2	1	Low	2	3	4	High	3	2	1	Low	4	2	1	Low	5	2	3	Med	6	2	1	Low	7	2	1	Low	8
Dissolution	3	4	High	1	2	1	Low	2	3	4	High	3	2	1	Low	4	2	1	Low	5	2	1	Low	6	2	1	Low	7	2	4	High	8
Degradation product	2	1	Low	1	2	1	Low	2	2	1	Low	3	3	4	High	4	3	2	Med	5	2	1	Low	6	2	1	Low	7	2	1	Low	8

*Refer to Table 4.9 for justification.

FIGURE 4.9 Risk assessment of impact of drug substance material attributes on product CQAs.

TABLE 4.9

Justification for the Risk Assessment of the Drug Substance Attributes for Ibuprofen

Ref. No.	Drug Substance Attributes	Drug Product CQAs	Justification
1	Particle size/bulk density	Physical attributes	PSD of drug substance is not expected to have any impact on physical attributes of the tablet; the risk is low.
		Assay Uniformity of dosage unit	Particle size will affect the flow properties of the blend and subsequently assay and uniformity of dosage unit, but since the percentage of ibuprofen in the formulation is high, the risk is medium.
		Dissolution	Ibuprofen is poorly soluble and shows pH-dependent drug solubility; hence, PSD of the drug substance may affect the dissolution of the tablet. The risk is high.
		Degradation product	The PSD of the drug substance is unlikely to affect degradation products. The risk is low.
2	Hygroscopicity/ light sensitivity/ oxidation	Physical attributes Assay Uniformity of dosage unit Dissolution Degradation product	Ibuprofen is nonhygroscopic in nature. Photo degradation and oxidation are also not reported in literature, API Drug Master File and so on; hence, the risk is low.
3	Assay	Physical attributes	Assay has no impact on tablet physical attributes. The risk is low.
		Assay Uniformity of dosage unit Dissolution	Assay of the drug substance will have a direct impact on the drug product assay, uniformity of dosage unit and dissolution. Hence, the risk is high.
		Degradation product	Assay of the drug substance is not related to degradation products of the drug product. Thus, the risk is low.
4	Related substances	Physical attributes Assay Uniformity of dosage unit Dissolution	Related substance is unlikely to have an effect on physical attributes, assay, uniformity of dosage unit and dissolution. Hence, the risk is low.
		Degradation product	The related substance of API will have a direct impact on the degradation product of the drug product. Hence, the risk is high.

(Continued)

TABLE 4.9 (CONTINUED)

Justification for the Risk Assessment of the Drug Substance Attributes for Ibuprofen

Ref. No.	Drug Substance Attributes	Drug Product CQAs	Justification
5	Residual solvents	Physical attributes Assay Uniformity of dosage unit Dissolution	The residual solvents in the drug substance are well below the ICH Q3C levels. As such, the risk of drug substance residual solvents that will affect the drug product CQAs is low.
		Degradation product	Generally, residual solvents can affect the degradation product. However, there are no known incompatibilities between the residual solvents and drug substances or commonly used excipients. Hence, the risk is medium.
6	Flow properties	Physical attributes	Flowability of the drug substance is not related to physical attributes of tablet. Thus, it is at low risk.
		Assay Uniformity of dosage unit	Ibuprofen has poor flow properties. Poor flow will affect uniformity of dosage unit and assay, but as the percentage of ibuprofen in the formulation is high, the risk is medium.
		Dissolution Degradation product	The flowability of the drug substance is not related to its degradation pathway or dissolution. Therefore, the risk is low.
7	Solid state form	Physical attributes	Drug substance does not exhibit any solid state form/polymorphic form. Thus, all CQAs are at low risk.
		Assay Uniformity of dosage unit Dissolution Degradation product	
8	Solubility	Physical attributes	Solubility does not affect tablet assay, uniformity of dosage unit and degradation products. Thus, the risk is low.
		Assay Uniformity of dosage unit	
		Dissolution	Ibuprofen exhibited low solubility. Drug substance solubility strongly affects dissolution. The risk is high. The formulation and manufacturing process will be designed to mitigate this risk. This is at high risk.
		Degradation product	Solubility does not affect tablet degradation products. This is at low risk.

CQA	Formulation composition variables															
	Disintegrant level				Binder level				Surfactant level				Lubricant level			
	U	S	Risk	Ref.*	U	S	Risk	Ref.*	U	S	Risk	Ref.*	U	S	Risk	Ref.*
Physical attributes	2	2	Low	1	3	2	Med	2	2	1	Low	3	2	1	Low	4
Assay	2	1	Low	1	2	1	Low	2	2	1	Low	3	2	1	Low	4
Content uniformity	2	1	Low	1	2	1	Low	2	2	1	Low	3	2	1	Med	4
Disintegration time	3	4	High	1	3	3	High	2	2	1	Med	3	3	3	High	4
Dissolution	3	4	High	1	3	4	High	2	3	4	High	3	3	4	High	4

*Refer to Table 4.10 for justification.

FIGURE 4.10 Risk assessment of impact of formulation variables on CQAs.

TABLE 4.10

Justification of Risk Assessment for Formulation Variables

Ref. No.	Formulation Component	Drug Product CQAs	Justification
1	Disintegrant level	Physical attributes Assay Content uniformity	Disintegrant level is not expected to have any effect on physical attributes, assay and content uniformity based on prior experience and general pharmaceutical development understanding; hence, the risk is low.
		Disintegration time Dissolution	Amount of disintegrant has direct impact on disintegration time and subsequently dissolution; hence, the risk is high.
2	Binder level	Physical attributes	Although the binder level has no direct impact on physical attributes, the binder quantity may have an impact on granules characteristics, which may subsequently affect the physical attributes such as hardness, friability and so on. Hence, the risk is medium.
		Assay Content uniformity	Binder level is not expected to have any effect on assay and content uniformity based on prior experience and general pharmaceutical development understanding; hence, the risk is low.
		Disintegration time Dissolution	Amount of binder has direct impact on disintegration time and subsequently dissolution; hence, the risk is high.
3	Surfactant level	Physical attributes Assay Content uniformity	Surfactant level is not expected to have any effect on physical attributes, assay and content uniformity based on prior experience and general pharmaceutical development understanding; hence, the risk is low.
		Disintegration time	Surfactants are added to improve the wettability of the drug. Thus, level of surfactant would have impact on the disintegration time (DT). As the level of surfactant does not have a direct impact on DT, it is at medium risk.
		Dissolution	Level of surfactant will have a direct impact on the drug release. It is at high risk.
4	Lubricant level	Physical attributes Assay	Lubricant level in the formulation will be low and is not expected to affect physical attributes and assay. The risk is low.
		Content uniformity	Lubricant level may have an impact on the flow property of the blend and subsequently may affect the uniformity of dosage unit. Hence, it is at medium risk.
		Disintegration Dissolution	Magnesium stearate is a hydrophobic lubricant and acts by particle coating. At higher concentrations, the formed hydrophobic coating may inhibit the wetting and consequently tablet disintegration and dissolution. Hence, the risk is high.

understanding. FMEA methodically breaks down the analysis of complex processes into manageable steps. It is a powerful tool for summarising the important modes of failure, factors causing these failures and the likely effects of these failures. FMEA is a risk prioritisation tool and is used to monitor the risk mitigation measures being developed for any product or process. FMEA can be applied to equipment and facilities and might be used to analyse a manufacturing operation and its effect on product or process. It identifies elements/operations within the system that render it vulnerable. The output/results of FMEA can be used as a basis for design or further analysis or to guide resource deployment.

Initial Risk Assessment of Manufacturing Process: As explained in Section 4.4.4.1, in this risk assessment tool, the severity and uncertainty/probability is evaluated along with detectability as an additional criterion. Different companies have different approaches to defining this scale. Table 4.11 is an example of one such scale that could be used. Assignment of Uncertainty (U), Severity (S) and Detectability (D) scores is done on the basis of the below-mentioned understanding of formulation development

TABLE 4.11

Scale for Uncertainty/Probability, Severity and Detectability

Uncertainty/Probability Risk Scale

1	**Full knowledge.** The parameter set-point/range/limit has already been properly evaluated on this particular product.	**Negligible.** Less than 1 in 200 chance of failure.
3	**Prior experience.** The parameter set-point/range/limit has been properly evaluated on sufficiently similar product (same or related API, dosage form, formulation, process), or the evaluation is based on proven mechanistic or statistical models.	**Low.** Less than 1 in 50 chance of failure.
7	**Published literature.** The influence of the parameter can be inferred from literature (articles, patents, general knowledge of pharmaceutical technology).	**High.** Less than 1 in 20 chance of failure.
10	**Insufficient knowledge.** The effect of this parameter is not properly understood.	**Very high.** More than 1 in 20 chance of failure.

Severity Risk Scale

1	The parameter has no impact on patient	No impact
3	Some impact on patient, but reversible	Minor
7	Impact on patient, but not life threatening	Major
10	High impact on patient, which is irreversible and potentially life threatening	Critical

Detectability Risk Scale

1	Failure can be detected during unit operations	Very high
3	Failure can be detected after unit operation and before end product testing	High
7	Failure can be detected by end product testing	Low
10	No ability to detect	None

and general knowledge of science. The risks are identified for each process and prioritised using the Risk Priority Number (RPN). The cumulative risk for the identified process for each failure mode is ascertained by the RPN. RPN is the product of severity × probability × detectability. Each process can have more than one failure mode, and for each failure mode, there can be more than one cause (ICH 2005).

RPNs are a relative risk scoring system and it is the relative size of the RPN score that indicates the priority that should be given to reducing risk. For ease of visualisation, RPN scores are often grouped into high, medium and low risk. The boundaries for differentiation between high, medium and low should be established by the risk assessment team, and for this exercise, the criteria as depicted in Figure 4.11 have been used. RPN scores should not be considered absolute; rather, it is the relative ranking of the scores that enables resources to be focussed on the areas of greatest risk.

A list of the potential failure modes, their cause and effect should be discussed by the risk assessment team and presented in a table such as the example given in Figure 4.12. Measures for further study and minimisation of risk should also be included in this table for future reviews and references. As an example, the proposed process for ibuprofen tablets has been assessed for risks related to product CQAs and presented in the table.

Based on the above risk analysis and assessment, dry mixing, granulation and compression unit processes have been assigned a high-risk category. Some of the other unit operations of blending, coating and packing have been analysed to be medium-risk category. The process parameters for these unit operations will be studied in the experimental plan of work so that they can be thoroughly understood and appropriate parameters are optimised and selected for the manufacturing process to be followed on a large scale.

While evaluating unit processes for any manufacture, the scalability of the process from laboratory scale to commercial scale should be considered and built into the development plan. Some process parameters are scale independent and so those can be optimised very well on a laboratory scale and applied directly during scale-up to a commercial scale. However, the parameters that are scale dependent need to be optimised, qualified and verified on a commercial scale on the basis of the equipment available at the manufacturing site. Using this example of ibuprofen tablets and the proposed process as given in Section 4.4.2.2, the scale dependency of the unit operations could be defined as given in Table 4.12.

4.4.5 PRECLINICAL AND CLINICAL STUDY STRATEGY

The strategy development team needs to discuss the expected and desired therapeutic outcomes of the development with internal and external stakeholders such as

RPN	Severity	Level of risk
>50	>5	High
>50	<5	Medium
<50	>5	Medium
<50	<5	Low

FIGURE 4.11 Risk ranking chart.

Source	Failure mode(s)	Effect(s)	Sev	Cause(s) of failure	Prob	Current controls	Det	RPN	Action plans risk controls/ reductions
Selection of API	API distribution	• Assay • Content uniformity (CU)	7	Non uniform distribution of API	7	• Micronised grade of material will be used • Particle size and PSD to be monitored • Particle size specification $(D_{0.9})$ for Ibuprofen below 250 μm • Wet granulation process selected	1	49	Grade of API will be evaluated at the laboratory and pilot scale for impact on CU, assay and dissolution
Dry mixing of the blend	Selection of API particle size	• Assay • Content uniformity	10	Improper distribution of API and excipient	7	Tentative range of mixing for 5–15 mins	1	70	To be optimised
Granulation	Kneading time Impeller and chopper speed	• PSD • Bulk density • Tapped density • Compressibility • Weight variation • Hardness • Dissolution	10	Granulation time and speed would have an impact on formation of hard or soft granules which will impact the dissolution profile	7	Granulation to be done at slow speed with slow binder addition Amperage or torque reading to be monitored	1	70	Kneading time and speed to be optimised during scale up

(Continued)

FIGURE 4.12　Potential failure modes: their cause and effect for a coated tablet.

Process	Effect		Cause		Action	RPN	Remarks
Drying	Less drying: sticking, picking, chipping Overdrying: capping during compression	LOD Hardness Disintegration time	10	Inadequate drying time and fluidisation	7	LOD limit to be established observing the LOD of dry mix. Tentative limit NMT 1% above the equilibrium moisture content	1	70	Fluidisation and drying time to be optimised
Sizing/milling	Selection of screen size	Particle size distribution Flow properties	3	Improper sieve selection	7	#20 sieve and retentions mill through 1.2/1.00 mm screen	1	21	Screen size impact to be evaluated and finalised
Blending and lubrication	Mixing time, occupancy and type of blender	Blend uniformity Dissolution	10	Low mixing time may cause nonuniformity and overlubrication Magnesium stearate may cause retarded dissolution profile	3	Blending and lubrication time need to be optimised; blend assay and blend uniformity test to be performed. Limit for blend uniformity NLT 90.0%–NMT 110% of the label claim. RSD NMT 5%	1	30	Need to be optimised
Compression	Machine speed Compression pressure	• Weight variation, hardness, assay and content uniformity	10	Very high machine speed	7	Speed can be adjusted for the range of 10–30 RPM in the laboratory scale	1	70	Need to be optimised at the laboratory and pilot scale

(Continued)

FIGURE 4.12 (CONTINUED) Potential failure modes: their cause and effect for a coated tablet.

Target parameter (hardness)	• Impact on the dissolution profile	7	Hardness variation	3	Hardness to be adjusted within the range of 100–200 N	1	21	Need to be optimised at the laboratory and pilot scale
Coating								
Improper spray pattern	• Loss of solution on spray drying and roughness of tablet surface	7	Lower spray rate and atomisation pressure, distance of gun	3	Spray rate: 5–10 g/min 1.0 bar pressure gun position – fixed	1	21	Need to be optimised at the laboratory and pilot scale
	• Overwetting, twining, sticking, peeling	7	Higher spray rate and atomisation pressure, distance of gun	3	Spray rate: 5–10 g/min 1.0 bar pressure gun position – fixed	1	21	Need to be optimised at the laboratory and pilot scale
High product bed temperature	• Weight variation, degradation, poor film formation, spray drying, rough tablets	3	Operational error/ setting of inlet temperature	3	Selected range for product bed temperature: 30°C–45°C	1	9	Need to be optimised at the laboratory and pilot scale
Low product bed temperature	• Overwetting • Twining • Sticking							
High pan RPM	• Overwetting • Sticking • Nonuniformity of coating	3	Setting of pan RPM	3	Selected range of pan RPM: 5–20 RPM	1	9	Need to be optimised at the laboratory and pilot scale
Low pan	• Edge erosion • Chipping							

FIGURE 4.12 (CONTINUED) Potential failure modes: their cause and effect for a coated tablet.

(Continued)

	RPM		• Roughness • Overwetting of tablets				
	Improper product bed temperature for drying	7	• LOD, related substances	3	Operational error	1	Initial and final LOD of tablets to be monitored as per CQA specifications	21	Need to be optimised at the laboratory and pilot scale
	Inadequate drying time	7	• LOD, related substances	3	Monitoring error	1	Initial and final LOD of tablets to be monitored	21	Need to be optimised at the laboratory and pilot scale
Product failure on in-process storage	Moisture uptake by the product	3	• Related substances	3	Improper selection of in-process storage and bulk pack	1	Store in double polybag with silica gel bag in between two bags and sealed with fastener	9	Based on the hold time study IP packing recommendations to be proposed
Packaging	Inadequate pack profile	10	• Physical attributes • Related substances	1	Improper selection of packing material	1	Leak testing/stress testing on packed formulation	10	Leak test checking to be done at frequent intervals
Stability	Moisture and temperature	10	• Product failure w.r.t. specification	1	Improper selection of packing material	1	Packing as per tentative trade dress and continuous stability monitoring	10	Finalisation of trade dress based on actual formula

FIGURE 4.12 (CONTINUED) Potential failure modes: their cause and effect for a coated tablet.

TABLE 4.12

Scale Dependency of Unit Processes

Unit Process	Parameters	Tentative Range to Be Studied		Critical/ Noncritical	Scale Dependent/ Independent
Dry mixing	Dry mixing time	10–20 min		Critical	Dependent
	Impeller speed/ chopper speed	To be finalised based on the development trials		Critical	
Granulation parameters	Binder addition time	5–10 min		Critical	Dependent
	Kneading time	1–2 min		Critical	
	Impeller speed/ chopper speed/binder quantity	To be determined based on the development trials		Critical	
Drying	Drying/product bed temperature	40°C ± 10°C		Critical	Independent
	LOD of dried granules	NMT 1.0%		Critical	
Sizing/milling	Milling speed	Medium		Noncritical	Independent
	Mill screen size	0.5–2.5 mm		Critical	
Prelubrication	Blender geometry	Double cone/bin blender		Noncritical	Dependent
	Blending time	10–20 min		Critical	
	Blending RPM	10–20 RPM		Critical	
	Blend uniformity	90.0%–110.0%		Critical	
Lubrication	Lubrication time	2–3 min		Critical	Dependent
	Lubrication RPM	10–20 RPM		Critical	
	Loss on drying	NMT 1.0%		Critical	
	Bulk density/tapped density	To be finalised based on the development data		Critical	
	Blend uniformity	90.0%–110.0%		Critical	
	Assay	95.0%–105.0%		Critical	
Compression	Machine speed	10–25 RPM		Critical	Dependent
		400 mg	200 mg		Independent
	Average weight	700–800 mg	400–500 mg	Critical	
	Hardness	100–200 N	30–150 N	Critical	
	Thickness	6.2 ± 0.4 mm	4.2 ± 0.4 mm	Noncritical	Independent
	Friability	NMT 1.0%		Critical	

(Continued)

TABLE 4.12 (CONTINUED)
Scale Dependency of Unit Processes

Unit Process	Parameters	Tentative Range to Be Studied		Critical/ Noncritical	Scale Dependent/ Independent
Coating		400 mg	200 mg		
	Spray rate	5–8 g/min		Critical	Dependent
	Inlet/exhaust/product bed temperature	20°C–40°C		Critical	Dependent
	Pan RPM	5–20 RPM		Critical	Dependent
	Appearance	White to off-white smooth film-coated tablets		Critical	Independent
	Weight gain	2%–3%		Critical	Independent
	Thickness	6.4 ± 0.4 mm	4.4 ± 0.4 mm	Noncritical	Independent

physicians, regulatory authorities and patient groups very early into the development so that the clinical studies can be designed in line with expectations of all interested parties. Clinical studies are the most resource intensive amongst all development processes and also carry a very high liability for pharmaceutical companies as they deal with human subjects. Depending on the therapy area under consideration, SMEs from the medical and research fraternity and regulatory bodies are often consulted regularly during the development so that the design of the studies is appropriate and all factors such as safety considerations, variability expected, statistical treatment, bias elements and so on are considered. The clinical study strategy often evolves and undergoes change as the product is developed and data from earlier studies start trickling in. This topic is a very vast area and readers are requested to refer to other reference books and articles on this subject for more in-depth understanding (Gibson 2009; EMEA 2014; UK-BDISR 2014).

4.4.6 COMPILATION OF THE STRATEGY DOCUMENT

As discussed in this chapter, the process for devising a strategy and design for a pharmaceutical product should be very systematic and science based and very often we find that if the team brainstorms intensely during this phase and is able to identify all the challenges ahead, then the development work execution is smooth and straightforward. In most companies, the strategy document is the start of knowledge generation for any product. This documents the prior knowledge that is being considered, the proposed approaches for the formulation and manufacturing process for the product, the proposed analytics, the initial risk assessment, the clinical strategy and the key challenges and risks for the project. It is usually initiated by the research and development team and approved by quality and regulatory team members. Once this document is signed off, it sets the tone for the development process and ensures a strong acceptance and buy-in from all internal stakeholders for the product. As the development progresses, the knowledge base for the product continues to be expanded and built.

4.5   CONCLUDING REMARKS

This chapter emphasises the role of the design and strategy phase for a new product development. Some of the pharmaceutical and biopharmaceutical considerations vital to decisions regarding the dosage form selection for a drug substance are explained in brief. It also discusses and explains how the QbD and risk management concepts can be integrated into the product development process and the advantages of doing so. A systematic approach for preparing the strategy/design for any product is invaluable and is very often the driver for the success of the new product development. This strategy document can be a very useful document and helps the team focus on developing the right product, for the right market, meeting all regulatory requirements and at the right time. This document should be reviewed at important milestones in the project and should be updated at regular time points. Some information may change with time and a review will help refocus the product to the actual market and patient needs.

REFERENCES

Amidon G. L., Lennernäs H., Shah V. P., Crison J. R. 1995. A theoretical basis for a biopharmaceutic drug classification: The correlation of *in vitro* drug product dissolution and *in vivo* bioavailability. *Pharm Res.* 12(3):413–20.

EMEA. 2014. Guideline on process validation for finished products – Information and data to be provided in regulatory submissions. http://www.ema.europa.eu/docs/en_GB/document_library/Scientific_guideline/2014/02/WC500162136.pdf (accessed June 19, 2015).

FMC Biopolymer. 2015. http://www.fmcbiopolymer.com/Pharmaceutical/Products/Avicelfor soliddoseforms.aspxto (accessed June 19, 2015).

Gibson M. 2009. *Pharmaceutical preformulation and formulation – A practical guide from candidate drug selection to commercial dosage form*, second edition, Drugs and The Pharmaceutical Sciences, Volume 199. New York: Informa Healthcare USA Inc.

Hassan B. A. R. 2012. Overview on pharmaceutical formulation and drug design. http://www.omicsonline.org/overview-on-pharmaceutical-formulation-and-drug-design-2153-2435.1000e140.pdf (accessed June 19, 2015).

ICH. 2005. International conference on harmonisation of technical requirements for registration of pharmaceuticals for human use, ICH harmonised tripartite guideline, quality risk management Q9. http://www.ich.org/fileadmin/Public_Web_Site/ICH_Products/Guidelines/Quality/Q9/Step4/Q9_Guideline.pdf (accessed June 19, 2015).

ICH. 2008. International conference on harmonisation of technical requirements for registration of pharmaceuticals for human use, ICH harmonised tripartite guideline, pharmaceutical quality system Q10. http://www.ich.org/fileadmin/Public_Web_Site/ICH_Products/Guidelines/Quality/Q10/Step4/Q10_Guideline.pdf (accessed June 19, 2015).

ICH. 2009. International conference on harmonisation of technical requirements for registration of pharmaceuticals for human use, ICH harmonised tripartite guideline, pharmaceutical development Q8(R2). http://www.ich.org/fileadmin/Public_Web_Site/ICH_Products/Guidelines/Quality/Q8_R1/Step4/Q8_R2_Guideline.pdf (accessed June 19, 2015).

ISPE. 2010. ISPE Product Quality Lifecycle Implementation® (PQLI®) Guide: Overview of product design, development and realization: A science- and risk-based approach to implementation. http://www.ispe.org/pqli-guides/product-design-development-realization (accessed June 19, 2015).

ISPE. 2011a. ISPE Guide Series: Product Quality Lifecycle Implementation (PQLI®) from concept to continual improvement part 1 – Product realization using QbD, concepts and principles. http://www.ispe.org/pqli-guides/product-realization-qbd-concepts (accessed June 19, 2015).

ISPE. 2011b. Product Quality Lifecycle Implementation® (PQLI®) from concept to continual improvement part 2: Product realization using QbD, illustrative example. http://www .ispe.org/pqli-guides/product-realization-qbd-example (accessed June 19, 2015).

Juran J. M. 1992. *Juran on quality by design: The new steps for planning quality into goods and services*, revised edition. New York: Free Press.

PhMRA. 2014. 2015 Profile bipharmaceutical research industry. http://www.phrma.org/sites /default/files/pdf/2014_PhRMA_PROFILE.pdf (accessed June 19, 2015).

Signet. 2015. http://www.signetchem.com/Signet-The-Complete-Excipients-Company-Product -Kollidon (accessed June 19, 2015).

Smith C. G., O'Donnell J. T. 2006. *The process of new drug discovery and development.* New York: Informa Healthcare USA Inc.

Tufts CSDD. 2014. Tufts Center for the study-briefing-cost of developing a new drug. http://csdd.tufts.edu/files/uploads/Tufts_CSDD_briefing_on_RD_cost_study_-_ Nov_18,_2014.pdf (accessed June 19, 2015).

UK-BDISR. 2014. UK Government, Department of Business Innovations and Skills Report, focus on enforcement regulatory reviews, review of the pharmaceutical manufacturing and production sector. http://discuss.bis.gov.uk/focusonenforcement/files/2014/06 /Pharma-report-final-publication.pdf (accessed June 19, 2015).

USFDA. 2011. FDA guidance for industry, process validation: General principles and practices. http://www.fda.gov/downloads/Drugs/Guidances/UCM070336.pdf (accessed June 19, 2015).

Van Buskirk G. A., Asotra S., Balducci C. et al. 2014. Best practices for the development, scale-up, and post-approval change control of IR and MR dosage forms in the current quality-by-design paradigm. *AAPS PharmSciTech.* 15(3):665–93.

Yu L. X. 2008. Pharmaceutical quality by design: Product and process development, understanding, and control. *Pharm Res.* 25(4):781–91.

Yu L. X., Amidon G., Khan M. A. et al. 2014. Understanding pharmaceutical quality by design. *AAPS J.* 16(4):771–83.

5 Design of Experiments
Basic Concepts and Its Application in Pharmaceutical Product Development

Amit G. Mirani and Vandana B. Patravale

CONTENTS

5.1   INTRODUCTION

Pharmaceutical products are formulated as specific dosage forms such as oral tablets, solutions, suspensions, creams, gels, parenterals and so on for effective drug delivery to the patient ensuring safety and efficacy. Today, patient safety, which is directly linked to product quality, has become an essential part of pharmaceutical product development. In the past, a traditional approach, that is, decision based on univariate approach (i.e. changing one variable at a time approach), or a trial and error-based approach was used to develop quality pharmaceutical products (Armstrong and James 2006; Lewis et al. 1999). However, execution of the traditional approach requires a large number of experiments without the assurance of achieving the optimum parameters. Also, it is associated with several other drawbacks such as inability to predict interaction, nonreproducibility and being strenuous, expensive and time consuming. To overcome the aforementioned problems associated with the traditional approach, the newer concept of quality by design (QbD) approach for product development was introduced (Figure 5.1) (Lindberg and Lundstedt 1995; Yu et al. 2014).

QbD is an essential tool to ensure quality in the finished product. Implication of the QbD approach starts with defining quality target product profile (QTPP) and critical quality attributes (CQAs), prioritising critical process and product parameters

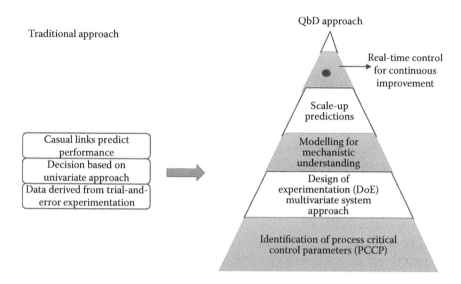

FIGURE 5.1   Translation from a traditional approach to the QbD approach.

(based on risk assessment studies), statistical optimisation of critical parameters using design of experiment (DoE), generating design space and finally assigning a control strategy to ensure the development of a quality product.

DoE is one of the essential elements of the QbD approach, which was discovered by British statistician Sir Ronald Fisher in 1925 (Fisher 1925). DoE represents a statistically organised experimental plan that provides information with higher precision on the variation of product/process response as a function of change in the input variables. The DoE technique involves the application of experimental design based on suitable variables, followed by its statistical evaluation and the search for an optimum using either a graphical or a mathematical approach. Thus, it depicts the information such as effect of variables at each level, their interaction and so on from systematically organised experiments (Cochran and Cox 1992; Doornbos 1981; Lewis et al. 1999). DoE primarily has three basic objectives, namely, screening (identification of critical variables and their levels), optimisation (identification of optimum input variable composition to achieve optimum response) and robustness (identification of sensitivity of response to small changes in the factor) (Singh et al. 2004). DoE exhibits wide applications in screening of important factors, optimisation of formulation variables, optimisation of manufacturing process, analytical instruments, robustness testing of method or product and so on. DoE proves to be the most efficient, economical and expedient approach as compared to the conventional approaches. Hence, currently it has gained wide acceptance as a pivotal developmental tool in pharmaceutical product development. In this chapter, we shall discuss the common terminologies used in DoE to provide the readers with an in-depth understanding of the topic. The chapter will focus on how the current pharmaceutical industry uses the DoE approach along with the sequential stages to perform DoE such as selection of variables, selection of levels, selection of experimental design and postulation of mathematical model for various chosen responses, fitting experimental data, mathematical modeling, generating optimum and design validation. The chapter also aims at simplifying the DoE concept with suitable case studies.

5.2 FUNDAMENTALS OF DESIGN OF EXPERIMENTATION

5.2.1 VARIABLES

Pharmaceutical product development is associated with a number of processing and manufacturing parameters that may affect the quality of the finished product. These parameters are designated as variables (Doornbos 1981; Schwartz and Connor 1996; Singh et al. 2011; Stetsko 1986). Variables in pharmaceutical product development can be further classified as input and output variables.

5.2.1.1 Input Variables

The variables that have a direct influence on the output or response of the finished product and that can be controlled by the formulation scientist are termed *input variables* or *independent variables*. The input variables can be determined using a risk assessment analysis or with prior knowledge about the product and process. Input variables can be further classified as qualitative or quantitative input variables.

Qualitative Input Variables: Qualitative input variables are the parameters that qualitatively affect the finished product and are directly under the control of the formulation scientist. For example, type of disintegrant, type of surfactant, grades of polymer, granulation method and so on.

Quantitative Input Variables: Quantitative input variables are the parameters that quantitatively affect the finished product and are directly under the control of the formulation scientist. For example, concentration of disintegrant, ratio of surfactants, percentage of coating, compression speed, compaction force and so on.

5.2.1.2 Output Variables

These variables are the measured properties of the process (e.g. disintegration time, hardness of tablet, drug release, particle size, assay, etc.) that need to be controlled to obtain the desired quality of the finished product. As they are dependent on input variables, they are also termed *dependent variables* or *response variables*. The measured response (y) for an experiment (i) will be written as y_i. The output variables can be decided based on QTPP and CQAs.

5.2.2 LEVEL

The value assigned to the variable is termed *level* and can be identified from factors influencing studies or preliminary studies (Lewis 2002).

5.2.3 EFFECT

Effect is the extent of change in output/response obtained by changing the input variable (Anderson et al. 2002; Stack 2003). The effect obtained can be further explained in terms of orthogonality, interaction and confounding.

5.2.3.1 Orthogonality

Orthogonality is the effect that is linearly proportional to changes in the input variables; that is, the response is mainly attributed to the factor of interest.

5.2.3.2 Interaction

Interaction is the effect that is not linearly proportional to the changes in the input variables and thereby represents a lack of orthogonality. Interaction represents the inherent quality of input data and can be assessed quantitatively.

5.2.3.3 Confounding

Confounding is the lack of orthogonality; that is, the effect obtained is not directly related to changes in input variables. Confounding effect represents the improper selection of factors, level, experimental design and data analysis. This effect needs to be controlled to avoid the misinterpretation of data. The degree of confounding can be assessed qualitatively by the resolution method (Myers and Montgomery 1995).

5.2.4 Coding of Variables

Coding involves transformation of a natural variable into a nondimensional coded variable to facilitate ease in mathematical calculation, orthogonality of effect, interpretation of effect and interaction among the factors. It also circumvents any anomaly in factor sensitivity to changes in levels (Armstrong and James 2006). Coding to the level of variables can be denoted as −1, 0, 1 where '−1' represents a low level, '0' represents a medium level and '1' represents a high level.

5.2.5 Experimental Domain

It is the dimensional space defined by the coded variables that is investigated experimentally (Anderson et al. 2002).

5.2.6 Design Space

It is an imaginary area bound by the extremes of the tested factors (Anderson et al. 2002; Singh et al. 2011).

5.3 DESIGN OF EXPERIMENTATION METHODOLOGIES

DoE, an integral part of QbD, implies statistical screening and optimisation of drug product and process (Armstrong and James 2006; Lewis 2002). It involves systematic planning of experiments in such a way that the optimum region in the form of design space can be obtained in an unbiased manner. The DoE methodologies are further classified into two methods on the basis of its implementation in product development: simultaneous optimisation and sequential optimisation (Figure 5.2) (Schwartz and Connor 1996; Stetsko 1986).

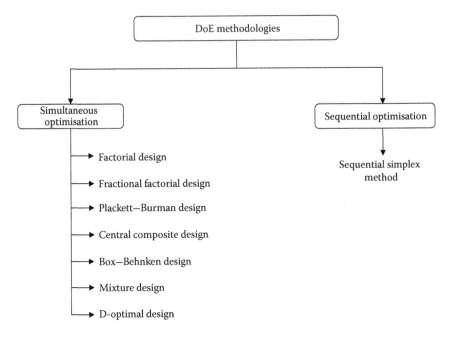

FIGURE 5.2 Design of experimentation methodologies.

5.3.1 SIMULTANEOUS OPTIMISATION METHOD

Simultaneous optimisation is a method in which statistical modeling and search for an optimum are performed after the experimental part is concluded (Armstrong and James 2006; Lewis and Chariot 1991). Different experimental designs under the simultaneous optimisation method (Figure 5.2) are briefly described in Sections 5.3.1.1 through 5.3.1.7.

5.3.1.1 Factorial Design

Factorial design was first introduced by Fisher in 1946 and is the simplest method for simultaneous optimisation of a pharmaceutical product. Factorial design is primarily based on the first-degree mathematical model for linear responses. It involves studying the effect of each factor at each level, thereby identifying the main effect as well as the interaction between factors at different levels. The number of experiments in factorial design is designated as X^n, where X represents the number of level and n represents the number of factors. In general, the factorial design can be performed either at two levels or at three levels. The selection of factor and their level is critical because as the number of factors and level increases, the number of experiments simultaneously increases and its mathematical interpretation also becomes difficult. The factorial design using different factor–level combinations is depicted in Figure 5.3 and the 2^3 factorial design is described briefly in the following paragraph.

2^3 *Factorial Design*: The simplest form of factorial design is the 2^3 factorial design where the three factors are studied at two levels. The experimental design for

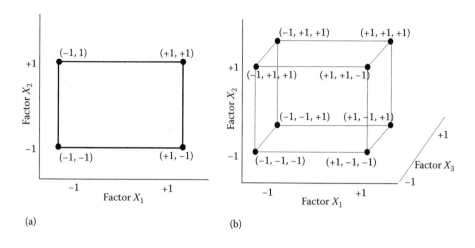

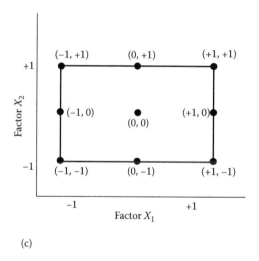

FIGURE 5.3 Factorial design: (a) 2^2 factorial design, (b) 2^3 factorial design and (c) 3^2 factorial design.

the 2^3 factorial design is explained using an example of sustained-release metformin tablet, where microcrystalline cellulose (diluent), ethyl cellulose (sustained-release polymer), PVP-K30 (binder), magnesium stearate (lubricant) and Aerosil (glidant) were taken as inactive ingredients. Among all inactive ingredients, microcrystalline cellulose, ethyl cellulose and PVP-K30 were taken as the independent factors based on preliminary risk assessment studies and prior knowledge, while the drug release profile was taken as the dependent variable considering the desired QTPP. The independent factors and their levels are given in Table 5.1.

TABLE 5.1

Independent Variables and Levels

Sr. No.	Notation	Independent Factors (mg/tab)	Levels	
			−1	+1
1	X_1	Microcrystalline cellulose	80	100
2	X_2	Ethyl cellulose	5	10
3	X_3	PVP-K30	3	5

The experimental plan for a three-factor, two-level (2^3) design is shown in Table 5.2. This design is graphically depicted as a cube (Figure 5.3b), where the experiments with factor X_1 are represented on the right-hand side for high level and left-hand side for low level. Similarly, factor X_2 is represented on the top (high level) and bottom (low level) faces of the cube, while factor X_3 is on the back (high level) and front (low level) faces.

In general, the variables and their levels are transformed into a coded term to facilitate ease in mathematical calculation, determining the main effect and interaction among the factors. Coding of the variables involves natural transformation and is expressed as an experimental unit. For two-level experiments, the low level is coded as '−1' and high level is coded as '+1', while for three-level experiments,

TABLE 5.2

Experimental Plan for the 2^3 Full Factorial Design

Experiment	Microcrystalline Cellulose (mg/tab)	Ethyl Cellulose (mg/tab)	Polyvinyl Pyrrolidone (mg/tab)	Drug Release (%) (12 h)
1	80	5	3	80
2	100	5	3	78
3	80	10	3	65
4	100	10	3	64
5	80	5	5	72
6	100	5	5	71
7	80	10	5	62
8	100	10	5	60

an intermediate term is added to improve the precision of data and designated as '0' for the intermediate level. The 2^3 factorial design in coded form and its effects are shown in Table 5.3.

The 2^3 factorial design exhibits seven effects, namely, three individual factor effects, three two-way interactions (X_1X_2, X_1X_3 and X_2X_3) and one three-way interaction ($X_1X_2X_3$). The magnitude of the main effect can be calculated by taking the mean of all experiments with a high level of factor X_1 minus the mean of all those with a low level of the same factor. However, the magnitude of interaction can be calculated by taking the mean of all results of interaction terms with a positive value minus the mean of all results with a negative value. Taking this information from Table 5.3, the magnitude of the effect of factor X_1 is as follows:

$$\text{Effect of factor } X_1 = 1/4\{(78+64+71+60)-(80+65+72+62)\}$$
$$= 1/4\{273-279\} = -1.5$$

(5.1)

Similarly, the effect of factors X_2, X_3, X_1X_2, X_1X_3, X_2X_3 and $X_1X_2X_3$ is calculated and the values are shown in Table 5.4.

Factors X_1, X_2 and X_3 are found to have a negative effect, which indicates that as the level of factors decreases, the % drug release increases. The interaction among factors X_1X_2 and X_1X_3 is nil while factor combinations X_2X_3 and $X_1X_2X_3$ reveal an interaction wherein the combination of factors increases the drug release.

TABLE 5.3

Sign to Calculate the Main Effect and Interaction Effect of the 2^3 Factorial Design

Experiments	Notation	X_1	X_2	X_3	X_1X_2	X_2X_3	X_1X_3	$X_1X_2X_3$	Drug Release (%)
1	(−1,−1,−1)	−1	−1	−1	+1	+1	+1	−1	80
2	(+1,−1,−1)	+1	−1	−1	−1	+1	−1	+1	78
3	(−1,+1,−1)	−1	+1	−1	−1	−1	+1	+1	65
4	(+1,+1,−1)	+1	+1	−1	+1	−1	−1	−1	64
5	(−1,−1,+1)	−1	−1	+1	+1	−1	−1	+1	72
6	(+1,−1,+1)	+1	−1	+1	−1	−1	+1	−1	71
7	(−1,+1,+1)	−1	+1	+1	−1	+1	−1	−1	62
8	(+1,+1,+1)	+1	+1	+1	+1	+1	+1	+1	60

TABLE 5.4

Magnitudes of the Main Effects and Interactions of the Factors

Factor			Interaction			
X_1	X_2	X_3	X_1X_2	X_2X_3	X_1X_3	$X_1X_2X_3$
−1.5	−15	−22	0	+8.0	0	+5.0

5.3.1.2 Fractional Factorial Design

The full factorial design provides information about the interaction of each variable at each level. However, as the number of variables increase, the experimental runs also increase and even the higher-order interactions have no significant effect. Hence, to overcome these issues in a methodical approach, fractional factorial design (FFD) or partial factorial design is introduced. The experimental runs in FFD are expressed as X^{n-x}, where X represents the number of levels, n represents the number of factors and x represents degree of fractionation. The selection of degree of fractionation is like playing with fire; if not selected properly, it may lead to confounding/aliasing of effects (Armstrong and James 2006; Lewis 2002; Loukas 1997). Confounding is the effect that falsely claims the presence of interaction between variables. The degree to which the main effect is confounded can be estimated by resolution. Resolution of design is represented by Roman numerals such as resolution III (main effect confounded with two factor interactions), resolution IV (two-factor interaction aliased with each other) and resolution V (two-factor interaction aliased with three factor interactions). The lower the resolution, the more will the alias become a problem. Hence, resolution V design is used for optimisation where both the main effect and interaction information are required while resolution III design is used for screening study where the purpose is to find only the main effects. The experimental plan for FFD based on resolution is shown in Figure 5.4.

The FFD can be further simplified by comparing the 2^5 full factorial design and resolution V 2^{5-1} FFD using an illustrative case study of a sustained-release tablet. The critical factors and their levels are shown in Table 5.5.

Experimental	Number of Factors					
Runs	2	3	4	5	6	7
4	2	2^{3-1} (Res III)				
8		2^3	2^{4-1} (Res IV)	2^{5-2} (Res III)	2^{6-3} (Res III)	2^{7-4} (Res III)
16			2^4	2^{5-1} (Res V)	2^{6-2} (Res IV)	2^{7-3} (Res IV)
32				2^5	2^{6-1} (Res V)	2^{7-2} (Res IV)

FIGURE 5.4 Experimental plan for FFD based on resolution.

TABLE 5.5

Critical Factors and Levels

Factor	Name	Low Level (−)	High Level (+)
X_1	Hydroxy propyl methyl cellulose (%)	10	20
X_2	Ethyl cellulose (%)	5	10
X_3	Polyvinyl pyrrolidone (%)	5	10
X_4	Microcrystalline cellulose (%)	30	60
X_5	Magnesium stearate (%)	0.5	1

Full factorial design would require 32 runs for five factors at two levels, which would reveal 31 effects (i.e. 5 main effects, 10 two-factor interactions, 10 three-factor interactions, 5 four-factor interactions and 1 five-factor interaction). The principle of sparsity of effect states that only 20% of the main effects and two-factor interactions are likely to be significant. If this is true, then only 3 effects, out of 31, will be significant and the rest of the effects will be an experimental error. Therefore, a full factorial design will be extensively time consuming.

A properly constructed 2^{5-1} FFD requires 16 runs estimating 15 effects (5 main effects and 10 two-factor interactions). The experimental plan for FFD is shown in Table 5.6. Designing of the experimental plan starts with a negative sign in the first row and ends with a positive sign (high level). The first four factor columns form the full factorial design with 16 runs, and the fifth column (X_5) is the product of the first four columns ($X_1X_2X_3X_4$).

5.3.1.3 Plackett–Burman Design

In 1946, Plackett and Burman invented an orthogonal two-level FFD, commonly known as the Plackett–Burman (PB) design. The PB design is also known as the Hadamard design (Armstrong and James 2006; Plackett and Burman 1946; Singh et al. 2011). The number of experimental runs in the PB design is in multiples of four, that is, 12, 16, 20, 24, 28 and higher runs. In each case, the number of experimental runs is one more than the factor and is expressed as $N = K + 1$, where K represents the factor and N represents the number of experimental runs (Table 5.7). The designing of the experimental plan starts with the first column, which is defined for different experimental runs as given in Table 5.8. The experimental design is generated from the first column by moving the elements down by one row and placing the last element in the first position. Subsequent columns are prepared in the same way. Lastly, the design is completed by adding a minus sign to the last line. A PB design for a study of 11 factors in 12 experiments is given in Table 5.9. The PB design exhibits a high degree of confounding/aliases and hence cannot be used to evaluate main effects and interactions. They are mainly used in screening and robustness testing (Anderson et al. 2002; Loukas 2001; Singh et al. 2004, 2011).

TABLE 5.6

Experimental Plan for the 2^3 Full Factorial Design

Experiments	X_1	X_2	X_3	X_4	X_5	Drug Release (%)
1	−	−	−	−	+	80
2	+	−	−	−	−	78
3	−	+	−	−	−	65
4	+	+	−	−	+	64
5	−	−	+	−	−	72
6	+	−	+	−	+	71
7	−	+	+	−	+	62
8	+	+	+	−	−	60
9	−	−	−	+	−	78
10	+	−	−	+	+	72
11	−	+	−	+	+	68
12	+	+	−	+	−	64
13	−	−	+	+	+	75
14	+	−	+	+	−	67
15	−	+	+	+	−	63
16	+	+	+	+	+	56

This two-level design is unable to predict the curvature and hence it is advised to augment it to the response surface method (RSM) by adding a central point. The most common RSMs are detailed in Sections 5.3.1.4 and 5.3.1.5.

5.3.1.4 Central Composite Design

Central composite design (CCD) was first proposed by Box and Wilson and is mainly used for factors with nonlinear responses. CCD is an amalgamation of the (2^k) factorial design with ($2k$) star design and central point. The experimental runs in this design can be expressed as $2^k + 2k + n$ (Table 5.10), where k is the number of factors

TABLE 5.7

Experimental Runs in the PB Design at Different Factors

Number of Experimental Run	Number of Factors
4	3
8	7
12	11
16	15
20	19
24	23

TABLE 5.8

Coded Variables for the PB Design

N	First Line of Design
4	+ + −
8	+ + + − + − −
12	+ + − + + + − − − + −
16	+ + + + − + − + − − + − − −
20	+ + − − + + − + − + − + − − − − + + −
24	+ + + + + − + − + + − − + + − − + − + − − − −

and n is the number of centre points (Armstrong and James 2006; Box and Wilson 1951; Lewis et al. 1999; Singh et al. 2004).

The CCD can be simplified by adding central points to the factorial design. For maximum efficiency, the 'axial' (or 'star') point at distance $\pm\alpha$ of the centre point should be added outside the original factor range (Figure 5.5). The value of axial point ($\pm\alpha$) will be the square root of the number of the factors. For a two-factor study, $\alpha = 1.414$; for a three-factor study, $\alpha = 1.682$; and for a four-factor study, $\alpha = 2.00$ (Armstrong and James 2006; Lewis et al. 1999). The additional centre points provide more power to estimate second-order effects, that is, the quadratic effect, which is needed for curvature characterisation. The experimental runs using CCD for two factors are given in Figure 5.6.

TABLE 5.9
PB Design for the Study of 11 Factors in 12 Experiments

					Factor					
X_1	X_2	X_3	X_4	X_5	X_6	X_7	X_8	X_9	X_{10}	X_{11}
+1	−1	+1	−1	−1	−1	+1	+1	+1	−1	+1
+1	+1	−1	+1	−1	−1	−1	+1	+1	+1	−1
−1	+1	+1	−1	+1	−1	−1	−1	+1	+1	+1
+1	−1	+1	+1	−1	+1	−1	−1	−1	+1	+1
+1	+1	−1	+1	+1	−1	+1	−1	−1	−1	+1
+1	+1	+1	−1	+1	+1	−1	+1	−1	−1	−1
−1	+1	+1	+1	−1	+1	+1	−1	+1	−1	−1
−1	−1	+1	+1	+1	−1	+1	+1	−1	+1	−1
−1	−1	−1	+1	+1	+1	−1	+1	+1	−1	+1
+1	−1	−1	−1	+1	+1	+1	−1	+1	+1	−1
−1	+1	−1	−1	−1	+1	+1	+1	−1	+1	+1
−1	−1	−1	−1	−1	−1	−1	−1	−1	−1	−1

TABLE 5.10
Experimental Runs at Different Factors

Factors	Number of Experimental Runs
2	13 (5 centre point runs)
3	20 (6 centre point runs)
4	30 (6 centre point runs)
5	48 (6 centre point runs)

Depending on the type of domain and α value, CCD is further explained in terms of Rotatable CCD and Face-Centred CCD (Anderson et al. 2002).

5.3.1.5   Box–Behnken Design

The Box–Behnken design (BBD) is a second-order quadratic model for nonlinear responses. It is mainly used when the factor space is within the experimental domain, which differentiates BBD from CCD. BBD does not contain factorial design or FFD and also overcomes the limitations associated with a CCD, which requires experiments to be performed at five levels (Armstrong and James 2006; Box and Behnken 1960; Lewis et al. 1999). As BBD uses only three levels for each factor, lesser numbers of experiments are run as compared to CCD. BBD is applied in three factors, where the design is represented as a cube and its experimental points are at the midpoints of the edges of the cube with a domain volume of 6. This differentiates BBD from the conventional three-factor cube design where the experimental points

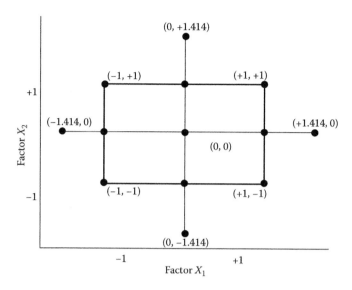

FIGURE 5.5 Central composite design.

Experiment	Component	Factor X_1	Factor X_2
1	Centre point	0	0
2		−1	−1
3	Factorial point	−1	+1
4		+1	+1
5		+1	−1
6	Star points	−1.414	0
7		+1.414	0
8		0	+1.414
9		0	−1.414

FIGURE 5.6 Experimental run using CCD.

are from corners and has a domain volume of 8. BBD is popularly used for response surface optimisation studies in drug delivery (Singh et al. 2011). CCD and BBD are compared in Table 5.11.

5.3.1.6 Mixture Design

Several drug delivery systems such as emulsions, solutions, syrups, tablets and so on involve the use of multiple components for their development in the final dosage form (Armstrong and James 2006; Cornell 1979, 1990; Doornbos 1981; Snee and Marquardt 1976). In such systems, each component cannot be varied at each level and hence factorial design, CCD, BBD and so on may not be applicable for their optimisation. For optimisation of such drug delivery systems, a systematic approach using mixture design is introduced, where the sum total of the proportions of all variables is restricted to unity and where no fraction is negative; that is, $0 \leq X_i \leq 1$, where X_i is the sum of all variables. This differentiates mixture design from the

TABLE 5.11

Comparison between the CCD and the BBD

Sr. No.	CCD	BBD
1	Experiments are run at five levels	Experiments are run at three levels
2	Contains embedded factorial design (FD) and star points	FD with central point form cube
3	Works when the factor space is outside the experimental domain	Works when the factor space is within the experimental domain
4	Domain volume is 20	Domain volume is 6

Factors	Number of Runs	
2	13 (5 centre point runs)	–
3	20 (6 centre point runs)	15
4	30 (6 centre point runs)	27
5	48 (6 centre point runs)	46

general factorial design, CCD and BBD. In mixture design, the factor level is varied independently up to $n - 1$ where n represents the number of mixture components. For instance, in a two-component mixture, only one factor can be independently varied, while in a three-component mixture, only two factors can be independently varied and the other factor is chosen to complete the sum to unity. Mixture design is mainly classified into two types: *standard mixture design* and *constrained mixture design*.

1. *Standard Mixture Design*: This is the design where the experimental region exhibits a simple $n - 1$ dimensional space and is also termed *simplex mixture design* (Scheffe's design). The simplex mixture design (Scheffe's design) is used for a single constraint that must add up to unity for its application in mixture design. This design can be either a centroid or a lattice design because the component forms a lattice within the factor space. The simplex mixture design can be further explained in terms of mixture components, that is, two-component system, three-component system or more-than-three-component system.

   a. *Two-Component System*: The two-component mixture can be represented by a straight line. For example, the % drug release of a tablet is mainly affected by two components, that is, X_1 and X_2; hence, the factor space will be generated by joining the point representing these two components (Figure 5.7a). The vertices represent the drug release

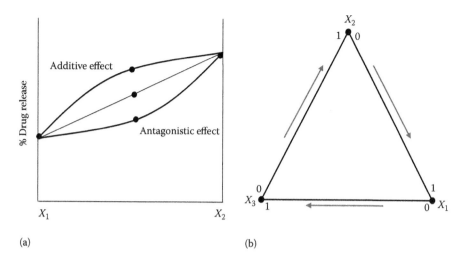

FIGURE 5.7 (a) Two-component mixture system. (b) Three-component mixture system.

of compound X_1 and X_2. The upward concave line represents low drug release while the downward convex line represents high drug release.

b. *Multi-Component System*: The three component mixture can be represented by a two-dimensional, three-cornered figure, that is, an equilat-
eral triangle. Consider the three components X_1, X_2, X_3 at each corners
of the triangle and the proportion of that component at that point as
unity (Figure 5.7b). The interior of the triangle represents a mixture
in which all three components are present while the boundaries of the
triangle (the straight lines) represent two-component systems. The scale
for all three sides must be the same and usually increases clockwise.

Similarly, the four-component system can be represented by a regular tetrahedron. However, for the five-or-more-component systems, it is
not possible to represent the design space and their properties can be
represented by plotting response surfaces on ternary diagrams using
three components to vary while holding the remaining components
constant.

2. *Constrained Mixture Design*: The constrained design is used when each of
the mixture components is bound by upper and lower limits. For example,
compression of blend containing moisture content less than 0.5% causes
capping while more than 5% causes sticking. In such a design, the observations are made at the corners, middle and centre of the design space by
keeping the lower limit 0.5% and the upper limit 5%. The constrained
design mainly used for this type of study is extreme vertices design.

5.3.1.7 D-Optimal Design

D-optimal design is a computer algorithm-based nonorthogonal design, particularly
used when the classical designs do not apply (Chariot et al. 1988; De Aguiar et al.

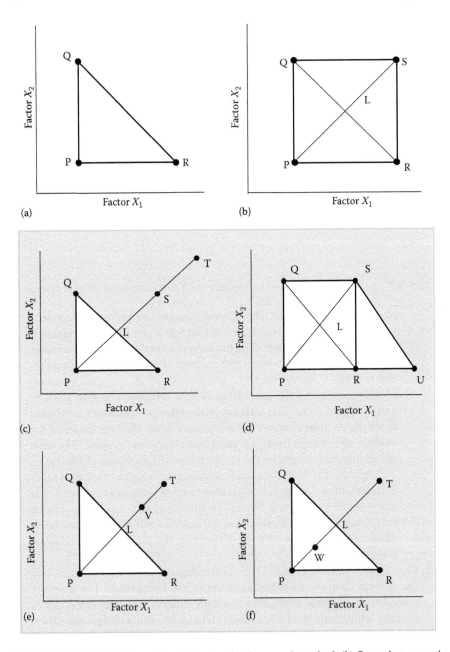

FIGURE 5.8 (a) First-stage optimisation by simplex search method. (b) Second-stage optimisation by simplex search method. (c) Expansion: When D is better than A, B or C. Hence, the next point E is located along this line such that PD = DE. (d) When D is better than A and B but worse than C, then vertex D is retained, and the next point F is located by moving away from B, reflecting triangle BCD about axis CD. (e) If D is worse than B and C, but better than A, the next experiment (G) is located along the AD axis at (P + 0.5AP). (f) When D is worse than A, B or C, then point H is located along the same axis at (P − 0.5AP).

1995; Lewis and Chariot 1991). The D-optimal design is generated based on specific optimality criteria, which are ultimately dependent on minimisation of variance of parameters estimated for the model. This indicates that the 'optimality' of a given D-optimal design is model dependent. Hence, one needs to specify a model before a computer algorithm can generate the specific treatment combination using a candidate set of treatment runs. This candidate set of treatment runs usually consists of all possible combinations of various factor levels that one wishes to use in the experiment. The D-optimal design is useful when resources are limited or there are constraints on factor settings.

5.3.2 SEQUENTIAL OPTIMISATION

Sequential optimisation is an approach where the experimentation proceeds consequently with optimisation (Armstrong and James 2006; Lewis 2002). This can be further simplified using the sequential simplex optimisation method as detailed in Section 5.3.2.1.

5.3.2.1 Sequential Simplex Optimisation

Sequential simplex optimisation was first proposed by Spendley et al. (1962). As the name indicates, the experimentation and its optimisation are analysed one by one as each is carried out. The sequential simplex method is a convex geometric figure where the number of vertices is equal to one more than the number of variables in the space. Thus, a simplex of two variables is a triangle while that of three variables is a tetrahedron. In this method, the experimentation is started at an arbitrary point in the experimental domain in a direction directly opposite to the 'worse' point of the simplex, that is, simplex expanding. The worse point is discarded and a new simplex is obtained, and the process is repeated till the optimum response is achieved, that is, simplex contracting (Doornbos 1981; Schwartz and Connor 1996).

The sequential simplex method can be further simplified by taking a suitable case, where there are two independent variables, that is, X_1 and X_2. The simplex design for these two variables would be in a triangle shape representing three experiments (Figure 5.8a). The responses of these experiments are designated P, Q and R, respectively. The optimum response can be identified by moving opposite to the point that showed the worse response. Let us assume that the response at P is worse than those at Q and R. The values of the independent variables for the next experiment (S) are therefore chosen by moving away from point L. This is achieved by reflecting the triangle PQR about the QR axis (Figure 5.8b). Hence, PL = SL. The experiment at point S is performed and the response of D is compared with the responses at points P, Q and R. The next experiment will be based on the response from four experiments (Figure 5.8c through f).

5.4 STAGES IN OPTIMISATION METHODOLOGY

Systematic optimisation/DoE is accomplished in several sequential stages (Figure 5.9). They are discussed in Sections 5.4.1 through 5.4.6.

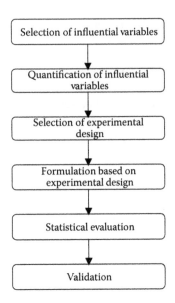

FIGURE 5.9 Flowchart of optimisation methodology.

5.4.1 SCREENING OF INFLUENTIAL VARIABLES

Development of a pharmaceutical product is associated with numerous process- and formulation-related variables that may affect the quality of the finished product. Variables that demonstrate a significant impact on the quality of the finished product are considered as active or influential variables and those having no significant impact are considered as inactive or noninfluential variables. The identification of criticality among variables is the preliminary step in optimisation methodology, which can be identified using a process termed *screening of the influential variables* (Lewis et al. 1999; Singh et al. 2011). The screening process utilises prior knowledge, reported literature and experimental designs such as screening design, PB design, mixture design and so on for the identification of critical variables. The screening design is also known as the main effect design as it provides only the factors having the main effect rather than main effect interaction. The screening process prevents the false identification of critical variables and thereby minimises the cost of product development.

5.4.2 QUANTIFICATION OF INFLUENTIAL VARIABLES

The study aims to quantify the lowest and highest level of critical variables that may affect the quality of the finished product (Armstrong and James 2006). Quantification of a critical variable is the second important step in optimisation methodology. This study utilises similar experimental designs as used for screening.

5.4.3 EXPERIMENTAL DESIGN

Experimental design is a statistically organised experimental plan to get the optimum responses (Araujo and Brereton 1996). The number of experiments mainly depends on the type of experimental design. It is essentially based on the principles of randomisation (i.e. random treatment of experimental units), replication (number of replicas of experiment) and error control (experiment grouping to increase precision) (Araujo and Brereton 1996; Armstrong and James 2006; Lewis et al. 1999). Experimental design helps in estimating the main effects, interaction effects and nonlinear effects, that is, quadratic effects (Lewis et al. 1999; Singh et al. 2004). The experimental design can be classified on the basis of their application and degree of model (Figure 5.10) (Armstrong and James 2006). The experimental design, which appropriately optimises critical factors using the minimum number of experiments with detailed information on effect and interaction, should be selected for optimisation in pharmaceuticals. Different experimental designs are summarised in Table 5.12.

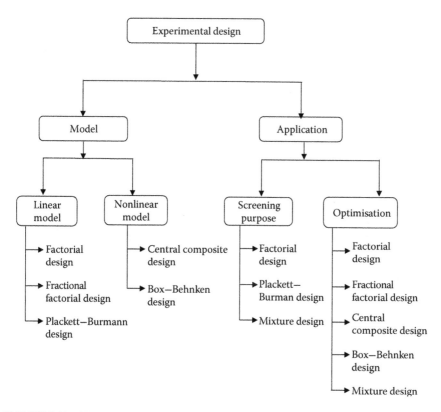

FIGURE 5.10 Types of experimental design.

TABLE 5.12

Summary of Experimental Designs

Type of Experimental Design	Notation	Factors	Levels	Type of Model	Advantages	Disadvantages
Factorial design	FD	≤3	2 or 3	Linear model, quadratic model, interaction model	Main effect and interaction can be estimated Useful for screening study and optimisation study	Large number of experimentation required Prediction outside region not advisable No curvature at two level
Fractional factorial design	FFD	>6	2	Linear model, interaction model	Suitable for large number of factors Useful for screening and optimisation study	Confounding/aliasing effect Effect cannot be uniquely estimated Limited degree of fractionation possible
Plackett–Burman design	PBD	≥7	2	Linear model, interaction model	Suitable for a large number of factors where FFD cannot be applicable Used for preliminary screening	Fixed design in which runs are predetermined Suitable for two levels only Confounding effect
Mixture design	MD	≥7	≥3	Mixture model	Large number of parameters can be screened with minimum experiments Useful for mixture components Useful for preliminary screening	Interaction of factor at each level cannot be estimated

(*Continued*)

TABLE 5.12 (CONTINUED)

Summary of Experimental Designs

Type of Experimental Design	Notation	Factors	Levels	Type of Model	Advantages	Disadvantages
D-optimal design	DOD	<6	≥3	Custom-made model	Minimise the variance of parameters Used for product and process optimisation (large level)	Complex model
Central composite design	CCD	<6	5	Quadratic model, interaction model	Requires fewer experiments compared to factorial design Useful for product and process optimisation Curvature can be estimated Main effect and interaction can be estimated	Design works at five levels only Large number of experiments required Difficult to implement for variables with fractional value
Box–Behnken design	BBD	<6	3	Quadratic model, interaction model	Minimum experiments compared to CCD Nonlinear effect can be obtained Used for product and process optimisation	Nil

5.4.4 Formulation Based on Experimental Design

On the basis of the selected experimental design, formulation batches are prepared and evaluated for their response variables. The experimental data of response variables are then subjected to computer-assisted mathematical modeling and statistical optimisation.

5.4.5 Computer-Assisted Statistical Data Analysis

Computer-assisted statistical data analysis of formulation batches can be carried out in sequential steps, which starts with experimental data input, calculating the coefficient of equation, mathematical model selection, statistical evaluation of the model using analysis of variance (ANOVA) or Student's *t* test, model diagnostic plots, response mapping and finally search for an optimum using either graphical optimisation or numerical optimisation (Bolton and Bon 1997; Box and Wilson 1951; Fisher 1925; Lindberg and Lundstedt 1995; Myers and Montgomery 1995). The layout of sequential steps of data analysis is depicted in Figure 5.11.

5.4.5.1 Experimental Data Input

The resultant data of the experimental batches, which are performed as per the experimental design, are fed in the software for statistical evaluation. The software used for experimental design is further explained in Section 5.6.

5.4.5.2 Transformation of Experimental Data

It is an important component of any data analysis. Transformation of experimental data is needed when there is an assumption of error in the magnitude of response.

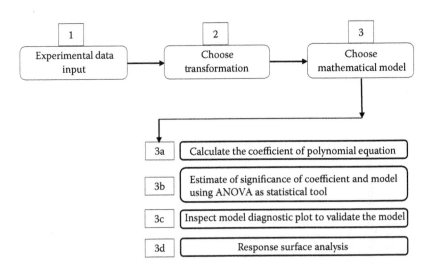

FIGURE 5.11 Layout of sequential steps of data analysis.

Several mathematical transformations are square root, natural log, base 10 log, inverse square root, inverse and power. The need for mathematical transformation of experimental data is further confirmed by diagnostic plot, such as Box–Cox plot, normal plot of the residuals and so on.

5.4.5.3 Mathematical Model Selection

Mathematical model is an expression that defines the dependence of the response variable on the independent variables. Mathematical models can be either theoretical or empirical. A theoretical model is often a nonlinear model, where transformation of a linear function is generally not possible. This model is rarely used in pharmaceutical product development. The empirical model is generally employed to describe the factor–response relationship. Most frequently, it is the set of polynomials of a given order or degree. The models mostly employed are first-order, second-order and, very occasionally, third-order polynomials. The first-order model is commonly used to describe the response, but if it fails, then the high-order model is followed. Implication of an effective mathematical model is the most critical part as inappropriate selection of the model results in confounding or high variance in the prediction and low reliability in the prediction of the optimum. Hence, special consideration should be given for model selection.

The most popularly used mathematical models are expressed as follows:

$$\text{Linear } Y = \beta_0 + \beta_1 X_1 + \beta_2 X_2 \tag{5.2}$$

$$\text{2FI } Y = \beta_0 + \beta_1 X_1 + \beta_2 X_2 + \beta_{12} X_1 X_2 \tag{5.3}$$

$$\text{Quadratic } Y = \beta_0 + \beta_1 X_1 + \beta_2 X_2 + \beta_{12} X_1 X_2 + \beta_1 X_1^2 + \beta_2 X_2^2 \tag{5.4}$$

where Y represents the measured responses, X_1 represents the value of the factor, β_0 represents the mean and β_1 and β_2 are the constants representing coefficients of first-order terms and coefficients of the second-order quadratic terms, respectively. The $X_1 X_2$ term indicates the interaction term among the variables while X_1^2 and X_2^2 represent the presence of curvature. Once the model is hypothesised, the coefficients of the polynomial are calculated using regression analysis or several other methods.

1. *Calculation of Coefficient of Polynomial Equation:* Regression is the most commonly used method for quantitative factors and should be performed in coded fashion. Generally, the least-squares regression method is used for estimating the coefficient in a linear model (Equation 5.2). This method is suitable for one or two factors. For multiple factors, multiple linear regression analysis is used (Equation 5.4). In certain cases wherein the factor–response relationship is nonlinear, multiple nonlinear regression analysis may also be performed.

In case the number of variables is high, such as in multivariate studies, the methods of partial least squares or principal component analysis is used for regression analysis (Dumarey et al. 2015). This method is an extension to multiple linear regression analysis and used when there are fewer observations than the number of predictor variables.

2. *Estimation of Significance of the Coefficient and Model:* Estimation of significance of the coefficient and the mathematical model is described in the following paragraph.

   a. *Significance of Coefficient:* ANOVA is the statistical tool used for estimation of the significance of coefficients as well as lack of fit for models using Yates algorithm. It is advisable to retain only significant coefficients in the final model equation. The significance of coefficient can be determined by comparing Pearsonian coefficient determination (r^2) and their adjusted values for degrees of freedom $\left(r^2_{adj}\right)$ of the polynomial equation. According to the model, the value of r^2 is the proportion of variance explained by the regression and is the ratio of the explained sum of squares to that of the total sum of squares (Equation 5.5).

$$r^2 = \frac{SS_{TOTAL} - SS_{RESIDUAL}}{SS_{TOTAL}} \tag{5.5}$$

$$SS_{TOTAL} = \sum_{i=0}^{n}(y_i - \bar{y}_i)^2 \tag{5.6}$$

$$SS_{RESIDUAL} = \sum(y_i - \hat{y}_i)^2 \tag{5.7}$$

where y_i is the experimental response value and $\hat{y}_i$ is the predicted response value. The residual sum of squares therefore represents deviation of experimental data from the model.

The closer is the value of r^2 to unity, the better is the fit and, apparently, the better is the model. However, this parameter has limitations to its use in multiple linear regression analysis. Hence, an alternative parameter, that is, $\left(r^2_{adj}\right)$, is preferred, whose value is usually less than r^2. The value of $\left(r^2_{adj}\right)$ is calculated using equivalent mean squares (MS) in place of the sum of squares (SS), as described in Equation 5.8.

$$r^2_{adj} = \frac{MS_{TOTAL} - MS_{RESIDUAL}}{MS_{TOTAL}} \tag{5.8}$$

$$MS_{TOTAL} = \frac{SS_{TOTAL}}{Degree\ of\ Freedom} \tag{5.9}$$

$$MS_{\text{RESIDUAL}} = \frac{SS_{\text{RESIDUAL}}}{\text{Degree of Freedom}} \qquad (5.10)$$

3. *Predicted Residual Error Sum of Squares (PRESS):* The PRESS statistics indicate how well the model fits the data. The PRESS for the chosen model should be small in comparison to the other models under consideration. PRESS is calculated as the sum of squared differences between the observed values (Y_i) and the predicted values ($\hat{Y}_i$), using the leave-one-out method. Ideally, the value should be zero or close to it (Armstrong and James 2006; Lewis et al. 1999)

$$\text{PRESS} = \sum (Y_i - \hat{Y}_i) \qquad (5.11)$$

   a. *Significance of Model:* The significance test for the selected model can be varied depending upon the types of variables to be investigated and type of experimental design; for example, for a two-level factorial design and general factorial design, a half normal plot is used, while for RSM, a mixture and crossed design normal plot is used.

4. *Half Normal Plot:* Significance of an effect can be identified using a half normal plot. The effect of different variables, either positive or negative, is expected to be statistically significant. If none of the effects are significant, it would be expected that they would be all normally distributed about zero. Considering this, the half normal plot can be plotted using the absolute value of the effect on the *x*-scale and the cumulative probability of getting results on the *y*-scale. In the half normal plot, the points that fall linearly represent normal scatter and can be used as an estimate of error for ANOVA. The points that outlie from the line are considered as a significant effect in the statistical sense (Figure 5.12a).

5. *Normal Plot:* The normal plot is used for estimation of the significance of the model for RSM design, mixture design and so on. The normal plot can be plotted using the actual algebraic value of the sample versus the expected ordered value from a true population on a two-way probability scale, where the effect from the normal population represents a straight line in a normal probability plot while few effects are represented as outliers in the plot (Figure 5.12b). The outliers in the plot can be considered to have a significant effect.

6. *F test:* The *F* test, which is used to compare the variance among the treatment means versus the variance of individuals within the specific treatments, was named after Sir Ronald Fisher. A high *F* value indicates that one (or more) of the means differs from another. The *F* test is a vital tool for any kind of DoE and can be calculated using Equation 5.12.

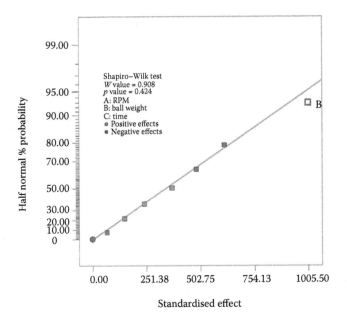

(a)

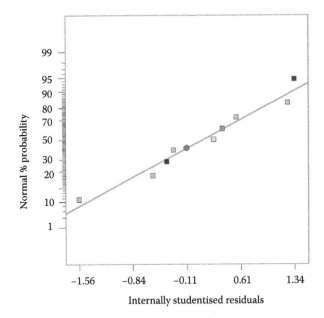

(b)

FIGURE 5.12 (a) Half normal plot. (b) Normal probability plot.

$$F \equiv \frac{s_1^2}{s_2^2} \tag{5.12}$$

where s_1 and s_2 represent variances.

This formula assumes that all samples are of equal size. The equation represents F as a ratio of signal versus noise. The F ratio increases as the treatment difference becomes larger. Thus, if the treatment has no effect, then the F ratio will be closer to a value of 1. As the F ratio increases, it becomes less and less likely that this could occur by chance. The probability of effect can be further confirmed statistically by the p value, a bottom line indicator of significance. When the F ratio becomes so high that the p value falls below 0.05, then it can be concluded with 95% confidence that one or more treatment is having an effect on the measured response.

7. *Lack of Fit:* The lack-of-fit tests compare the residual error to the pure error from the replicated design points. A lack of fit error significantly larger than the pure error indicates that there is some residual retention that can be removed by selecting a more appropriate model. If a significant lack of fit (Prob > F value 0.10 or smaller) is observed, then the model should not be used as a predictor of response.

8. *Model Diagnostic Plots:* To investigate the goodness of fit of the proposed model, the diagnostic plots shown in Table 5.13 are used.

Finally, all these parameters are assessed to aid in choosing the most appropriate model for a particular response.

9. *Response Surface Analysis:* Similar to the mathematical interpretation of the impact of input variables at different levels of various responses, graphical interpretation using a two-dimensional (2-D) plot, that is, a contour plot, and a three-dimensional (3-D) plot, that is, a response surface plot, can also be performed (Doornbos 1981; Singh et al. 2004).

   a. *Contour Plot or 2-D Plot:* A contour plot is the geometric illustration of a 3-D relationship in two dimensions, with X_1 and X_2 (independent variables) plotted on x- and y-scales and response values represented by contours (z-scale) (Figure 5.13a). Contour plots are useful for establishing the response values and operating conditions as required.

   b. *Response Surface Plot (3-D plot):* Response surface plot (3-D plot) is the graphical illustration of the potential relationship between three variables. It is obtained by plotting two independent variables on the x- and y-scales, and the response (z) variable is represented by a smooth surface (3-D surface plot) (Figure 5.13b). Similar to contour plots, 3-D surface plots are useful for establishing the response values but in a more concise manner.

5.4.6 SEARCH FOR AN OPTIMUM

On completion of data analysis of all variables, the next important step in the optimisation method is the search for an optimum. Optimisation generally aims at determining the variables that lead to the best value of the response. This response could

TABLE 5.13
Model Diagnostic Plot

Model Diagnostic Plots	Application and Plot Inferences
Predicted vs. actual	• The plot of predicted vs. actual response value indicates whether the observed values of responses are analogous to those predicted. • Ideally, this plot exhibits linearity with r^2 value close to unity.
Residual vs. predicted	• Residual is the difference between actual and predicted response. • The plot of residual vs. predicted response indicates an assumption of a constant variance. • Ideally, this plot has a random and uniform scatter with the points close to zero axes. • Wide distribution of points suggests the need for suitable data transformation such as logarithmic, exponential, square root, inverse and so on.
Residual vs. run	• The plot of residual vs. experimental run indicates the presence of lurking variables. • Ideally, this plot should show random and uniform scatter.
Normal probability plot	• The plot of normal% probability vs. residual indicates whether the model transformation is required or not and also predicts the significance of the effect. • Model transformation: When this plot shows an S-shaped curve, model transformation is suggested. • Effect: When the effect is substantially farther from the straight line fit into the normal plot, it is considered significant.
Box–Cox plot	• This plot of leverages vs. run number determines whether the transformation is required or not.

be kept as a target value, or in range or within maximum and minimum limits. Optimum composition of variables to attain the desired responses is achieved using four different methods, as depicted in Figure 5.14 and briefly discussed in Sections 5.4.6.1 through 5.4.6.4 (Armstrong and James 2006; Doornbos 1981; Lewis et al. 1999; Singh et al. 2004).

5.4.6.1 Graphical Method

The graphical methods are widely preferred for systems that possess less than four factors and one response. The different types of graphical methods are briefly described in Table 5.14.

5.4.6.2 Mathematical Method

The mathematical model is the preferred method for a system that possesses multiple factors (two to six) and multiple responses. The commonly used mathematical method for optimum search is the desirability function.

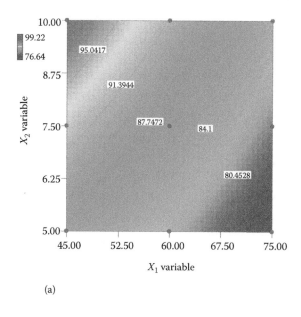

(a)

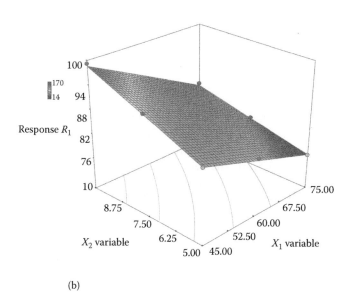

(b)

FIGURE 5.13 **(See colour insert.)** (a) Contour plot. (b) 3-D plot.

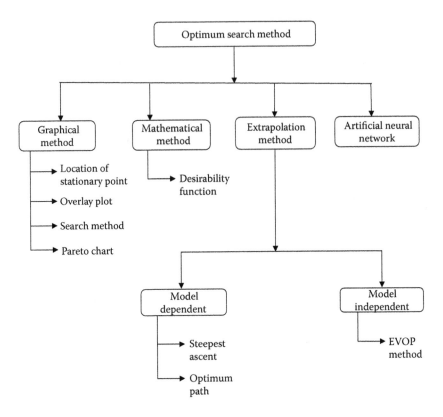

FIGURE 5.14 Optimum search methods.

Desirability Function: The desirability function is the most commonly used optimum search method for multiple responses. In optimisation, each response (i) is associated with its own partial desirability function (d_i). The overall desirability function (D) can be calculated by multiplying the partial desirability function and taking its root (Equation 5.13). The desirability function value varies from 0 to 1. The nearer the value is to 1, the better will be the response map to its target value. This numerical optimisation by desirability function gives a single point, and thus, to complete the determination of optimum, the overall function is graphically mapped over the domain using the contour plot.

$$D = \left[\prod_{i=1}^{r} d_i \right]^{\frac{1}{r}}$$
(5.13)

5.4.6.3 Extrapolation Method

The extrapolation method is used when the experimental domain is unknown. The different types of extrapolation methods are briefly listed in Table 5.15.

TABLE 5.14

Graphical Method of Optimum Search

Method	Salient Features
Location of stationary point	Graphical interpretation of optimum by identifying the stationary point location. Stationary point could be minimum, maximum or target value. The target value stationary point is known as *saddle point*.
Search method	Used for choosing the upper and lower limits of the response of interest. Two methods: feasibility search and grid search. *Feasibility search:* locates the set of responses that exhibit feasible attributes within the given constraints. *Grid search:* where the selected experimental range is further divided into a grid of a smaller size and searched methodically for the response value.
Overlay plot	Overlay plot or combined contour plot: where the response surface of all variables are superimposed over each other to get the optimum. Minimum and maximum values need to be set. Use of overlay plot is limited only to 3–4 response variables.
Pareto optimality charts	Pareto chart is used when multiple criteria need to be considered to get an optimum, and hence it is also known as multiple criteria decision-making plot. The graph is plotted between the predicted value of the objective and variables to determine the magnitude and the importance of an effect. The graph displays the absolute value of the effects and draws a reference line on the chart. Any effect that extends beyond this reference line is potentially important.

5.4.6.4 Artificial Neural Network

The Artificial Neural Network (ANN) is a model-independent neural computing design that was first proposed by Aleksander and Morton. This design mimics the working principle of the brain where the interconnecting adaptable nodes collect bunches of information, store it in the form of knowledge like neurons and make this knowledge available to forecast the response. A typical ANN consists of one input layer, one output layer and one or more hidden layers. The information is passed from the input layer to the output layer through the hidden layers by the network synapses. The modelling of ANN involves iterative training of networks with random sets of synaptic weight. The synaptic weight is the strength of connection between any two units. The training of a network is continued till the connection weight is adopted to become the memory unit and learning is achieved. The completion of learning will be further confirmed by δ error. The δ error is the difference between the output of the network and the measured value of the response. Once learning is completed, the memory unit will be then used to predict the output from the sets of new input values (Armstrong and James 2006; Singh et al. 2004).

TABLE 5.15
Extrapolation Method for Optimum Search

Method	Salient Features
Steepest ascent method	The steepest ascent method is the direct optimisation method for first-order design.
	Used when the optimum is outside the domain.
	Method: Response is set to maximise and is determined as a function of coded variables X_1, X_2.
	As depicted in the figure, steepest ascent (Δx) is obtained by taking the partial derivative of the maximum response and taking it as zero. A straight line is drawn from the central region of interest, followed by experimentation at suitable spacing along this line and subsequent measurement of response. This is continued until an optimum is achieved.
	For single response.
Optimum path method	The optimum path method is mainly used for second-order design.
	This method is parallel to the steepest ascent method.
Evolutionary operation (EVOP) (Chattopadhyay et al. 2015)	EVOP is an alternative methodology for optimisation of industrial process.
	In this method, two to three variables that were found to be critical during scale-up study are to be taken in a systematic manner and statistically analysed to search for an optimum. The process is continued until the optimum peak is obtained.
	Generally, two-level factorial or simplex designs are most commonly employed to search for an optimum.
	This method has limited application in pharmaceutical technology.
	Advantages
	Useful when prior knowledge about the factor and level does not exist.
	Disadvantages
	Impractical method for implementation in pharmaceutical product development.
	Expensive method.

Application of ANN in product development is increasing day by day (Bourquin et al. 1997). ANN represents superiority over multilinear regression methodologies in the case of many complex systems such as fluidised bed granulation, tableting process and so on. However, this design is difficult to implement when a higher number of factors come into consideration.

5.5 VALIDATION OF THE OPTIMISATION METHOD

As DoE is performed using computer-assisted software, the reliability of the selected software needs to be assured in order to confirm the reproducibility of predicted data (Armstrong and James 1990; Singh et al. 2004, 2011). To validate the design, check-point batches should be conducted, which provide the desired value of the dependent variables after taking an appropriate ratio of the independent variables. The experimental results of all checkpoint batches should be compared to predicted values from point-to-point prediction. Furthermore, the linear regression plots between predicted and experimental values of all the response variables should be determined along with the percentage error in prognosis. Also, the model fitness parameters such as r^2, r^2_{adj} and PRESS guide the prognostic ability of the model.

5.6 SOFTWARE USED FOR DoE

Use of software for optimisation makes the task a lot easier, faster, elegant and economical. Software used for DoE can be selected based on several features such as presence of design for screening and RSM optimisation, choice of suitable model fitness, amenities to randomise the order of experimental runs, simple and user-friendly graphic user interface and so on. Several software packages commonly used in DoE are given below:

Design Expert (http://www.statease.com)
Minitab (http://www.minitab.com)
MODDE (http://www.umetrics.com)
STAVEX (http://www.aicos.com/front_content.php?idart=36&changelang=2)
DoE Fusion PRO (http://www.sigmazone.com)
STASTISTICA (http://www.statsoftinc.com)
DoE WISDOM (http://www.launsby.com)
JMP (http://www.jmp.com)
OPTIMA (http://www.optimasoftware.co.uk)

5.7 CASE STUDY

The utility of DoE in quality pharmaceutical product development can be simplified using a case study conducted by Patravale et al. (data unpublished). The case study describes the development of the sustained-release floating matrix tablet of verapamil hydrochloride (HCl) prepared using the direct compression method. Herein, DoE was employed as a systematic tool for optimisation of the floating matrix tablet. The product development as per DoE is summarised in sequential stages discussed in Sections 5.7.1 through 5.7.7.

5.7.1 SCREENING INFLUENTIAL VARIABLES

Considering the target of the developing sustained-release floating tablet, the criticality among formulation and process parameters was screened using risk assessment analysis based on prior knowledge. Based on a risk assessment study (Figure 5.15), the critical formulation variables for the development of the floating matrix tablet were identified as sustained-release polymers (e.g. hydroxypropyl methylcellulose [HPMC]), which control drug release, and gas-generating agents (e.g. sodium bicarbonate), which maintain the integrity of the tablet for longer periods in the stomach. As the process used was direct compression, process parameters such as

Drug Product CQAs	Concentration of Swellable Polymer	Concentration of Gas-Generating Agent	Compaction Force	Compression Speed
% Friability	Low	Low	Medium	Medium
Floating lag time	High	High	Medium	Medium
Drug content	Low	Low	Low	Low
Drug release (Q_{t8h})	High	High	Medium	Medium

FIGURE 5.15   Risk assessment analysis.

compaction force and compression speed were considered as medium risk and could be controlled.

5.7.2   QUANTIFICATION OF INFLUENTIAL VARIABLES

The identified critical variables were further screened to quantify the upper and lower levels that affect the target profile of the product.

5.7.3   EXPERIMENTAL DESIGN SELECTION

From screening studies, two variables were identified as high risk. Considering this, the 3^2 full factorial design was chosen for systematic optimisation of the floating matrix tablet. The influential factors HPMC (X_1) and sodium bicarbonate (X_2) and their influencing concentrations were selected as independent variables based on risk assessment analysis. The independent variables and their levels are depicted in Table 5.16. The formulation design of 3^2 full factorial batches of nine different combinations and their responses is depicted in Table 5.17.

5.7.4   COMPUTER-ASSISTED STATISTICAL ANALYSIS OF DATA

The results of 3^2 factorial batches were initially fitted into the Design Expert version 9.0 software for mathematical modeling and response surface mapping to get an optimum composition.

TABLE 5.16
Independent Variables and Their Levels

Sr. No.	Independent Variable	Coded Value		
		0	1	2
1	HPMC (X_1) (mg)	45	60	75
2	Sodium bicarbonate (X_2) (mg)	5	7.5	10

TABLE 5.17

3^2 Full Factorial Batches with Their Results

Sr. No.	Batches	X_1	X_2	Floating Lag Time (Y_1)	Q_{t1h} (Y_2)	Q_{t8h} (Y_3)
1	F1	45	5	170	33.73	86.77
2	F2	45	7.5	98	37.99	92.93
3	F3	45	10	45	41.07	99.22
4	F4	60	5	110	28.31	82.18
5	F5	60	7.5	50	31.21	86.41
6	F6	60	10	25	36.12	90.26
7	F7	75	5	65	23.45	76.64
8	F8	75	7.5	35	26.91	80.52
9	F9	75	10	14	32.89	84.31

5.7.4.1 Mathematical Modeling

Mathematical relationships generated using multiple regression analysis for the studied response variables are expressed in Equations 5.14 and 5.15.

Floating Lag Time (Y_1): The results show that the amount of swellable polymer HPMC (X_1) and sodium bicarbonate (X_2) plays an important role in floating lag time. HPMC being a swellable polymer rapidly hydrates in water to form a gel layer where the CO_2 bubbles are trapped. The trapped CO_2 bubbles decrease the density of the tablet and thereby decrease the floating lag time. Similarly, sodium bicarbonate, when it comes in contact with 0.1 N HCl, results in increased generation of CO_2. The generated gas trapped within the gel layer formed by the HPMC polymer decreases the density of the tablet and thereby decreases the floating lag time. The quadratic equations for the reduced model in coded units are as follows:

$$Y_1 = +54.67 - 33.17X_1 - 43.50X_2 + 18.50X_1X_2 + 9.50X_1^2 + 10.56X_2^2 \quad (5.14)$$

$$Y_2 = +84.57 - 6.22X_1 + 4.68X_2 - 1.16X_1X_2 + 0.54X_1^2 + 0.033X_2^2 \quad (5.15)$$

Equation 5.14 shows that the coefficient X_1 bears a negative sign, indicating that Y_1 decreases with an increase in X_1 concentration. Also, X_2 bears a negative sign, indicating that Y_1 decreases with an increase in concentration of sodium bicarbonate. Similarly, the mathematical relation between two variables X_1 and X_2 against Q_{t8h} (Y_2) was generated.

5.7.4.2 Model Fitting and Statistics of the Measured Responses

The significance of the model applied to the design was further validated statistically using ANOVA. The ANOVA computation was performed using Yates algorithm to find the significance of each coefficient based on F statistics. The F statistics for the responses Y_1 and Y_2 are detailed in Table 5.18. The data in Table 5.18 show that the model F values for the Y_1 and Y_2 variables are 196.15 and 170.26, respectively, implying that the model is significant. There is only a 0.02% chance that a model F value this large could occur because of noise. Further, the significance of the model can also be predicted from the r^2 value (Table 5.18). Statistically, high values of r^2 for all dependent variables indicate a good fit. Adj r^2 and Pred r^2 values were also in reasonable agreement, particularly for reduced models, signifying a good model fit (Table 5.18). Better Pred r^2 values obtained for the reduced model might be attributed to the elimination of insignificant terms.

5.7.4.3 Model Diagnostic Plot

The model diagnostic plots, that is, normal plot of residuals, residuals versus predicted, residuals versus run, predicted versus actual and Box–Cox plot for floating lag time and Q_{t8h}, are depicted in Figure 5.16a through e. The normal plot of residuals for responses reveals that the residuals appear to follow a straight line and thus the existence of nonnormality, outliers, skewness or unidentified variables can be ruled out. From the plot of residuals versus run and residuals versus predicted of all responses, it can be stated that residuals appear to be randomly scattered and the existence of missing terms and nonconstant variance can be ruled out. The Box–Cox plot of all the responses reveals that the confidence interval of the lambda value is around 1 and hence no specific power transformation is required.

5.7.5 Response Surface Analysis

The mathematical relation was further expressed graphically using response surface analysis. Response surface analysis is the graphical mapping of the effect of independent variables on dependent variables. Figure 5.17a and b portrays the contour plot and 3-D response surface plots for the studied response properties, namely, Y_1 and Y_2. Both the plots show an inverse relation between factor and response variables, wherein as X_1 and X_2 values increased from 0 level to 2 levels, Y_1 and Y_2 were found to decrease linearly.

TABLE 5.18
ANOVA Results – Effect of Independent Variables on the Measured Responses

Mass Response	Sum of Square	DF	F Value	Prob $> F$	PRESS	r^2	Adj. r^2	Pred r^2
FLT	19,723.67	5	196.15	0.0006	502.67	0.9970	0.9919	0.9746
Q_{t8h}	372.58	5	170.26	0.0007	16.00	0.9965	0.9906	0.9572

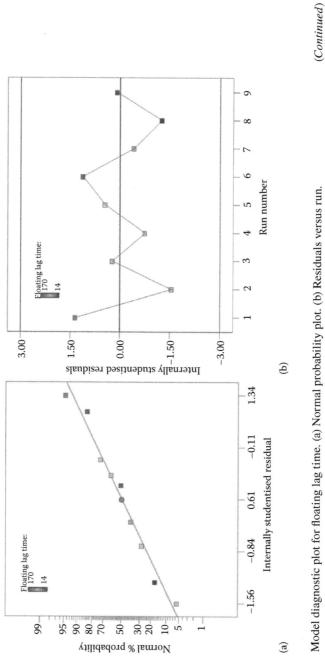

FIGURE 5.16 Model diagnostic plot for floating lag time. (a) Normal probability plot. (b) Residuals versus run.

(Continued)

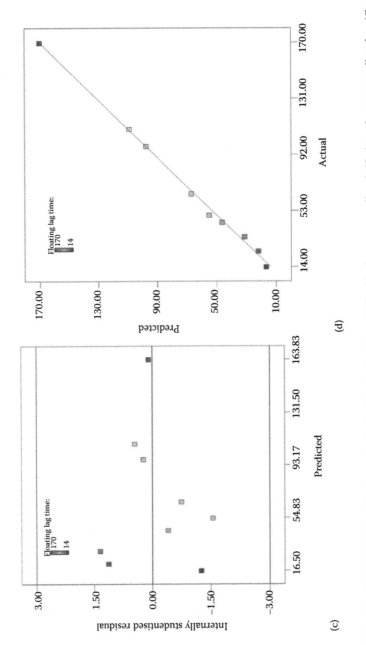

FIGURE 5.16 (CONTINUED) Model diagnostic plot for floating lag time. (c) Residuals versus predicted. (d) Actual versus predicted. *(Continued)*

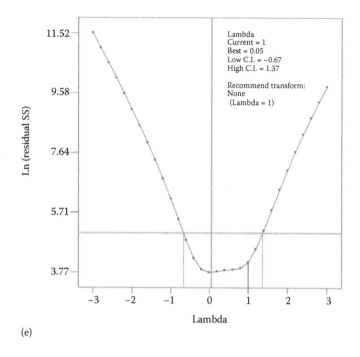

Lambda
Current = 1
Best = 0.05
Low C.I. = −0.67
High C.I. = 1.37

Recommend transform:
None
(Lambda = 1)

(e)

FIGURE 5.16 (CONTINUED) Model diagnostic plot for floating lag time. (e) Box–Cox plot.

5.7.6 Optimisation Using Desirability Function

The desirability function was calculated for floating lag time and Q_{t8h}. Based on the acceptance limit for the response variable (Table 5.19), the software provided several combinations of both factors that show a desirability value of 1.

5.7.7 Validation of Optimised Formulations

The three checkpoint formulations taken from the overlay plot (Figure 5.18) were evaluated to validate the design. The predicted and experimental values of all the response variables and percentage error in prognosis are depicted in Table 5.20. The linear correlation plots between the observed and the predicted value of floating lag time and Q_{t8h} are shown in Figure 5.19. The linear correlation plots drawn between the predicted and the observed responses demonstrated a higher value of r^2, indicating that the software is validated and can be used further to develop a control strategy for the verapamil HCl-based floating matrix tablet. Comparison of the observed response with that of the anticipated responses reveals the prediction error as well as the significant value of r^2 in the current study indicates a high prognostic ability of floating matrix tablet formulation of verapamil hydrochloride using response surface method of optimisation validation.

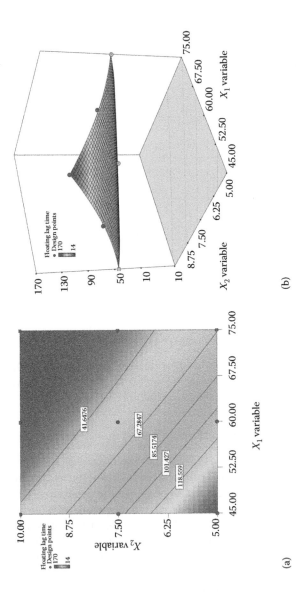

FIGURE 5.17 **(See colour insert.)** Response surface plot. (a) 2-D contour plot of floating lag time. (b) 3-D response surface plot of floating lag time.

TABLE 5.19

Acceptance Criteria of the Response Variable

Code	Dependent Variable	Acceptance Limit
Y_1	Floating lag time	Less than 60 s
Y_2	Quantity released in 8 h (Q_{8h})	80%–100%

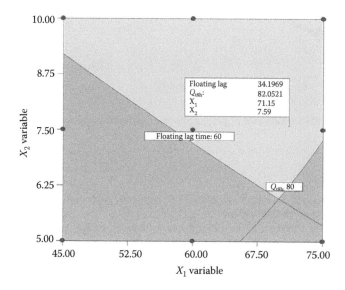

FIGURE 5.18 Overlay plot for the verapamil HCl floating tablet.

TABLE 5.20

Checkpoint Analysis

Formulation	Composition		Response Variable	Predicted Value	Experimental Value	% Error
	X_1	X_2				
V1	60.08	7.51	Floating lag time	58.55	60.7	3.67
			Q_{8h}	96.07	95.92	0.15
V2	64.34	7.44	Floating lag time	56.28	58.27	3.53
			Q_{8h}	94.27	93.56	0.75
V3	66.45	7.48	Floating lag time	55.36	57.23	3.37
			Q_{8h}	90.87	90.91	0.04

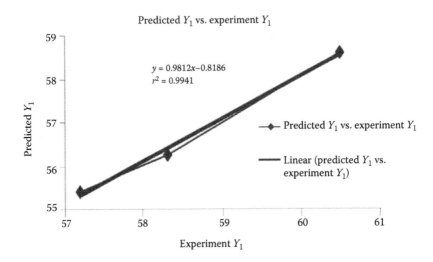

Predicted Y_1 vs. experiment Y_1

$y = 0.9812x - 0.8186$
$r^2 = 0.9941$

⬥ Predicted Y_1 vs. experiment Y_1

— Linear (predicted Y_1 vs. experiment Y_1)

FIGURE 5.19 Correlation plot of predicted and experimental values for floating lag time (Y_1).

REFERENCES

Anderson, M., Kraber, S., Hansel, H. et al. 2002. *Design Expert® Software Version 6 User's Guide*. Minnesota: Statease Inc.

Araujo, P. W. and Brereton, R. G. 1996. Experimental design II. Optimization. *Trends in Analytical Chemistry* 15:63–70.

Armstrong, N. A. and James, K. C. 1990. *Understanding Experimental Designs and Interpretation in Pharmaceutics*. London: Ellis Horwood.

Armstrong, N. A. and James, K. C. 2006. *Pharmaceutical Experimental Designs and Interpretation*, 2nd edition. London: Ellis Horwood.

Bolton, S. and Bon, C. 1997. *Pharmaceutical Statistics: Practical and Clinical Applications*, 3rd edition. New York: Marcel Dekker.

Bourquin, J., Schmidli, H., Hoogevest, P. et al. 1997. Application of artificial neural networks (ANN) in the development of solid dosage forms. *Pharmaceutical Development Technology* 2:111–121.

Box, G. E. P. and Wilson, K. B. 1951. On the experimental attainment of optimum conditions. *Journal of Royal Statistical Society Series B* 13:1–45.

Box, G. E. P. and Behnken, D. W. 1960. Some new three-level designs for the study of quantitative variables. *Technometrics* 2:455–475.

Chariot, M., Lewis, G. A., Mathieu, D. et al. 1988. Experimental design for pharmaceutical process characterization and optimization using an exchange algorithm. *Drug Development and Industrial Pharmacy* 14:2535–2556.

Chattopadhyay, J., Banerjee, S., and Bhattacharyya, A. K. 2015. Comparative optimization study for chemical synthesis of hydroxypropyl starch from native corn and taro starch through evolutionary operation (EVOP) factorial design technique. *Starch* 67(1–2):54–162.

Cochran, W. C. and Cox, G. M. 1992. *Experimental Design*, 2nd edition. New York: Wiley.

Cornell, J. A. 1979. Experiments with mixtures: An update and bibliography. *Technometrics* 21:95–106.

Cornell, J. A. 1990. *Experiments with Mixtures: Design, Models, and the Analysis of Mixture Data*, 2nd edition. New York: Wiley.

De Aguiar, P. F., Bourguignon, B., Khots, M. S. et al. 1995. D-optimal designs. *Chemometric and Intelligent Laboratory System* 30:199–210.

Doornbos, D. A. 1981. Optimization in pharmaceutical sciences. *Pharmaceutisch Weekblad, Scientific Edition* 3:33–61.

Dumarey, M., Goodwin, D., and Davison, C. 2015. Multivariate modelling to study the effect of the manufacturing process on the complete tablet dissolution profile. *International Journal of Pharmaceutics* 486(1–2):112–120.

Fisher, R. A. 1925. *Statistical Methods for Research Workers*. Edinburgh: Oliver and Boyd.

Lewis, G. A. 2002. Optimization methods. In: *Encyclopedia of Pharmaceutical Technology*, eds. Swarbrick, J., Boylan, J. C., 2nd edition. New York: Marcel Dekker.

Lewis, G. A. and Chariot, M. 1991. Non classical experimental designs in pharmaceutical formulations. *Drug Development and Industrial Pharmacy* 17:1551–1570.

Lewis, G. A., Mathieu, D., and Phan-Tan-Luu, R. 1999. *Pharmaceutical Experimental Design*, 1st edition. New York: Marcel Dekker.

Lindberg, N. O. and Lundstedt, T. 1995. Application of multivariate analysis in pharmaceutical development. *Drug Development Industrial Pharmacy* 21:987–1007.

Loukas, Y. L. 1997. A 2 (k-p) fractional factorial design via fold over: Application to optimization of novel multicomponent vesicular system. *Analyst* 122:1023–1027.

Loukas, Y. L. 2001. A Plackett–Burman screening design directs the efficient formulation of multicomponent DRV liposomes. *Journal of Pharmaceutical and Biomedical Analysis* 26:255–263.

Myers, R. H. and Montgomery, D. C. 1995. *Response Surface Methodology: Process and Product Optimization Using Designed Experiments*. New York: Wiley.

Plackett, R. L. and Burman, J. P. 1946. The design of optimum multifactorial experiments. *Biometrica* 33:305–325.

Schwartz, J. B. and Connor, R. E. 1996. Optimization techniques in pharmaceutical formulation and processing. In: *Modern Pharmaceutics*, eds. Banker, G. S., Rhodes, C. T., 3rd edition. New York: Marcel Dekker.

Singh, B., Kumar, R., and Ahuja, N. 2004. Optimizing drug delivery system using systematic design of experiments. Part I—Fundamental aspects. *Critical Review in Therapeutics Drug Delivery System* 22(1):27–105.

Singh, B., Kapil, R., Nandi, M., and Ahuja, N. 2011. Developing oral drug delivery system using formulation by design: Vital precepts, retrospect and prospects. *Expert Opinion in Drug Delivery* 8(6):1341–1360.

Snee, R. and Marquardt, D. 1976. Screening concepts and designs for experiments with mixtures. *Technometrics* 18(1):19–29.

Spendley, W., Hext, G. R., and Himsworth, F. R. 1962. Sequential application of simplex designs in optimization and evolutionary operations. *Technometrics* 4(4):441–461.

Stack, C. B. 2003. Confounding and interaction. In: *Encyclopedia of Biopharmaceutical Statistics*, ed. Chow, S. C. New York: Marcel Dekker.

Stetsko, G. 1986. Statistical experimental design and its application to pharmaceutical development problem. *Drug Development and Industrial Pharmacy* 12:1109–1123.

Yu, L., Amidon, G., and Khan, M. 2014. Understanding pharmaceutical quality by design. *AAPS Journal* 16(4):771–783.

6 Preformulation Studies
Role in Pharmaceutical Product Development

Swati S. Vyas, Amita P. Surana
and Vandana B. Patravale

CONTENTS

6.1 INTRODUCTION

Preformulation testing entails all studies conducted on a drug substance towards generating practical data for the successful fabrication of a stable and biopharmaceutically apt drug dosage form. Pharmaceutical preformulation investigations can have an enormous impact on drug discovery and product development as they reveal important data with respect to the intrinsic and extrinsic properties of a drug substance and its suitability to be formulated into a medicinal dosage form. Moreover, such studies at an earlier phase can bring down drug developmental costs and duration. Before the development of any dosage form of a drug substance, it is crucial that particular fundamental physical and chemical properties of a drug substance are determined. Interestingly, the application of high-throughput technologies in previous years has expedited drug substance screening and selection, permitting useful compilation of physicochemical properties (Chen et al. 2013). These properties form the basis for fabricating a drug's combination with appropriate pharmaceutical excipients for producing the right dosage form. This information may also influence many of the ensuing processes in formulation development, such as manufacturing methods and conditions. This preliminary stage of evaluation is described as the preformulation phase and is at the interface between the drug entity and formulation development. Pharmaceutical preformulation spans across disciplines and is construed differently by different groups in an industrial research setting. For instance, medicinal chemists can manipulate structural traits of potential drug candidates on the basis of information received on the stability, solubility and cell penetrability from preformulation studies, whereas pharmacologists can use the same information to determine drug therapeutic and toxic levels, and pharmaceutical analysts can optimise assay procedures for determination of the drug substance/product. After drug candidate selection, further along the developmental stages, preformulation studies provide insight to large-scale manufacturing, dosage form development and clinical investigation processes. Drug substance manufacturers utilise information on salt form, purity and yield to control drug production and formulators integrate preformulation data such as drug physicochemical characteristics for designing a suitable dosage form for drug delivery by appropriate excipient selection and opting for the right processing conditions. Information on physicochemical properties such as solubility and permeability of drug substances, obtained from preformulation studies, can provide insights into pharmacokinetic and pharmacodynamic behaviour of the drug substances. Preformulation begins after extensive literature survey of identical or similar types of compounds. It provides an understanding of the relationship between physicochemical properties of the drug substance and its bioavailability, potential degradation, side effects/toxicity of formulation components, and helps select the appropriate dosage forms.

Preformulation dictates the choice of the form of the drug substance, formulation components, manufacturing process for drug substance and drug product, suitable container closure system, analytical methods and storage conditions for the drug substance and drug product. This chapter provides a brief account about how applicative principles of physicochemical and structural characteristics of drug substance/active pharmaceutical ingredient (API) can be of great importance in designing and developing a pharmaceutical product.

6.2 PROCESS OF PREFORMULATION

Before pursuing a proper plan of preformulation, it is imperative that researchers consider available information pertaining to various properties of the drug, therapeutic category and anticipated dose, cost and amount of drug available. The outline of preformulation activities is detailed in Figure 6.1.

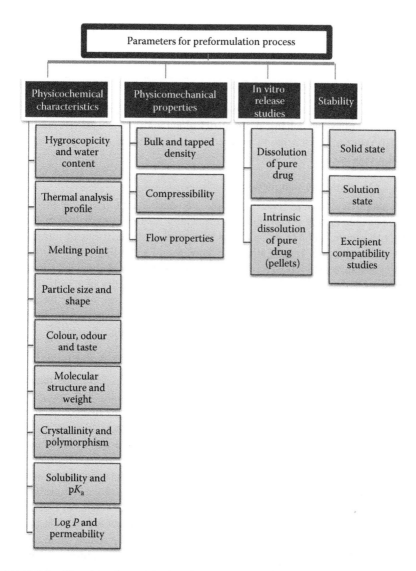

FIGURE 6.1 Flowchart for a typical preformulation process.

6.2.1 Preliminary Information

At this stage, available information about the drug substance and its desired dosage form under development is compiled from literature. Most of the information on a drug substance can be obtained from its manufacturer as a technical data package. The technical information may contain manufacturing processes employed in the production of the raw material and data pertaining to the quality and safety (exposure limits) of the raw material. Specifics such as molecular formula and structure, physical and chemical properties including degradation pathways, handling hazards, stability information and analytical methods can be obtained from the technical data package.

6.2.2 Parameters for Primary Evaluation

The preformulation process gives a detailed understanding of physicochemical characteristics and physicomechanical properties along with in vitro release studies and stability (Figure 6.2). Depending on the dosage form to be developed, desired or required properties will be studied.

6.2.2.1 Organoleptic Properties

Quality parameters such as the organoleptic and aesthetic properties of the product are indispensible during preformulation especially for oral drug delivery. A drug's molecular structure defines its colour, which relates to the extent of unsaturation present in the structure – especially the occurrence of a chromophore. Additionally, the drug substance may have an innate odour typical of chief chemical functionalities present. This property can directly affect the palatability of the final dosage form. On a related note, if taste is found unpalatable, a lesser soluble form of the drug is employed. Also, the odour and taste can be masked by inclusion of the right flavours and excipients or by coating the final product or drug substance.

6.2.2.2 Crystal Properties (Particle Size, Shape and Crystallinity)

Crystal properties have a great impact on physicochemical and biopharmaceutical properties of drugs. The solid state is important as it affects the flow and processes of filtration, tableting and dissolution. Compounds have many diverse habits based on the crystallinity (external appearance and intrinsic molecular arrangement). There may not be major variations in the bioavailability of drugs with diverse crystal habits; however, from a technological perspective, study of crystal properties is essential. For instance, the injectability of a suspension comprising a drug in crystal form is affected by its habit. Crystal properties also modulate tablet compression and powder flow characteristics of the drug in solid state. For instance, Li Destri et al. found that using tolbutamide in its plate-like crystal form caused particle bridging in the hopper of the tablet press and also precipitated capping issues while tableting (Li Destri et al. 2013). Bristol-Myers Squibb retracted antihypertensive formulation of hydrochlorothiazide and irbesartan (Avapro tablets) in January 2011, because of problems associated with dissolution of irbesartan (Wang et al. 2008). Irbesartan formed secondary crystals in some batches, which led to poor solubility. The formation of the less-soluble crystal form was attributed to tablet manufacturing changes. Thorough investigations disclosed that longer granulation times caused production of secondary crystals.

Particle size affects critical parameters such as solubility, dissolution rate, uniform distribution, suspendability, lack of grittiness and absorption; it can also be a factor for instability. Finer particles are comparatively more susceptible to atmospheric oxygen, humidity and interacting excipients than their coarser counterparts. Commonly, particle size is determined by sieving, microscopy, sedimentation rate method, laser techniques and light scattering.

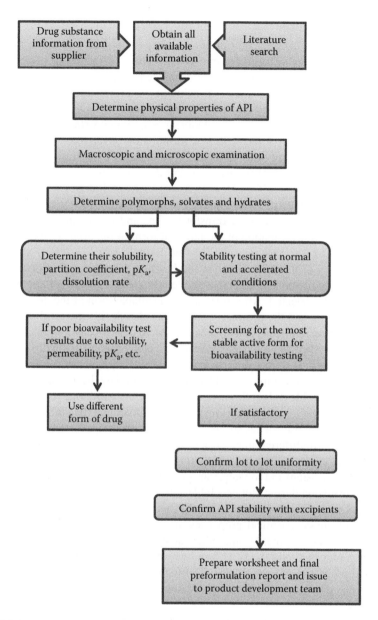

FIGURE 6.2 General parameters considered during preformulation studies.

6.2.2.3 Polymorphism

Polymorphism is the capacity of a drug substance to crystallise in more than one particular crystal type with diverse intrinsic structures. Creation of diverse polymorphs is controlled by factors such as solvents, temperature, pressure and cooling rate. Polymorphs vary considerably in solubility, thermal profile, density, hardness, crystal type, optical and elemental characteristics (e.g. vapour pressure). The selected solid form of a drug for development must remain stable in formulation throughout the shelf life of the API. For formulation development, generally the crystalline form of a drug with defined and reproducible physical characteristics is preferred. Polymorphic transitions can occur during milling, granulation, drying and compressing processes. Transformations between solid phases in dosage forms often affect the drug release process and, therefore, it is important to understand the mechanisms and kinetics of phase transformations and factors that may influence them. The risk of transformation of the physical form of the drug substance can be reduced by selecting a stable physical form of the API; however, it may also affect the bioavailability of the API. Thus, on the basis of the collective information on the bioavailability of stable, metastable and least stable forms of API, an appropriate form should be selected. After selection of a suitable form of the drug, the process of formulation and the prevailing conditions also decide the solid state of the drug in the formulation. This phenomenon is referred to as processing-induced phase transition. For finished tablet formulations, in situ conversion additionally influences characteristics such as porosity, particle–particle interactions and hardness of the tablet. In situ phase transitions of API in formulation have been exemplified with thiamine hydrochloride and erythromycin. Thiamine hydrochloride transforms from its nonstoichiometric hydrate (NSH) form to a hemihydrate (HH) form in the presence of water. The HH form has stronger molecular assembly compared to the NSH form, and both have different microstructures and physical properties. In one study by Chakravarty et al. (2012), tablets of thiamine hydrochloride (NSH form) made by dry (direct compression) and wet (aqueous granulation) techniques were observed for phase transformation during storage. Tablets prepared by wet granulation showed a faster rate of phase transformation to the HH form, which was evidenced by major differences in physicochemical properties and an increase in disintegration time, tablet hardness and volume. On the other hand, in tablets of NSH made by direct compression (dry method), the NSH form was retained during storage as the physicochemical properties did not change. Eliminating exposure of NSH to water molecules during tablet manufacture reduced the incidence of phase transformation significantly.

Abbott Laboratories documented failure of erythromycin tablet dissolution as a result of creation of isomorphic dehydrates in situ (Bauer et al. 1999). Failure of pharmacopoeial specifications for dissolution was seen in particular batches of 1-year stability samples of erythromycin dihydrate tablets (label claim 250 mg). Batch no. 15, which displayed excellent preliminary dissolution (>99% at 60 min), failed uniformly at the 1-year time point (71% release at 1 h). Milling and drying processes employed during the tablet manufacture caused variations in crystal properties and water content. During tablet manufacture, erythromycin dihydrate loses water of crystallisation, slowly interacts with hydroxyl-containing excipients such as magnesium hydroxide and consequently leads to prolonged dissolution. Various techniques used for studying polymorphism are hot stage microscopy, infrared spectroscopy, single crystal x-ray and powder x-ray diffraction (PXRD) and thermal analysis.

6.2.2.4 Thermal Analysis Profile

Melting point is defined as the temperature at which the solid and liquid phases are in equilibrium and is characteristic to a drug substance. There are a number of interrelated thermal analytical techniques that can be used to characterise the salts and polymorphs of drug molecules. Giron has reviewed thermal analytical and calorimetric methods used in the characterisation of polymorphs and solvates (Giron 1995). Endothermic processes are fusion, boiling, vapourisation, chemical degradation and solid–solid transition. Exothermic processes are crystallisation and degradation. From amongst the thermal methods available for investigating polymorphism and related phenomena, differential scanning calorimetry (DSC), thermal gravimetric analysis (TGA) and hot stage microscopy are the most widely used methods.

6.2.2.5 Hygroscopicity

Hygroscopicity is the interaction of a material with moisture in the atmosphere (Newman et al. 2007), that is, the tendency of a solid to take up water from the atmosphere as it is subjected to a controlled relative humidity (RH) program under isothermal conditions. The propensity of a compound to absorb moisture is an important aspect to consider during the selection of a drug. Moisture absorption can depend upon humidity, temperature, surface area and mechanism for moisture uptake. Properties such as the crystal structure, powder flow, compaction, lubricity, dissolution rate and permeability of polymer films can be affected by moisture absorption (Ahlneck and Zografi 1990). As a paradigm of its significance in inhalation dosage forms, Young et al. (2003) studied the influence of humidity levels on the aerosolisation performance of micronised sodium cromoglycate, salbutamol sulfate and triamcinolone acetonide. They reported that discharge of fine particles and delivered dose for both salbutamol sulfate and sodium cromoglycate decreased with higher RH levels. Moisture is an important factor that can also affect the stability of drugs and their formulations. Absorption of water molecules by a drug (or an excipient) can frequently cause hydrolysis. The effect that moisture has on stability relates to the binding strength, that is, whether the water molecules are bound or unbound. Normally, degradation results as a function of unbound moisture, which could be attributed to its capacity to change the pH of the surfaces of drug and excipient (Wang et al. 1984). The interaction of water molecules with drug–excipient blends may cause surface dissolution and changes in pH, which in turn could induce degradation. For instance, remacemide hydrochloride salt mixed with lactose seemed stable even at storage conditions of 90°C for 12 weeks. However, the experiment conducted in the presence of moisture resulted in general degradation owing to Maillard reaction. As specified by the European pharmacopoeia, the degree of hygroscopicity can be performed by using a loss on drying bottle (50 mm external diameter and 15 mm high), which is placed in a desiccator at 25°C containing a saturated solution of ammonium chloride or ammonium sulfate or in a climatic cabinet set at 25°C/80% RH and allowed to equilibrate for 24 h after which the percentage increase in mass is calculated. While performing this method, maintenance of controlled hygroscopicity conditions is essential. This technique gives a hint about the degree of hygroscopicity instead of a true value. Halogen or infrared moisture analysers are also used for moisture determination at fixed temperature and fixed times.

6.2.2.6 Partition Coefficient

Lipophilicity of a drug substance is typically described in terms of the partition coefficient (log P). Log P is the ratio of concentration of the unionised drug substance partitioned between the organic and aqueous phases at equilibrium. Notably, the scale is logarithmic; hence, a log P value of zero indicates that the drug substance has equivalent solubility in water and the lipophilic partitioning medium. If the compound has a log P of 5, then the compound is 100,000 times more soluble in the partitioning solvent. A value of log $P \leq 2$ implies that the drug substance is 100 times more soluble in water and is therefore hydrophilic. Log P values can be determined by the octanol–water system, which is commonly adopted as a model of the aqueous–lipid biphasic system (Leo et al. 1971).

As per Lipinski's rule, log P <5 is optimal for a drug substance to show drug-like properties. By and large, drugs with log P values between 1 and 3 display excellent absorption properties, but those with log P >6 or <3 display less efficient transport. Highly lipophilic drug substances preferentially exist in the lipidic regions of cell membranes, and highly hydrophilic drugs demonstrate low bioavailability because of their failure to cross membrane barriers. Therefore, there is a synchronous relationship between log P and transport of drug substances.

6.2.2.7 Permeability

Permeability of a drug substance is defined as the capability of drug molecules to cross cell membrane barriers, such as the intestinal membranes or the blood–brain barrier. Determination of permeability of a drug substance is key to predicting absorption, distribution, metabolism, excretion and toxicity profile of the drug substance. In general, lipophilic drugs show higher permeability than hydrophilic molecules since they can easily cross the lipid bilayer membranes of cells and are therefore absorbed faster. Steroids such as morphine, for instance, have very high lipophilicity and can therefore rapidly cross cell membranes, thereby displaying a faster onset of action. According to the US Food and Drug Administration guidance document for oral absorption studies of solid dosage forms, permeability can be determined by calculating the extent of absorption (in situ animal and in vitro epithelial culture methods – Caco 2 cells) of the drug substance and determination of the rate of mass transfer of the drug substance across the human intestinal membrane. A drug is considered to be highly permeable if the extent of absorption is >85% of the administered dose in comparison to a reference dose given intravenously.

6.2.2.8 Solubility

The amount of substance that passes into solution in order to establish equilibrium at constant temperature and pressure to produce a saturated solution is defined as solubility. The solubility of a drug is an important physicochemical property because it affects the rate of drug released into dissolution medium and, therefore, the bio-availability of the drug. The solubility of materials is usually determined by the equilibrium solubility method, which employs a supersaturated solution of the mate-rial obtained by stirring an excess of material in solvent for a prolonged time at a particular temperature until equilibrium is achieved. The solubility should ideally be measured at (i) 4°C because the minimum density of water occurs at 4°C (minimum aqueous solubility), (ii) 25°C ambient temperature and (iii) 37°C. After stirring or shaking, the solvent is separated from the suspension by centrifugation or filtra-tion using polytetrafluoroethylene filters or other suitable filters. The filtrate is then assayed preferably by high-performance liquid chromatography (HPLC), although ultraviolet (UV)–visible spectroscopy can also be used to determine the solubility, if the degradation products or impurities do not interfere with the estimation. On the basis of predefined protocol, data need to be generated (initial solubility followed by other time points) till maximum equilibrium solubility is reached (Figure 6.3). It is recommended that the experiment is conducted at least overnight, but longer periods may be required if the compound has a very low solubility; for example, saturation of morphine in water at 35°C was obtained only after approximately 48 h (Roy and Flynn 1989). It may also be useful to measure the pH of the filtrate (if the experiment is conducted in water) and to analyse any undissolved material by DSC or PXRD for detecting any phase changes that may have occurred during the course of the experiment.

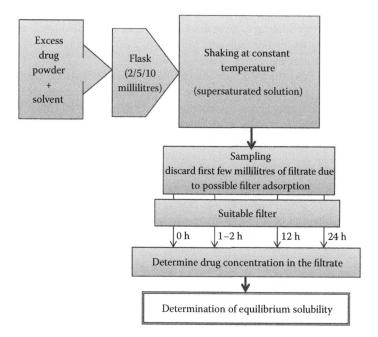

FIGURE 6.3 Equilibrium solubility determination.

Determination of solubility is very critical and hence the following points should be considered:

- The solvent and solute must be pure.
- No sample should be removed for solubility determination before super-saturated solution is achieved. The technique of sample separation from undissolved solute in a saturated solution should be acceptable.
- The method of analysis of the solution must be reliable.
- Temperature should be sufficiently controlled and recorded.

Plot of solubility against temperature is commonly used for solubility determination. Solubility data of a drug substance aids in deciding the suitable media for dissolution testing and for selecting the correct solvent or pH for the function of granulation and coating. The maximum absorbable dose can also be determined from the solubility data (Johnson and Swindell 1996). The biopharmaceutical classification system (BCS) is very helpful in understanding the relationship of solubility and permeability of a drug substance to its potential bioavailability. Rate and extent of drug absorption from dosage form depend on solubility, permeability of the API and dissolution of the product. BCS segregates drugs into four classes on the basis of their solubility and permeability: Class I (high permeability, high solubility), Class II (high permeability, low solubility), Class III (low permeability, high solubility) and Class IV (low permeability, low solubility). According to BCS, a drug substance is highly soluble when the highest dose strength is soluble in <250 ml of water over a pH range of 1 to 7.5 and highly permeable when the extent of absorption in humans is >90% of the dose administered to humans. Additionally, there are interclass boundaries that define terms such as *rapidly dissolving* (>85% of labelled amount of drug substance dissolves within 30 min) and *very rapidly dissolving* (>85% of labelled amount of drug substance dissolves within 15 min).

General media used for solubility studies have pH ranging from 1 to 7.5 (pH solubility profiles), and if required, solubilising agents such as cosolvents, surfactants, complexation agents and a combination of techniques are included. For BCS Class II drugs, which are usually weak bases and lipophilic, the drug release rate is the rate-limiting step in absorption. Hence, biorelevant gastrointestinal media that simulate the fasted and fed state in the gastrointestinal tract (media containing bile salts and lecithin, pharmacopoeial-simulated media containing gastrointestinal tract enzymes) are also considered appropriate for dissolution testing. This permits in vitro testing to mimic the conditions in vivo as closely as possible. Furthermore, for poorly soluble drugs, solubility needs to be also performed in these biorelevant media such as simulated gastric fluid–fasted state, simulated gastric fluid–fed state, simulated intestinal fluid–fasted state and simulated intestinal fluid–fed state (Dressman and Reppas 2000). Solubility data generated in different media will help determine the biorelevant dissolution media for in vitro testing.

The solubility of acidic and basic drugs will show a difference in solubility with changes in pH. Solubility is affected by the solid-state form of drugs such as amorphous, crystalline and polymorphs, temperature and different salts. If the solubility of a compound is accompanied by degradation, the solubility figure obtained is inaccurate and problematic. In this case, it is preferable to quote a solubility figure, but with the stipulation that a specified amount of degradation was found. Evidently, higher quantities of degradation will cause the solubility to be insignificant. Several drug substances are ionizable organic molecules, and hence, there are numerous factors that will establish the solubility of a drug substance, such as molecular size and functional groups on the molecule, degree of ionisation, ionic strength, physical form, temperature, crystal habit and complexation.

6.2.2.8.1 Solubility Modifiers

Solubility modifiers may increase or decrease the solubility of a solute in a given solvent. Salts that augment the solubility are supposed to 'salt in' the solute and those that diminish it are said to 'salt out' the solute. The effect of the solubility modifier depends on the interaction it has with water molecules or on its ability to compete with solvent water molecules. An additional feature regarding the influence of electrolytes on the solubility of a salt is the theory of solubility product for poorly soluble drug substances. The experimental effects are that if the common ion concentration is elevated, then the concentration of the other ion must reduce in a saturated solution of the drug substance; that is, precipitation should take place. On the other hand, the effect of external ions on the solubility of poorly soluble salts is contrary, and the solubility enhances. This is described as the salt effect. Setschenow constants have been computed for eight hydrochloride salts of some α-adrenergic agonists and β-adrenergic agonist/blocker drugs (Thomas and Rubino 1996). The constant was determined by computing the solubility of the salts in sodium chloride salt solution, and the outcomes demonstrated that they were highest for those with the least water solubility and greatest melting point. Moreover, the amount of aromatic rings and ring substituents seemed to add to the salting out constant values. Larsen et al. (2007) have researched the application of the common ion effect as a tool to prolong the bupivacaine release from mixed salt suspensions.

6.2.2.9 pK_a

pK_a is the dissociation constant of a drug, that is, the pH at which 50% of the drug is ionised. For any drug substance, the unionised form exhibits better permeability across the biological membrane as compared to the ionised form owing to the lipophilic nature of the unionised species.

Most drugs are either weak acids (pK_a 2–8) or weak bases (pK_a 7–11). Thus, for weakly acidic drugs, a substantial concentration of the unionised species will exist at pH below the pK_a, making the stomach and the upper intestinal region major sites of permeation. For weakly basic drugs, a substantial concentration of the unionised species will exist at pH above the pK_a; thus, these drugs are present significantly as the ionised species throughout the gastric and upper intestinal region. In the case of weakly basic drugs, the degree of ionisation reduces as the pH increases and therefore results in the intestinal region (high pH and high surface area) being a major site of permeation. Having understood that highly ionised drugs cannot permeate the biological membrane, permeation can only occur through the active transport mechanism via specialised transporters. pK_a can give an understanding of drug substance solubility – it is a very important parameter for understanding permeability and solubility. Ionised drugs are much more soluble than their unionised form.

The pK_a for drug substances can be measured by spectroscopic and pH titration methods (Reijenja et al. 2013). For drug substances with a UV-detectable chromophore whose absorbance changes with pH (or the extent of ionisation), UV spectroscopy can be used for pK_a measurement. In this method, the UV spectra of the drug substance are recorded as a function of pH, and mathematical analysis of the spectral shifts are then used to calculate the pK_a. Large UV shifts are obtained for drug substances with ionising groups close to the aromatic chromophore or part of the chromophore itself upon ionisation, making this method most suitable for pK_a determination of such drug substances. Similarly, Kong et al. (2007) have described a pH indicator titration method for measuring pK_a. This method employs a universal indicator solution with spectrophotometric detection for the determination of the pK_a instead of a pH electrode and calculates the pH from the indicator spectra in the visible and UV regions. For drug substances without chromophores, pH electrodes are used for pK_a measurement. More recent techniques employ reverse-phase HPLC for pK_a measurement.

For drug substances with low aqueous solubility, the measurement of pK_a may be tricky. In such cases, the apparent pK_a of the drug substance is measured in organic solvent–water mixtures, after which the data are extrapolated by a Yasuda–Shedlovsky plot to the aqueous pK_a value. Methanol, ethanol, propanol, dimethylsulfoxide, dimethyl formamide, acetone and tetrahydrofuran are widely used organic solvents. Methanol is largely utilised as its properties are closely similar to water. Interestingly, Takács-Novák et al. (1997) have presented a validation study using water and methanol. The calculation of the pK_a values of ibuprofen and quinine in several different organic solvent–water combinations is provided by Avdeef et al. (1999).

6.2.2.10   Intrinsic Dissolution

The intrinsic dissolution is the rate of dissolution of the drug substance at constant surface area. Throughout the preformulation phase, knowledge of the dissolution rate of a drug substance is required, given that this property of the drug substance is identified as a considerable aspect influencing drug bioavailability. The chief factor that establishes the dissolution rate is the water solubility of the drug substance, but other factors such as particle size, crystal state, pH and buffer concentrations can influence the rate. Physical characteristics such as viscosity and wettability can moreover manipulate drug substance dissolution. Nonetheless, intrinsic dissolution of a drug substance dictates its dissolution rate and is an important parameter since it affects drug absorption. Intrinsic dissolution of a drug varies for different forms of drug substances (polymorphs, salt forms, cocrystals, etc.) and helps screen drug forms for optimal candidate selection.

The basis for employing a compressed disc of a pure substance is that the intrinsic propensity of the test substance to dissolve can be tested without excipients. The rotating disc method has been theoretically studied by Levich (1962). Under hydrodynamic conditions, the intrinsic dissolution rate is affected by rotational speed.

For performing the experiment, approximately 200 mg of the drug substance is compressed into a round disc using a hydraulic press – an infrared press is idyllic and produces a 1.3-cm-diameter disc. Notably, certain drug substances are not properly compressed and may display elastic compression; the disc is either fragile or undergoes polymorphic transitions because of high pressure. If the disc shows sound compression characteristics, it is connected to the holder and subjected to agitation in the dissolution medium, at a rotational speed of 100 rpm. Diverse analytical techniques such as UV spectrophotometry or HPLC are normally used for analysis. The intrinsic dissolution rate is given by the slope of the concentration versus time graph divided by the area of the disc and is expressed as mg/min·cm^2.

6.2.2.11 Stability

The rationale of stability testing is to give proof of how the drug substance or product quality changes with time owing to the effect of an array of environmental aspects such as temperature, light and humidity. The final objective of stability testing is the application of suitable evaluation to permit the setting up of recommended storage conditions, retest periods and shelf lives. Stability studies are conducted for all phases of development of new drug substances, formulations and new excipients. Nevertheless, the stability study design and type of evaluation are related to the phase of the development and the characteristics of the drug substance and product. Stability monitoring of a drug substance presents an outline of the drug substance stability in an assortment of pharmaceutical circumstances and thus aids to recognise possible problems that may influence the drug development process. Initial stability studies at the preformulation stage provide the first quantitative assessment of stability of a drug. These investigations include solid and solution state testing of a drug substance along with stability in combination with excipients. Several processes such as blending, processing and storage and biological environments such as the gastrointestinal tract affect drug substance stability.

A classic preformulation procedure includes studies to test chemical, physical and light stability. Chemical stability is conducted in solid and solution states. Solid- and solution-state stability profiles can vary quantitatively and qualitatively. Solid-state stability relates to several parameters and processes such as temperature, pH, humidity, hydrolysis and oxidation. Samples are weighed, kept in open screw cap vials and subjected to direct exposure (light, temperature and humidity) for 12 weeks.

On the whole, solid-state interactions are gradual processes and more difficult to deduce than solution-state reactions, and therefore stability under stress conditions is conducted to hasten degradation processes and to study the causative aspects of degradation in a smaller time span. The information collected from studies conducted under stress conditions is then plotted and extrapolation is made to envisage stability under suitable storage conditions. Stress conditions employed are high-temperature and high-humidity investigations, light and oxidative stability.

For calculating solution phase stability, the principal aim is recognition of conditions required to get a stable solution. This involves investigating the influence of pH, ionic strength, cosolvent, light, temperature and oxygen. Solution stability studies are performed at extremes of pH and temperatures (e.g. 0.1 N HCl, water and 0.1 N NaOH, all at 90°C). These purposely degraded samples may be utilised to corroborate assay specificity and provide approximates of highest degradation rates. This preliminary experiment should be conducted after the production of an absolute pH-degradation rate profile to recognise the pH of highest stability. The accessibility of pH-induced degradation rate profile is helpful in forecasting the drug substance stability in the presence of excipients (acidic and basic). In case the drug degrades in an acidic environment, a less-soluble or less-susceptible chemical derivative or an enteric formulation may be recommended. Similarly, if a drug substance is molecularly unstable upon exposure to moisture, dry processes such as direct compression or nonaqueous solvent granulation procedure should be adopted for tablet formulation. For certain cases, drug substances in solution are photosensitive; hence, photostability evaluation becomes crucial. Such formulations are packaged in light-resistant containers for shielding from light. ICH Q1B guidelines specify photostability testing requirements for such drugs. According to the guidelines, the drug substance is subjected to light exposure in four ways – direct exposure, exposure of drug product outside of pack, exposure of drug product in the immediate pack and exposure of drug product in the marketing pack. Oxidation studies are done for the solution samples in presence of excessive headspace of oxygen, headspace of an inert gas such as helium or nitrogen, inorganic antioxidant such as sodium metabisulfite and organic antioxidant such as butylated hydroxytoluene.

6.2.2.12 Particle Density and Compressibility

Particle density is defined as the ratio of the mass of the particle to its volume. Bulk and tapped densities differ according to the way the particle volumes are considered. For bulk density, the pore volume and particle volume are considered, and in general, bulk density is the least density when the powder volume is highest, whereas tapped density is the highest density when the powder volume is the least. In the context of powder densities, porosity is a popular and critical factor that relates to the fraction of powder bed inhabited by pores (packing efficiency). The packing efficiency of a powder also dictates compressibility, which is a measure of the arch formation of powders. Compressibility is given by the Carr index (5–12 for free-flowing powder, 12–16 for good flow, 18–21 for fair flow and >21 for poorly flowing powders), angle of repose (<30° for good flow) and Hausner ratio (1.2 for free-flowing powder and 1.6 for cohesive powders).

6.2.2.13 Flow Properties

Powder flow is an important characteristic in manufacturing of dosage forms and is affected by interparticulate cohesive forces, presence of moisture, particle size and morphology. Interparticulate forces that are mainly involved in controlling powder flow are electrostatic (conductivity of particles) and van der Waals forces. Van der Waals forces become stronger as particle size reduces below 50 microns and, therefore, restrict powder flow as particle size reduces. Surface roughness is another parameter of importance; particles with rougher surfaces exhibit poor flow as compared to smoother particles. Powder flow in hoppers and powder flow through the orifices are key to any formulator to guarantee smooth manufacturing, and hence, early studies are conducted on particulate properties to ensure good powder flow.

6.3 COMPATIBILITY STUDIES

Efficacious formulation of a stable dosage form relates to careful choice of excipients that are included to assist administration, ensure uniform release and bioavailability of the drug substance and prevent degradation. Incompatibilities in formulations can be detected by alterations in colour or visual appearance, mechanical properties (such as tablet hardness), dissolution profiles, transformations in physical form, loss through sublimation, reduction in strength and escalation of degradation by-products.

Excipient compatibility investigations are conducted to envisage any probable incompatibilities arising in the final formulation. Such investigations present rationalisation for choice of excipients and their concentration in the dosage form as documented in regulatory filing. On this basis, specific control strategies for dosage form development can be primed. These investigations are imperative in the drug development process, since the information obtained from excipient compatibility data is applied to choose the excipient; study stability of the drug substance; spot degradation by-product and its pathway of formation; investigate reaction mechanisms, which helps prevent unforeseen predicaments; and recognise stable storage factors for a drug substance in the solid/liquid phase. The road map to carry out compatibility studies is outlined in Figure 6.4.

Based on available information on API, excipients and their interactions, a study design is proposed (Figure 6.5). Depending on the number of excipients to be

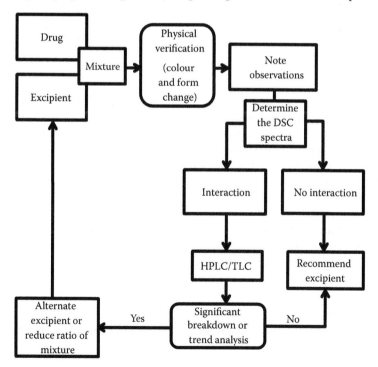

FIGURE 6.4 Flowchart for conducting compatibility studies.

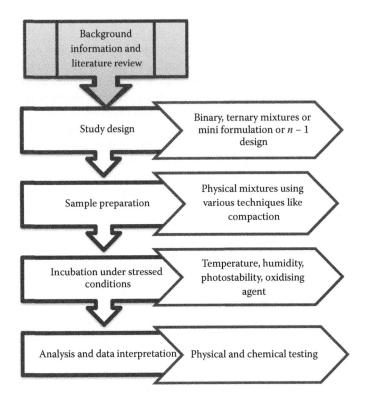

FIGURE 6.5 Drug–excipient compatibility study design.

investigated, compatibility tests can be expedited by using factorial or reduced factorial design experiments. The study design (stress conditions and exposure times) should be prepared based on information about the properties of the drug substance, potential degradation pathways, type of dosage form and proposed manufacturing process.

If the formulation contains multiple active ingredients, compatibility studies with individual active ingredients are carried out (Table 6.1). In addition to a mixture of drug with each excipient separately, sometimes a mixture of drug with multiple excipients is also studied simultaneously if interaction amongst the excipients is expected to affect the drug stability. An $n - 1$ design or mini-formulation study designs are performed with the omission of only a single component in every sublot to trace the source of incompatibility. Frequently, mini-formulations are made with

TABLE 6.1

Compatibility Study Design for Two Active Ingredients A and B

Active		1st	2nd	3rd
A	B	A + excipient	B + excipient	A + B + excipient

the omission of nonessential, quantitatively inconsequential or easily interchangeable excipients, such as colour, flavour from solutions and suspensions.

The choice of excipient depends upon the role it plays in the formulation. The active/excipient ratio should be based on the tentative anticipated concentrations of the excipient in the formulation and the dose of the drug. Table 6.2 gives indicative ratios that can be suitably changed as per dosage form requirement.

On the basis of the study design, samples are prepared and incubated at different conditions. It is essential to ensure that the mixtures are prepared carefully so that

TABLE 6.2
Suggested Ratios of Excipients for Preformulation Compatibility

No.	Excipients	Suggested Ratio of Active/Excipients	Indicative Ratio of Active/Excipients
1	Fillers (e.g. lactose, starch, microcrystalline cellulose, dibasic calcium phosphate)	1:1–1:20	1:5
2	Binders (e.g. PVP, HPMC, HEC, pre-gelled starch)	1:0.25–1.1	1:0.5
3	Disintegrants and super disintegrants (e.g. crospovidone, cross carmellose sodium, Primojel, sodium starch glycolate)	1:0.25–1.1	1:0.5
4	Lubricants, glidants and antiadherents (e.g. magnesium stearate, stearic acid, sodium stearyl fumarate, talc, colloidal silicon dioxide)	1:0.05–1:0.15	1:0.1
5	Syrup base constituents (e.g. sugar syrup, propylene glycol, sorbitol, glycerin)	1:1–1:20	1:10
6	Colours and flavours	1:0.1–1:1	1:0.5
7	Solubilisers, buffering agents and stabilisers	1:0.1–1:10	1:0.5
8	Suspending agents and thickeners	1:0.1–1:1	1:0.5
9	Cream and ointment oily bases	1:0.1–1:1	1:5
10	Controlled release polymers	1:0.1–1:1	1:0.5
11	Plasticisers	1:0.05–1:0.5	1:0.2
12	Film-coating materials	1:0.05–1:0.5	1:0.2

uniformity of the drug is maintained in the samples incubated for the study. Exposed samples are then evaluated using suitable techniques. Compatibility studies between the active and the proposed excipients are carried out by exposing the sample to stress conditions (under open and closed conditions). For open conditions, samples should be kept in open petri plates or in open glass vials, and for closed conditions, samples should be kept in closed and sealed glass vials.

For analysing samples towards identifying incompatibilities, the following methods are applied:

- Incompatibilities are physically traced by visual screening for colour or physical form alterations.
- Thermal analysis (DSC, TGA, etc.) are conducted.
- Monitoring of form changes by PXRD, nuclear magnetic resonance and near infrared is done.
- Rate of degradation by HPLC, gas chromatography and mass spectroscopy is determined.

Physical mixtures of active and excipients/powder blends/granules/compressed tablets at elevated temperatures and humidity are exposed for periods of up to 1 month at 40°C/75% RH and up to 2 weeks at 50°C/60°C with no humidity control or 25°C/60% RH. Simultaneously, individual drug and excipient samples are also separately tested. The drug substance itself is the positive control; when the compatibility testing results (amount of degradation products) of the drug–excipient mixtures show a similar trend as that of the drug, it means that the excipient does not accelerate the degradation of the drug and it is deemed to be compatible. If, however, the results show that the degradation trend is higher than that of the drug (e.g. appearance of more amounts of degradation products), the mixture is incompatible. On the other hand, the interactions of the excipient alone can give insights into the stability (formation of degradation products) of the excipient itself, and this serves as the negative control. Therefore, interpretation of compatibility data is based more on trends and not on any set point specifications and therefore the design of the study is very critical so as to ensure that appropriate positive and negative controls have been incorporated. The conditions for incubation should be selected, based on the reported heat and moisture sensitivity of the active and how quickly the results are required. For instance, for highly heat-sensitive actives, incubation at higher heat conditions such as 50°C/60°C may be excluded. For moisture-sensitive or hygroscopic actives, incubation under open condition may be excluded or the exposure time may be reduced as appropriate. Studies should be performed under closed conditions for actives known to be sensitive to oxidation due to atmospheric conditions. Control samples are usually stored at 2°C–8°C as backup in case some confirmation is needed at a later stage. Pack conditions may vary on the basis of the formulation to be designed.

Generally, excipients do not elicit a direct pharmacological effect, but they convey useful characteristics to the formulation. Nevertheless, they can also instigate unintended effects such as augmented drug degradation. Integration of additive(s) to an intravenous fluid may change the intrinsic properties of the drug substance, causing parenteral incompatibility. Physical compatibility of various components in parenteral

TABLE 6.3

Physicochemical Properties of Drug Substance for Specific Dosage Forms

Physicochemical Properties of Drug Substance for Evaluation during Preformulation[a]

Dosage Form Type	Flow Properties	Viscosity and Gelling Properties	Density	Particle Size and Morphology	Surface Area and Static Charge	Porosity	Hygroscopicity	Water Content
Tablets	●		●	●	●	●	●	●
Capsules	●		●	●	●	●	●	●
Suspensions	●	●	●	●	●	●		
Emulsions (otic, nasal, ophthalmic, oral)		●		●				
Solutions (otic, nasal, ophthalmic, oral)		●						
Inhalation dosage forms	●		●	●	●	●	●	●

(Continued)

TABLE 6.3 (CONTINUED)

Physicochemical Properties of Drug Substance for Specific Dosage Forms

Dosage Form Type	Physicochemical Properties of Drug Substance for Evaluation during Preformulation[a]							
	Flow Properties	Viscosity and Gelling Properties	Density	Particle Size and Morphology	Surface Area and Static Charge	Porosity	Hygroscopicity	Water Content
Transdermal				●	●		●	●
Topical		●		●	●		●	●
Parenteral		●		●	●			

Note: ●, Required.

[a] Physicochemical properties such as pH, log *P*, solubility and intrinsic dissolution; physical form properties; thermal properties; and degradation are required for drug substances formulated as any dosage form.

nutrition solutions has been a concern, particularly solubility of minerals such as calcium and phosphate. In another case, therapeutic incompatibility was seen in 5-mg Coumadin tablets, which were recalled by Bristol-Myers Squibb because of efficacy issues. In another case, combination bilayer metformin tablets (Starlix), a product of Novartis, intended for type II diabetes, showed low percentage recovery during analytical method development. Metformin (positively charged) interacts with croscarmellose sodium (negatively charged), resulting in incompatibility. In solid state, the interaction between metformin and croscarmellose sodium does not occur owing to low moisture content in the tablet (<2%). However, in solution, metformin and croscarmellose sodium underwent a charge interaction, causing a 4%–8% loss in metformin recovery in the tablets. Interestingly, inclusion of arginine, which has stronger interaction with croscarmellose sodium than metformin, prevented metformin from interacting with the excipient, thus enabling full recovery of metformin in the tablets.

6.4 DOSAGE FORM–SPECIFIC STUDIES

Selection of a suitable dosage form for a drug substance can be influenced by preformulation data in addition to other preclinical toxicological and pharmacological data. Prior knowledge of drug substance properties can define the drug's suitability to be administered as a specific formulation type by a particular route. A number of factors, such as particle size, particle shape, water content, hygroscopicity and surface properties, can govern the drug's behaviour during further formulation development stages including the manufacturing process development. For instance, good flow properties are a prerequisite for the successful manufacture of both tablets and powder-filled hard gelatin capsules. Depending on the selected dosage form to be developed, there could be a need to study some specific characteristics of the drug substance so as to generate understanding of its behaviour to aid the dosage form development. Table 6.3 charts specific physicochemical properties of a drug substance that need to be studied as a requisite for determining the suitability of the drug to be formulated in that specific dosage form.

6.5 CONCLUSION

Preformulation should always be an integral part of all drug development projects to facilitate the progression of formulation optimisation on a sound scientific basis precept by precept. Needless to say, as a prelude to formulation, it should pave the way for the formulation scientist to successfully formulate stable dosage forms for clinical trials in record time. Preformulation is considered as a rate-limiting step in the overall planning of product development activities. Right from the early phases, physicochemical investigations have an important role in predefining possible problems during formulation development and in recognising a rational path in dosage form design. Therefore, it is imperative that preformulation should be performed as carefully as possible to enable rational decision making.

REFERENCES

Ahlneck, C. and G. Zografi. 1990. The molecular basis of moisture effects on the physical and chemical stability of drugs in the solid state. *Int J Pharm* 62:87–95.

Avdeef, A., B. J. Box, J. E. Comer, M. Gilges, M. Hadley, C. Hibbert, W. Patterson, and K. Y. Tam. 1999. pH-metric log P 11. pKa determination of water-insoluble drugs in organic solvent–water mixtures. *J Pharm Biomed Anal* 20(4):631–41.

Bauer, J. F., W. Dziki, and J. E. Quick. 1999. Role of an isomorphic desolvate in dissolution failures of an erythromycin tablet formulation. *J Pharm Sci* 88(11):1222–7.

Chakravarty, P., R. Suryanarayanan, and R. Govindarajan. 2012. Phase transformation in thiamine hydrochloride tablets: Influence on tablet microstructure, physical properties, and performance. *J Pharm Sci* 101(4):1410–22.

Chen, M., X. Zhang, J. Liu, and K. B. Storey. 2013. High-throughput sequencing reveals differential expression of miRNAs in intestine from sea cucumber during aestivation. *PLoS One* 8(10):e76120.

Dressman, J. and C. Reppas. 2000. In vitro-in vivo correlations for lipophilic, poorly water-soluble drugs. *B T Gattefosse* 93:91–100.

Giron, D. 1995. Thermal analysis and calorimetric methods in the characterisation of polymorphs and solvates. *Thermochem Acta* 248:1–59.

Johnson, K. C. and A. C. Swindell. 1996. Guidance in the setting of drug particle size specifications to minimize variability in absorption. *Pharm Res* 13(12):1795–8.

Kong, X., T. Zhou, Z. Liu, and R. Hider. 2007. pH indicator titration: A novel fast pKa determination method. *J Pharm Sci* 96(10):2777–83.

Larsen, S. W., J. Ostergaard, S. V. Poulsen, B. Schulz, and C. Larsen. 2007. Diflunisal salts of bupivacaine, lidocaine and morphine. Use of the common ion effect for prolonging the release of bupivacaine from mixed salt suspensions in an in vitro dialysis model. *Eur J Pharm Sci* 31(3–4):172–9.

Leo, A., C. Hansch, and D. Elkins. 1971. Partition coefficients and their uses. *Chem Rev* 71(6):525–616.

Levich, V. G. 1962. *Physicochemical hydrodynamics*. Englewood Cliffs, NJ: Prentice-Hall, Inc.

Li Destri, G., A. Marrazzo, A. Rescifina, and F. Punzo. 2013. Crystal morphologies and polymorphs in tolbutamide microcrystalline powder. *J Pharm Sci* 102(1):73–83.

Newman, A. W., S. M. Reutzel-Edens, and G. Zografi. 2007. Characterization of the "hygroscopic" properties of active pharmaceutical ingredients. *J Pharm Sci* 97(3):1047–59.

Reijenja, J., A. van Hoof, A. van Loon, and B. Teunissen. 2013. Development of methods for determining pKa values. *Anal Chem Insights* 8:53–71.

Roy, S. D. and G. L. Flynn. 1989. Solubility behavior of narcotic analgesics in aqueous media: Solubilities and dissociation constants of morphine, fentanyl, and sufentanil. *Pharm Res* 6(2):147–51.

Takács-Novák, K., K. J. Box, and A. Avdeef. 1997. Potentiometric pK_a determination of water-insoluble compounds: Validation study in methanol/water mixtures. *Int J Pharm* 151(2):235–48.

Thomas, E. and J. Rubino. 1996. Solubility, melting point and salting-out relationships in a group of secondary amine hydrochloride salts. *Int J Pharm* 130(2):179–85.

Wang, Y. J., T. C. Dahl, G. D. Leesman, and D. J. Monkhouse. 1984. Optimization of autoclave cycles and selection of formulation for parenteral products, Part II: Effect of counter-ion on pH and stability of diatrizoic acid at autoclave temperatures. *J Parenter Sci Technol* 38(2):72–7.

Wang, G., J. D. Fiske, S. P. Jennings, F. P. Tomasella, V. A. Palaniswamy, and K. L. Ray. 2008. Identification and control of a degradation product in Avapro film-coated tablet: Low dose formulation. *Pharm Dev Technol* 13(5):393–9.

Young, P. M., R. Price, M. J. Tobyn, M. Buttrum, and F. Dey. 2003. Effect of humidity on aerosolization of micronized drugs. *Drug Dev Ind Pharm* 29(9):959–66.

7 Formulation Development and Scale-Up

Amit G. Mirani, Priyanka S. Prabhu
and Maharukh T. Rustomjee

CONTENTS

7.1 INTRODUCTION

The primary objective of drug product development is to ensure that the product meets its quality target therapeutic profile and quality attributes, which reflect the patient needs relating to its efficacy and safety, and remains stable during its designated shelf life. Traditionally, product development was carried out using trial-and-error/empirical experimentation resulting in incomplete understanding of the input variables and process variables for any product, eventually leading to inconsistent quality of the product being manufactured and supplied to the market. This deficient product development process has led to many problems being faced by both pharmaceutical companies and regulators. On one hand, the companies faced problems related to variable quality leading to batch failures, product recalls and inability to meet market demand leading to lost opportunity, while on the other hand, regulators faced problems relating to severe drug shortages and enhanced risk to patients. Over the past two decades, the regulators and industry associations have worked together under the aegis of the International Conference on Harmonisation (ICH) to discuss, debate and arrive at solutions to these problems by applying scientific and technology advancements to the field of pharmaceutical development, manufacturing and supply chain to arrive at many initiatives. ICH came up with a new paradigm in product life cycle approach, which involves systematic, risk-based, scientific approach (i.e. quality by design [QbD] approach) for pharmaceutical product development (ICH 2009). The process development, risk management and quality systems strategies and tactics outlined in ICH Q8 Pharmaceutical Development (ICH 2009), Q9 Quality Risk Management (ICH 2005a) and Q10 Pharmaceutical Quality System (ICH 2008) have been discussed in depth in Chapter 2.

Pharmaceutical product development based on the QbD approach as outlined in the ICH Q8 guideline is a systematic approach to design a quality product that begins with predetermined objectives and emphasises on product and process understanding and determination of control strategy based on sound science and quality risk management (ICH 2009). It is a concept that states that 'quality should be built in by design'. The objective of QbD and its essential elements for drug product development have been discussed in Chapters 2 and 4. The current chapter will focus on how the QbD-based approach can be adapted for the development of a pharmaceutical product. The chapter also aims at simplifying product development life cycle into sequential phases, that is, strategy phase, laboratory development phase, technology transfer and regulatory submission phase, as depicted in Figure 7.1.

The product development starts with identifying a suitable dosage form on the basis of the properties (physical, chemical and biological) of the drug substance, patient need, market need, regulatory and intellectual property right consideration and so on, as discussed in Chapter 4. Dosage form selection is an important consideration as it ensures optimum utilisation of drug by the body and provides maximum therapeutic effect.

7.2 PRODUCT DEVELOPMENT STRATEGY

Once the dosage form is proposed, the drug product development strategy using the QbD approach is initiated. The product development strategy involves defining the drug product quality target product profile and its quality attributes, identifying probable formulation and manufacturing process and, based on that, identifying critical quality attributes (CQAs). After this, an initial risk assessment is done on the basis of prior knowledge to identify the critical material attributes (CMAs), composition variables and critical process parameters (CPPs). The process of arriving at a product design and strategy is explained in Chapter 4.

7.3 RAW MATERIAL SELECTION

The first step in product development is initiation of raw material source selection. The raw materials used in formulation composition are composed of a drug substance, excipients and packaging materials; all of these can have an impact on the product quality.

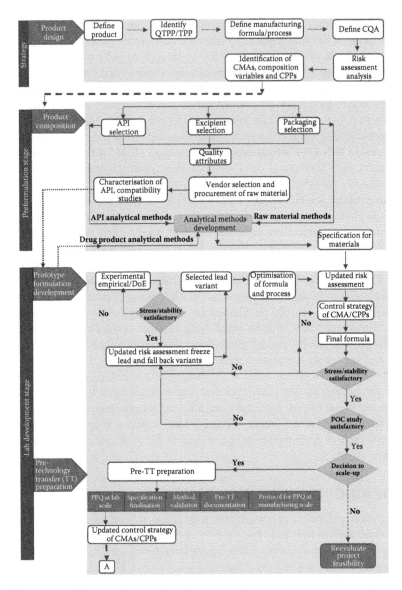

FIGURE 7.1 Product development life cycle. BMR, batch manufacturing record; BPR, batch packaging record; MFR, manufacturing formula record; MPR, manufacturing packaging record; PPQ, process performance qualification.   (*Continued*)

7.3.1 ACTIVE PHARMACEUTICAL INGREDIENTS

An active pharmaceutical ingredient (API) or a drug substance is the chemical entity that shows therapeutic effect when administered to the patient. The physicochemical properties (polymorphic form, particle size, surface area, impurity profile, etc.) of the APIs can have a significant effect on the safety and efficacy of drug products,

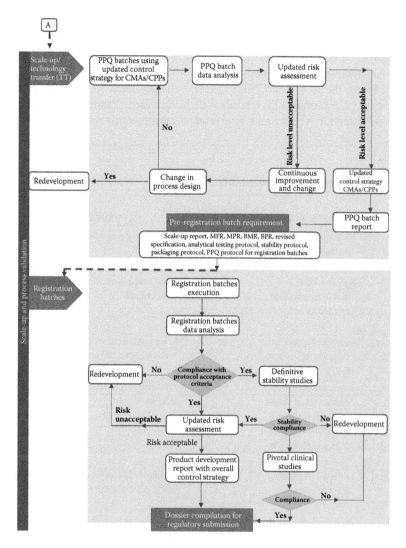

FIGURE 7.1 (CONTINUED) Product development life cycle. BMR, batch manufacturing record; BPR, batch packaging record; MFR, manufacturing formula record; MPR, manufacturing packaging record; PPQ, process performance qualification.

which ultimately affects the health of the patient. Hence, depending on the expected therapeutic requirements and CQAs of the drug product, the quality/material attributes of APIs are assigned during initial risk assessment. The quality attributes of the APIs ensure that the drug product meets its product CQAs reliably and reproducibly. The quality attributes of the APIs need to be controlled within the specified acceptance limit set as per official compendium, ICH guidelines or in-house specification. The initial risk assessment evaluates the impact of the quality/material attributes of the API on the CQAs of the drug product as explained in Chapter 4. The CMAs of the API are usually compiled as a draft specification to help the procurement team evaluate and shortlist the vendors. A hypothetical example of potential QAs of API with their justification is detailed in Table 7.1.

Considering the identified CMAs of the drug substance, the identification and qualification of potential API vendors will be initiated. Identification and qualification of potential API sources is a prudent step towards quality drug product

TABLE 7.1
Quality Attributes of Drug Substance

Quality Attributes	Acceptance Criteria	Criticality	Justification
Identity	Melting point, infrared, ultraviolet	Yes	Necessary for patient safety and efficacy
Assay	As per official monograph/vendor specifications	Yes	Demonstrates overall purity, hence considered to be critical
Appearance	Limit set for residual particles and colours	Yes	Must ensure freedom from particulate contamination or excessive colour
Impurities	Limit set for both unknown and known impurities (as per official monograph or ICH guideline Q3B)	Yes	Impurities in API affect the drug product stability
Crystal form	Different forms and their percentage	Yes	Polymorph affects solubility and stability, hence considered to be critical
Water content	As per official monograph	Yes	Water content may affect stability of drug, hence considered to be critical
Particle size distribution	Meets product requirement	Yes	Particle size affects the solubility, thereby affects the bioavailability, and hence needs to be considered as critical in case of BCS class II and IV molecules
Residual solvent	ICH acceptance criteria	No	Potential safety concern; however, it can be controlled, hence considered to be of low criticality
Microbial load	Pharmacopoeial requirements	No	Presence of organisms could be harmful to product safety; however, it can be controlled, hence considered to be of low criticality

TABLE 7.2

Parameters to Be Considered in Qualification of API Source

Supplier-Related Attributes	API-Related Attributes
cGMP certified manufacturing site and in compliance with customer quality audit	Source offering freedom to operate from enforceable patents
Complying with respective regulatory authorities of the intended market	In compliance with pharmacopoeial monograph and ICH guidelines
Availability of technical package containing information of the synthetic pathway employed for API manufacture, its purity profile, identification of impurities listed in the API specifications and their routes of formation, stability profile, pack profile and storage conditions	Drug product–specific requirements, for example, physical properties (particle size, particle size distribution, crystal polymorph, surface area, etc.) and chemical properties (tighter control on impurity levels or moisture content)
Capability to supply the material in bulk quantities (in conformance to quality agreement) reliably and consistently	
Economical	

development. The qualification of API supplier/source will be based on assigned CMAs and several other parameters as listed in Table 7.2.

It is preferable to evaluate at least three sources of API for qualification based on their compliance with the abovementioned criteria. The technical data package available from the open part of the drug master file, which includes information on the synthetic route, potential impurities and degradants, analytical methods and stability data, needs to be studied and evaluated to assess the suitability of the drug substance for the product development. The sample quantities of API from different lots that represent the production batch are requested from each supplier and evaluated for compliance with the official monograph (if applicable) and Certificate of Analysis (COA) provided by the supplier. Once compliance is confirmed, the vendor is approved and API is procured in bulk for further analytical method and product development. The development is usually undertaken with one/two of these approved sources all the way up to registration. It is preferable to register a drug product with at least two sources of drug substance so that during the commercial phase, reliability and consistency of the supply chain can be maintained. The analytical method development for a drug substance will be covered in detail in Section 7.4.

7.3.1.1   Specifications of APIs

The specifications for the quality attributes of APIs are set based on pharmacopoeial and ICH requirements (ICH Q6A) in addition to the specifications of the manufacturer (ICH 1999). These specifications would also include the identified CMAs as per the risk assessment. The specifications for the CMAs are decided on the basis of the experiments conducted during development using API with varying parameters associated with each attribute. As the development progresses, the specifications evolve and are fine-tuned (based on the subsequent risk analysis) to minimise the risk of failure in the drug product (Chen et al. 2009; DiFeo 2003; Yu 2008).

7.3.2 Excipients

Excipients are those chemical entities that do not have their own therapeutic effect but influence the effect of the drug substance when administered in the body. Hence, special consideration needs to be given to excipient selection. The criteria for selection of excipients will be based on intended dosage form and the proposed route of administration, safety profile, manufacturing process, regulatory aspects, intellectual property aspects and so on.

7.3.2.1 Intended Route of Administration

The excipients are selected based on the intended route of administration. For example, in parenteral products, it is mandatory for all the excipients to be sterile and pyrogen free. For ophthalmic, nasal, buccal, vaginal and rectal products, the excipients must not cause any mucosal irritation (Shargel and Kanfer 2014).

7.3.2.2 Safety Profile and Prior Use of Excipient

Excipients must have a favourable safety profile as they represent the major component of the drug product. The acceptable daily intake (ADI) of any nonactive ingredient is an important parameter to be considered when designing the dosage form as safety of use of the excipients in the particular quantity needs to be justified. Usually, this is justified based on historical information on usage of that particular excipient in a respective dosage form. The safety is also dependent on the route of administration; for example, the ADI of an excipient is lower for a parenteral dosage form compared to an oral dosage form. The Inactive Ingredient Database published by the US Food and Drug Administration (USFDA) is a good source of ADI based on historical usage in approved drug products. The excipient manufacturer also provides support for justifying the safety of the excipients for various routes of administration. In case a novel excipient is used for the first time in a particular dosage form, its safety has to be supported/established based on relevant toxicity studies, which could be provided by the excipient manufacturer. In cases where the excipient is used beyond the reported ADI, similar justification/support for their use is required. However, toxicological testing of novel excipients for European or US markets is a lengthy and tedious process.

7.3.2.3 Regulatory Aspects

It is a common practice to use compendial excipients (which are official in the main pharmacopoeias such as United States Pharmacopoeia, British Pharmacopoeia and Japanese Pharmacopoeia) for the development of pharmaceutical products. However, in some cases, noncompendial excipients such as co-processed excipients (Prosolv, Ludipress), flavours or colour mixtures are also used and in which case their regulatory status for the intended market needs to be considered before selection. The excipients that qualified as generally recognised as safe (GRAS) can be used for product development as per the Food, Drug and Cosmetics Act. Similarly, in the European market, the compliance with food additive requirement can be applicable for noncompendial excipients.

7.3.2.4 Compatibility with Drug Substance

Excipients can alter the stability of drug substances; hence, to understand the impact of excipient on drug product quality, drug–excipient compatibility studies are carried out. The selection of excipients for the compatibility study should be based on the mechanistic understanding of the drug substance and its impurities, degradation pathways and potential processing conditions for the drug product manufacture. A scientifically sound approach should be adopted while executing the compatibility studies. For more information on drug–excipient compatibility studies, see Chapter 6.

7.3.2.5 Manufacturing Process

The manufacturing process should be considered early while selecting excipients in product development to avoid problems at the stages of scale-up and process validation. The selection of a specific grade or type of the excipient is based on the manufacturing process. For example, if one intends to employ direct compression for tablet manufacturing, it necessitates the use of 'directly compressible' grades of diluents.

Once the excipients are selected based on the requirements of the drug product and its manufacturing process, a risk assessment is carried out to understand and identify the material attributes that can have an impact on the CQAs of the drug product. The next step is to assign the quality attributes that need to be considered while identifying and qualifying the excipients' vendors (DiFeo 2003; Yu 2008). A few examples of probable quality material attributes of excipients with their justifications are shown in Table 7.3.

TABLE 7.3
Quality Attributes of Excipients

Type of Product	Critical Material	Critical Material Attributes	Justification
Sustained-release tablet	Sustained-release polymer	• Viscosity • Degree of polymerisation • Polymer chain length • Swelling index • Water content • Particle size range	Viscosity, degree of polymerisation, polymer chain length, swelling index water content and particle size range all have an impact on drug release profile, hence considered as CMAs
Gel	Gelling agent	• Water sorption capacity • Chain length • Degree of polymerisation • Swelling index • Viscosity	Water sorption capacity, chain length of polymer, swelling index, degree of polymerisation and viscosity all have an impact on gelling property, hence considered as CMAs

Once the attributes are assigned, the identification and qualification of potential excipient vendors will be initiated. If possible, multiple sources of excipients should be qualified based on their compliance with compatibility studies and several other identified CMAs. The excipients from different lots must comply with the official compendium and COA provided by the supplier. Some other techno-commercial factors as elaborated in Table 7.2 should also be considered while selecting the excipient vendor. Excipients that are noncompendial are required to comply with the in-house specifications of the manufacturer. Once compliance is confirmed, the vendor is approved.

7.3.2.6   Specifications of Excipients

The quality attributes of excipients can be further controlled by setting acceptance criteria in terms of specifications. The specifications ensure that each batch represents reproducibility of quality attributes of excipients. The specifications are not static but dynamic and will continue to evolve as the development progresses. These are updated regularly on the basis of knowledge gained during their use in product and periodic revisions in the compendial monographs (DiFeo 2003; Yu 2008).

7.3.3   PACKAGING MATERIAL

Packaging materials are important input materials that ensure the integrity of products during transportation and distribution as well as stability of products during their shelf life. From a development point of view, the most important packaging components for pharmaceutical drug products are primary and secondary packaging. Primary packaging is most relevant to regulatory bodies owing to its direct contact with the product. Hence, the components used for primary packaging should be compatible with the drug product. The secondary packaging (e.g. carton) is not in direct contact with the drug product. However, it imparts some additional protection from light, moisture and so on (Shargel and Kanfer 2014). The selection of packaging and its material will be based on the quality target product profile (QTPP) and CQAs of the drug product. Similarly, several other factors such as type of dosage form, type of drug substance, regulatory aspects and intellectual aspects should be considered, the details of which are discussed in Chapter 10.

Once the pack profile and its material is selected based on QTPP and the above-mentioned parameters, the next step is to identify CMAs that need to be considered while identifying and qualifying packaging vendors. The initial risk assessment evaluates the impact of the quality/material attributes of the packaging material on the drug product CQAs (as explained in Chapter 4). A few examples of the probable quality material attributes of packaging materials with their justifications are illustrated in Table 7.4.

Considering the identified CMAs of the packaging material, the identification and qualification of potential packaging material source will be initiated. The packaging materials will be evaluated for the defined attributes to check for their compliance with supplier COA, reliability and consistency of supply and machinability on the packaging machine. Once the compliance is confirmed, the vendor is approved and the packaging material is procured in bulk for final packaging of the drug product.

TABLE 7.4

Quality Attributes of Packaging Material

Type of Product	Critical Material	Critical Material Attributes	Justification
Tablet	Alu-Alu blister pack	• Composition of the laminates • Thickness • Dimension • Moisture vapor transmission rate (MVTR) • Oxygen transmission rate (OTR) • Compatibility with drug product	All these attributes have an impact on drug product CQAs, hence considered as critical
Parenteral	Glass	• Type of glass • Dimension • Head space • Type of closure	All these attributes have an impact on drug product CQAs, hence considered as critical

7.3.3.1 Specifications of Packaging Materials

The specifications for the quality attributes of packaging materials are set based on compendial requirements and manufacturer's specifications. These specifications would also include the identified CMAs as per the risk assessment. The specifications for the CMAs will be decided based on the stability studies of experimental batches conducted during prototype development and performance on packaging machine during scale-up.

7.4 ANALYTICAL METHOD DEVELOPMENT

Analytical method development and its validation play an integral role in product development, which involves application of analytical tools for quality control (QC) testing and stability testing of drug substances and drug product. Analytical tools such as high-performance liquid chromatography (HPLC) and ultraviolet (UV) analysis are the most commonly used methods for identification and quantification of active drugs, known impurity and unknown impurities in drug substances and drug products. As the entire product development is based on the QbD approach, it is becoming imperative to apply QbD principles for the development of analytical methods as well (Shargel and Kanfer 2014; Vogt and Kord 2011). The critical steps for analytical method development based on the QbD approach are depicted in Figure 7.2. It starts with defining quality analytical target profile (QATP) in terms of accuracy, precision, repeatability, sensitivity and so on. Considering the QATP, the next step will be the selection of analytical techniques from different literature sources such as official monograph, research articles and the technical part supplied by the drug substance manufacturer. The selected method can then be assessed for critical method attributes using prior knowledge, that is, failure mode effect analysis

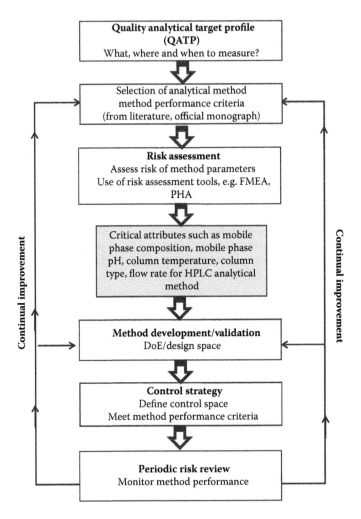

FIGURE 7.2 QbD-based approach for analytical method development.

(FMEA) method and screening design such as the Plackett–Burman design. The critical attributes in risk assessment analysis will be optimised further by experimental design to generate the design space. The design space further helps design a control strategy that ensures consistency in the performance of analytical methods. Analytical method parameters for testing of APIs, in-process materials and finished product will be finalised based on design space. QbD-based analytical method development not only results in shortening the development time but also ensures the development of robust methods, which builds confidence in the quality of the products being tested and released for human consumption. A well-developed analytical method is a prerequisite for an effective control strategy for a drug product. The analytical method development is broadly categorised into two sections, namely, method development for API and for drug product.

7.4.1 METHOD DEVELOPMENT FOR API

Analytical method development of API is an important tool in product development and is primarily used for API vendor evaluation and its release. Analytical method development is done based on compendial methods, manufacturer/supplier methods or in-house developed methods. For compendial methods, truncated validation is carried out, and for noncompendial methods, complete validation as per ICH guideline Q2 (R1) needs to be conducted. ICH guideline Q6A gives the Test Procedures and Acceptance Criteria for New Drug Substances (ICH 1999). Several test methods for the API release include identification, assay, chromatographic purity, impurities, residual solvent, inorganic impurities, microbial limits, particle size measurement, polymorph determination, water content (bound and unbound water), melting point, specific rotation, pH dependent solubility, heavy metals and so on, as shown in Table 7.5. Amongst all the parameters, solubility/intrinsic dissolution and purity profile are the most important as they may influence the drug release and stability of the final dosage form. Very often, the assay and chromatographic purity tests are conducted using HPLC procedures. However, sometimes semiquantitative thin-layer chromatography test methods are also used. Similarly, gas chromatography methods are used for residual solvent determination.

7.4.2 METHOD DEVELOPMENT FOR DRUG PRODUCT

Once the API method is developed, the next step is development of the analytical method for the drug product. Typically, this starts with the placebo product to identify the impact of excipients on the quantification of active ingredients or known impurities in dosage form. The analytical test parameters for drug products vary with the type of dosage forms and representative examples are as shown in Table 7.6.

The developed analytical method is required to show specificity, sensitivity, linearity, reproducibility, precision and accuracy for quantification of the drug in a dosage form to ensure that the method can be used for evaluation of dosage form stability.

7.4.2.1 Chromatographic Method for Purity Estimation

The chromatographic method developed and validated for estimation of the purity profile of the drug product should ensure sensitivity and specificity. For this method, limit of detection (LOD) and limit of quantification (LOQ) are very critical and should be appropriate based on the limit of impurity. For example, if the limit for the impurity is 0.1%, then usually it is desirable to have an LOD of 0.005% at least and an LOQ of 0.01% at least. The forced degradation studies under thermal, acid, base, oxidation and photodegradation conditions should be conducted to verify that the test method can reliably quantify degradation products. The degradation level obtained in forced degradation studies should be around 10% to 30% so as to be close to real-life situations. Higher degradation level during studies can provide misleading data. The forced degradation study also helps to understand the labile nature of the drug substance and supports understanding on the degradation products and degradation pathways of the drug substance. Similarly, the presence of impurity can be further confirmed by mass balance confirmation where a decrease in value for

TABLE 7.5
Test Methods for API Release

Test Parameters	Test Method
Description	Qualitative description of the physical state of the drug substance, solid/liquid and its colour.
Identification	To differentiate the given drug substance from compounds having a closely resembling structure. Identification tests ought to be specific for the new drug substance, for example, infrared spectroscopy. For salts, identification testing should be specific for the individual ions. Optically active drug substances require specific testing for chirality.
Assay	Specific stability-indicating methodology should be included to estimate the content of the new drug substance.
Impurities	Impurities are classified as organic, inorganic and residual solvents. ICH guideline Q3A gives the limits for impurities in drug substances.
Particle size	If the drug substance is supposed to be used in a solid or suspension product, then the particle size can affect the dissolution, bioavailability and stability. Particle size acceptance criterion is generally not needed in case of solutions.
Polymorphic forms	For drugs that are known to exist in different crystalline forms that can influence the performance and stability of the drug product. Polymorphic form of a drug substance studied using hot stage microscopy, solid-state infrared (IR) spectroscopy, solid-state nuclear magnetic resonance (NMR), differential scanning calorimetry (DSC), Raman spectroscopy, optical microscopy and x-ray powder diffraction.
Melting point	DSC.
Heavy metals	Specific tests for different heavy metals such as lead and iron.
Test for chiral new substances	If a new drug substance is mainly one enantiomer, then the opposite enantiomer is excluded from the qualification and identification thresholds mentioned in the ICH Guidelines on Impurities in New Drug Substances and Impurities in New Drug Products because of practical issues in quantification at those concentrations.
Water content	For drug substances known to be hygroscopic, sensitive to moisture-induced breakdown and stoichiometric hydrates. Loss on drying and Karl Fischer titration were employed to estimate water content.
Inorganic impurities	Tests for detection of impurities are decided based on the manufacturing process. Sulfated ash and residues on ignition are examples of pharmacopoeial tests whereas atomic absorption spectroscopy may be used to determine elemental impurities.

(Continued)

TABLE 7.5 (CONTINUED)
Test Methods for API Release

Test Parameters	Test Method
Microbial limits	Nature of drug substance, its manufacturing technique and final use decide the type of microbial limit tests and their acceptance criteria.
	Total aerobic microbial count, total fungal and mould count and absence of undesirable bacteria such as *Pseudomonas aeruginosa*, *Salmonella*, *Escherichia coli* and *Staphylococcus aureus*.
	Sterility tests are done for substances manufactured as sterile and pyrogen/endotoxin testing for drug substances intended to be used for parenteral formulation development.

the active is mass balanced with the amount of impurities observed (Gibson 2004; Shargel and Kanfer 2014).

7.4.2.2 Dissolution Method and Its Specification

In vitro dissolution is a pivotal tool in selection of the optimum formulation from various prototype formulations in the developmental stage. The in vitro dissolution testing methodology used for screening different formulations during product development need not be the same as the one used for QC of the finished drug product. The method used for QC testing should be able to show dissolution of the total amount of drug from the dosage form and so in some cases it may involve using surfactants that may compromise the biorelevance of the method. The dissolution methodology used for screening has to be able to differentiate between various formulations differing in their composition and other attributes such as hardness or disintegration time and be biorelevant; that is, it should be able to predict the in vivo performance of the product (Gao and Sanvordeker 2014). The biorelevant dissolution testing allows for the prediction of the in vivo performance of a drug product on a rank-order basis.

One cannot evaluate all possible product variants that are generated using the defined design space during product development in the clinic. Hence, it is important to link clinical/pharmacokinetic parameters to in vitro dissolution testing. This would result in usage of dissolution testing as a surrogate tool for evaluation of clinical efficacy of all possible variants during development. The products manufactured during development using variants of highest-risk variables (material and process) are evaluated using the discriminatory in vitro dissolution testing method (Dickinson et al. 2008). The selection of variable and variant formulation is dependent on an effective discriminatory in vitro dissolution method. The development of a dissolution testing method to differentiate between product variants involves selection of an appropriate dissolution medium and test conditions. The selected dissolution medium should meet sink requirements for in vitro release testing. Sink condition refers to utilisation of a large volume of dissolution medium (at least greater than threefold saturation volume of drug) so that the concentration of the drug in the medium is much lower than the saturation solubility of the drug (Kuwana 2007). The concentration of the

TABLE 7.6
Analytical Test Parameters for Various Drug Products

Parameters	Test
	Solid Dosage Form (Tablet, Capsule)
Dissolution	Single time point testing appropriate for immediate-release dosage forms. Multiple time point testing appropriate for extended-release dosage forms. Two-stage testing employing different dissolution media in series or simultaneously for delayed-release dosage forms. In vitro/in vivo correlation can be used to establish acceptance criteria for extended-release dosage forms if human bioavailability data are available for formulations showing different release rates.
Disintegration	Could be substituted for dissolution in case of rapidly dissolving products containing drugs highly soluble throughout the physiological pH range. Disintegration testing is most appropriate when a relationship to dissolution has been established or when disintegration has been shown to be more discriminating than dissolution.
Hardness/friability	Normally a part of in-process quality controls and not included as a part of product specification. Only if the hardness and friability would crucially influence product quality as in the case of chewable tablets and orodispersible tablets are their acceptance criteria to be considered in the product specification.
Uniformity of dosage units	Uniformity of weight and drug content in dosage form determined using pharmacopoeial procedures.
Water content	Done in the same way as discussed in specifications for new drug substances.
Microbial limits	Same as for drug substances.
	Oral Liquids and Powders Intended for Reconstitution as Oral Liquids
Uniformity of dosage units	Uniformity of mass is considered for powders intended for reconstitution as oral liquids.
pH	Acceptance criteria for pH must be specified and justified.
Microbial limits	Same as discussed for drug substances.
Antimicrobial preservative content	Acceptance criterion for antimicrobial preservative content decided based on the concentration of preservative needed to maintain the microbiological quality of the drug product throughout its usage period and proposed shelf life.
Preservative efficacy testing	Done to ensure effectiveness of the preservative system.

(Continued)

TABLE 7.6 (CONTINUED)

Analytical Test Parameters for Various Drug Products

Parameters	Test
Antioxidant and preservative content	Release testing and stability testing for antioxidant and preservative content is performed.
Extractables	Specification for this is needed in case there should be extraction from rubber closures, liners or dosage form containers.
	Not needed if extractables are consistently below the levels (from development and stability data) that are proven to be acceptable and safe.
	Once the packaging material is decided, the container/closure is listed and data are gathered on potential extractables.
Alcohol content	This is applicable for hydroalcoholic preparations such as elixirs, wherein the alcohol content needs to be mentioned on the label. Specifications for alcohol content are set and the amount is calculated by assay.
Dissolution	Applicable for suspensions and dry powders intended for reconstitution.
	Dissolution testing must be performed at release and in accordance with pharmacopoeial specifications.
Particle size distribution	Applicable for oral suspensions.
Redispersibility	Applicable for oral suspensions that settle on storage (formation of sediment), setting acceptance criteria for redispersibility is considered suitable.
	Mechanical or manual shaking must be specified. Time needed to resuspend the sediment must be clearly defined.
Rheological properties	Applicable for relatively viscous solutions such as syrups and elixirs and for suspensions and includes specifications for viscosity and specific gravity.
Reconstitution time	Applicable for powders intended for reconstitution as oral liquids.
Water content	Test and acceptance criterion for water content is specified.
Parenteral Drug Products	
Uniformity of dosage units	Same as for oral liquids.
pH	Acceptance criteria for pH must be specified and justified.
Sterility	Testing methodology and acceptance criterion to be established unless parametric release is employed for terminally sterilised drug products.

(Continued)

TABLE 7.6 (CONTINUED)
Analytical Test Parameters for Various Drug Products

Parameters	Test
Endotoxins/pyrogens	Limulus amoebocyte lysate test or test for pyrogens to be included in specifications.
Particulate matter	Acceptance criterion for visible and subvisible particulate matter to be included.
Water content	In case of nonaqueous injections and powders for reconstitution, determination of water content must be included.
Antimicrobial content	Same as for oral liquids.
Antioxidant content	Same as for oral liquids.
Preservative content	Same as for oral liquids.
Extractables	More important for parenterals than for oral liquids.
Functionality testing of delivery systems	Products packaged in prefilled syringes or autoinjector cartridges need to be tested for functionality of delivery system. Syringeability, pressure and seal integrity tip cap removal force and piston release force need to be monitored.
Container-closure integrity testing	To ensure tight seal between the opening of the glass container and the rubber closure.
Osmolarity	Osmolarity needs to be controlled in case of products where the tonicity is mentioned on the label.
Particle size distribution	For injectable suspensions. Must be performed at release.
Redispersibility	Same as for oral liquids.
Reconstitution time	Same as for oral liquids.

drug at the absorption site in the gastrointestinal tract is always greater than the concentration in plasma because the absorbed drug rapidly distributes into the large volume of body fluids; this provides sink for drug absorption in the in vivo condition. Thus, the sink condition ensures that there is no artefact in the testing owing to saturation of the dissolution medium. Initially, dissolution in aqueous buffers across the pH range of 1.2–7.4 (sometimes higher pH values are also used, in cases where the drug is not stable at a lower pH) would be carried out. Other dissolution media such as simulated gastric/intestinal fluids containing enzymes and bile salts can also be used for biorelevance. If the solubility of the drug substance is limited, then

TABLE 7.7

Dissolution Development as per the QbD Approach

Sr. No.	Step	Example
1	Conduct quality risk assessment (QRA)	QRA allows the most relevant risks (product and process variables) to in vivo drug release to be identified (ICH Q9).
2	Develop appropriate CQA tests	Develop in vitro dissolution test(s) with physiological relevance that is most likely to identify changes in the relevant mechanisms for altering in vivo drug release (identified in Step 1).
3	Understand the in vivo importance of changes	Determine the impact of the most relevant risks (from Step 1) to clinical pharmacokinetics based on in vitro dissolution data combined with prior knowledge including biopharmaceutics classification system (BCS), mechanistic absorption understanding or clinical 'bioavailability' data.
4	Establish appropriate CQA limits	Establish the in vitro dissolution limit that assures acceptable bioavailability.
5	Use the product knowledge in subsequent QbD steps	Define a control strategy to deliver product CQAs that assure that dissolution limits are met during routine manufacture (ICH Q10).

surfactants or solubilisers may be utilised to increase solubility so that the 100% drug will be released for discrimination in routine QC analysis. The medium showing the ability to distinguish between batches manufactured using different process variables and material attributes is selected. The rate of dissolution should be slow enough to discriminate between product variants, and the time scale for complete dissolution should be small enough to be used for routine testing (Dickinson et al. 2008). During development, an in vivo pharmacokinetic study is often carried out on the formulation variants with different in vitro profiles. On the basis of the in vivo pharmacokinetic data, an attempt can be made to develop in vitro–in vivo correlation (IVIVC), so that this information can be used for further development and even later during the life cycle of the drug product. However, if there is no effect of changes in those attributes and process variables (highest risk) on the pharmacokinetic profile, it can be concluded that the absolute risk of those variables to the clinical efficacy of the product is reduced. The discriminatory dissolution method can further be used to qualify changes in manufacturing site, scale, equipment, manufacturing process and so on. The key steps in utilising dissolution as a tool to confirm clinical performance in QbD are shown in Table 7.7 (Dickinson et al. 2008).

7.4.2.3 Setting Clinically Relevant Dissolution Specifications

This is an important consideration to ensure batch-to-batch consistency in clinical outcome (Bandi 2012). The different approaches for setting dissolution specifications are shown in Table 7.8.

TABLE 7.8

Approaches for Setting Dissolution Specification

Approach	Inferences
Approach 1 (no data linking in vitro dissolution to plasma concentration)	• No assurance regarding clinical relevance of the dissolution specification in all cases. • Specification is set solely on in vitro considerations. • If adopted in QbD approach, could result in design space with limited regulatory flexibility.
Approach 2 (establish range of release characteristics resulting in clinical data)	• Clinical relevance of the dissolution specification is assured within the established range. • When used in QbD approach, can result in design space having suitable regulatory flexibility.
Approach 3 (predictive and robust in vitro and in vivo correlation)	• Setting dissolution specifications based on targeted clinically relevant plasma concentration. • Mainly applicable to modified release formulations. • When used in QbD approach, can result in design space with increased regulatory flexibility.

7.5 PROTOTYPE FORMULATION DEVELOPMENT

The next important step towards product development is prototype formulation development. This can be initiated only if all the following activities have been successfully completed:

- Dosage form selection
- QTPP and CQAs of drug product
- Product design and strategy for development
- Raw materials vendor selection and its procurement
- Analytical method development of API

Prototype formulation development starts with preliminary experimental trials using prioritised CMAs and critical process attributes obtained from initial risk assessment analysis (Yu 2008). The preliminary experimental trials aim to evaluate formulation variables and process parameters at different levels using an empirical approach (changing one variable at a time) or using screening design (Plackett–Burman design, fractional factorial design, mixture design). The results of formulation batches from preliminary screening studies will be evaluated for required CQAs such as assay, drug release, disintegration time, impurities and so on. Simultaneously, as the development proceeds, few preliminary batches that qualify the CQAs will be subjected for developmental stability studies and stress studies. The developmental stress study involves exposure of formulation batches to different stress conditions to escalate their impact on product stability. The stress study results further help decide the impact of CMAs of packaging material and their components

TABLE 7.9
Stability Study and Stress Study Protocol

Stress Condition		0 Day	7 Days	15 Days	1 Month
40°C/75% RH (open)		√	√	√	√
30°C/65% RH (open)		√	√	√	√
60°C		√	√	√	√
Stability Condition	0 Day	1 Month	2 Months	3 Months	6 Months
40°C/75% RH	√	√	√	√	√
30°C/65% RH	√	√	√	√	√
25°C/60% RH	√	√	√	√	√

on drug product CQAs. For example, if the tablet is moisture sensitive and shows high impurity under the open condition, then this warrants to be packaged in a high density polyehtylene (HDPE) bottle with moisture absorber or in an Alu-Alu blister pack that has the lowest moisture vapour transmission rate (MVTR) to ensure product stability. The detailed stability studies conducted on a product as per ICH guidelines and trend analysis of the resulting data will be covered in Section 7.6. Stability studies will be conducted in all possible configurations (i.e. HDPE, blister, plastic, glass vial, etc.) intended for future commercialisation. However, stress studies will be conducted in a vial or in an HDPE bottle as per specified study protocol. The generalised protocol for stability studies and stress studies is shown in Table 7.9.

If the trend analysis of formulation composition and +ve control (reference product) shows comparative stability in stress condition as well as at 40°C/75% RH for at least 3 months or is complying with assigned CQAs, then that formulation variant will be taken for further development. Once the preliminary experimentation is done, the criticality of some material attributes and process attributes is reduced to a lower level; however, some attributes are still at high risk; hence, the risk priorities will be updated again (ICH 2005a,b). The formulation composition and manufacturing process attributes that show criticality in updated risk assessment will be further optimised using design of experiment (DoE) or experimental design (ICH 2009). For more information on DoE, see Chapter 5.

7.5.1 DESIGN OF EXPERIMENTS

DoE is a statistically organised experimental plan based on number of the CMAs/ CPPs as independent variables followed by its statistical evaluation and generation of design space to get an optimum. DoE is a tool to detect the cause–effect relationship as well as interaction amongst the variables. The type of DoE will be selected on the basis of the number of variables, type of response, mathematical model and so on. The experiments are performed as per the selected experimental design (i.e. central composite design, D-optimal design, fractional factorial design, etc.). The data obtained from experimental design are fed into the software that generates design space after sequential steps of statistical modeling. The generated design space provides the region that ensures the attainment of response as per the specification and hence facilitates the decline in criticality of material and process attributes from a high level to a low level (Yu 2008; Yu et al. 2014).

7.5.2 UPDATED RISK ASSESSMENT

The updated risk assessment for the critical formulation attributes and CPPs can be assessed based on experimental trials so as to ensure that all criticalities are moved down to the lower side (ICH 2005a,b). Hence, the product development with these CMAs and CPPs ensures reliability and consistency of the developed product.

7.5.3 CONTROL STRATEGY

Simultaneously, along with updated risk assessment, the control strategy for the drug product based on control of material attributes (raw and intermediate materials) and process parameters will be defined on the basis of the generated design space. The control strategy can be defined as a variable strategy (selection of range within the design space) or as a fixed strategy (selection of the fixed value within the design space). Both variable and fixed control strategies ensure that the CQAs of the drug product will be attained reliably and consistently at all times. The finalised formulation batches are then packaged in all possible configurations intended for future commercialisation and kept for stability studies as per the ICH guidelines, which are briefly discussed in Section 7.6.

7.6   STABILITY STUDIES

Stability of a pharmaceutical product is the ability of a given formulation in a particular container-closure system to retain its physical, chemical, microbiological, toxicological, protective and informational characteristics within the specified limits throughout its storage and use (Bajaj et al. 2012). It also helps to understand the risk to product quality, safety and efficacy in case of accidental exposure to conditions beyond the recommended storage conditions, on the basis of which risk mitigation strategies can be evolved. The data gathered from stability testing is employed to envisage the shelf life of the product, assign proper storage conditions and propose labeling instructions. Stability of drug products could be influenced by numerous factors such as stability of the drug substance, drug–excipient interactions, manufacturing process used, type of dosage form, container-closure system used for packaging and light, heat and moisture conditions encountered during shipment, storage and handling (Bajaj et al. 2012). The stability testing can be of two types, that is, developmental stability testing and definitive stability testing as summarised in Table 7.10.

ICH has come up with several stability guidelines to harmonise the stability testing requirements for pharmaceutical product registration. ICH guidelines for stability testing are summarised in Table 7.11 (ICH 1996a,b, 2002b, 2003a,b).

TABLE 7.10
Type of Stability Studies

Type	Description
Developmental stability studies	Performed during developmental stage
	Not governed by ICH guideline
	Mainly used to check cause-and-effect relationship at different temperatures and humidity
Definitive stability studies	Final product (as per ICH guidelines)

TABLE 7.11
ICH Guidelines for Stability Testing

Guideline	Description
ICH Q1A (R2)	Stability testing of new drug substances and products, gives stability testing protocols including temperature, humidity and trial duration.
ICH Q1B	Photostability testing of new drug substances and products, gives the basic protocol needed to evaluate light susceptibility and stability of new drugs and products.
ICH Q1C	Stability testing for new dosage forms extends main stability guideline for new formulations of already approved medicines and defines circumstances under which reduced stability data can be accepted.
ICH Q1D	Bracketing and matrixing designs for stability testing of new drug substances and products give general principles for reduced stability testing and provide examples of bracketing and matrixing designs.
ICH Q1E	Evaluation for stability data, explains possible situations where extrapolation of retest periods/shelf lives beyond real-time data may be appropriate and provides examples of statistical approaches to stability data analysis.

7.6.1 ICH Q1A (Stability Testing of Drug Products)

ICH guideline Q1A (R2) gives information on storage conditions and testing frequency for drug products under six storage conditions, namely, general case (room temperature) [A], drug products packaged in impermeable containers [B], drug products packaged in semipermeable containers [C], drug products intended for storage in a refrigerator [D], drug products intended for storage in a freezer [E] and drug products intended for storage below −20°C [F] as summarised in Tables 7.12 and 7.13 (ICH Q1A (R2) 2003). Similarly, the stability conditions for different climatic zones are shown in Table 7.14 (Bajaj et al. 2012; WHO 2009).

TABLE 7.12

Drug Product Stability Testing and Storage Conditions

Type of Product	Storage Condition for Stability Testing (Minimum Period to Be Covered by Stability Data at Time of Submission)		
A	Long term 25°C/60% RH or 30°C/65% RH (12 months)	Intermediate 30°C/65% RH (6 months)	Accelerated 40°C/75% RH (6 months)

*If significant change occurs at the accelerated testing condition, additional testing at the intermediate storage condition would be done.

B	Stability studies may be conducted at ambient humidity		

C	Long term 25°C/40%RH or 30°C/35% RH (12 months)	Intermediate 30°C/65%RH (6 months)	Accelerated 40°C/no more than 25% RH (6 months)

Besides the guideline mentioned above (*), aqueous-based products packaged in semipermeable containers have to be evaluated for water loss at conditions of low relative humidity apart from physical, chemical, biological and microbiological stability evaluation.

D	Long term 5°C (12 months)	Accelerated 25°C, 60% RH (6 months)

E	Long-term condition: −20°C (12 months)

F	Drug products that are to be stored in the freezer should be evaluated on a case-by-case basis.

Note: The change can be considered as significant change, if physical characteristics, appearance, functionality test, assay, degradation products and dissolution (12 dosage units) fail to meet the acceptance criteria.

TABLE 7.13

Frequency of Stability Testing

Testing Condition	Frequency of Testing (Months)
Long term	0, 3, 6, 9, 12, 18, 24, 36
Intermediate	0, 6, 9, 12
Accelerated	0, 3, 6

7.6.2 ICH Q1B (Photostability Testing)

ICH guideline Q1B gives the standard conditions for photostability testing. Photostability testing is done in order to ensure that the product does not undergo any unacceptable change on exposure to light. Photostability testing should be conducted on the drug product outside the primary pack. If this testing results in an

TABLE 7.14

Stability Testing and Storage Conditions at Different Climatic Zones

Zones	Major Regions	Long-Term Stability Condition	Accelerated Stability Condition
I—Temperate zone	Russia, United States, Europe and United Kingdom	21°C/45% RH	40°C/75% RH
II—Subtropical and Mediterranean zone	Japan and Southern Europe	25°C/60% RH	40°C/75% RH
III—Hot and dry zone	Iraq and India	30°C/35% RH	40°C/75% RH
IVa—Hot and humid zone	Iran and Egypt	30°C/65% RH	40°C/75% RH
IVb—Hot and very humid zone	Brazil and Singapore	30°C/75% RH	40°C/75% RH

unacceptable change in the product, then testing on the product in the immediate pack is conducted. If photostability testing of the product inside the immediate pack gives an unacceptable change, then the product is subjected to testing in the marketing pack, which would include the secondary package such as carton along with the primary pack. If this yields an unacceptable change in the product, then the packaging may be redesigned or the product formulation may be modified followed by photostability testing on the new product or product in the new package (Velagaleti 2008). The guideline recommends the light sources to which the drug product needs to be exposed for photostability evaluation. The generalised considerations for photostability testing are shown in the following list (ICH 1996a).

- Total three batches data (i.e. one development batch and two pilot batches).
- If primary package is impermeable to light, testing can be done by direct exposure of product.
- If product is unstable to oxidation, then it should be kept in a protective inert transparent container.
- Sample should be placed horizontally with respect to the light.
- Samples packed in aluminium foil are kept along with authentic sample as dark control to study the contribution of heat-induced change.

7.6.3 STABILITY TESTING PARAMETERS FOR VARIOUS DOSAGE FORMS

Appendix 2 of WHO Annex 2 (Stability testing of APIs and finished pharmaceutical products) gives the testing parameters for various dosage forms, as shown in Table 7.15 (WHO 2009).

TABLE 7.15
Stability Testing Parameters for Different Dosage Forms

Dosage Form	Parameter
Tablets	Dissolution, disintegration if justified, water content, hardness and friability.
Hard gelatin capsules	Dissolution, disintegration if justified, brittleness, water content and microbe bioburden.
Soft gelatin capsules	Dissolution, disintegration if justified, microbe bioburden, pH, leakage and pellicle formation.
Emulsions	Separation of phases, pH, viscosity, microbe bioburden, mean globule size and globule distribution.
Oral solutions and suspensions	Precipitate formation, clarity in case of solutions, pH, viscosity, microbe bioburden, extractables, conversion to polymorphic forms if applicable, redispersibility, mean size, particle distribution and rheological properties.
Powders for oral solutions and suspensions	Water content, reconstitution time and the reconstituted solution and suspension tested as specified in oral solutions and suspensions.
Topical, ophthalmic and otic preparations	Clarity, uniformity, pH, resuspendability (lotions), consistency, viscosity, microbe bioburden and water loss. Additionally, for ophthalmic and otic products, sterility, particulate matter and extractables would be tested.
Suppositories	Softening range, dissolution at 37°C.
Transdermal patches	In vitro release rate, leakage, microbe bioburden and peel and adhesion force.
Metered dose inhalers and nasal aerosols	Uniformity of content, aerodynamic particle size distribution, microscopic evaluation, water content, leak rate, microbe bioburden, valve delivery, extractables, leachables from plastic and elastomeric components.
Nasal sprays	Clarity, microbe bioburden, pH, particulate matter, unit spray medication content uniformity, droplet and particle size distribution, weight loss, pump delivery, microscopic examination of suspensions, particulate matter, extractables, leachables from plastic and elastomeric components of container closure and pump.

(Continued)

7.6.4   ICH Q1D (BRACKETING AND MATRIXING)

ICH guideline Q1D defines specific situations in which bracketing and matrixing can be applied (ICH 2002b). Bracketing and matrixing refers to reduced designs in which samples for every factor combination are not tested at all time points. Bracketing is a design for stability testing in which only samples on the extreme of particular design

TABLE 7.15 (CONTINUED)
Stability Testing Parameters for Different Dosage Forms

Dosage Form	Parameter
Small-volume parenterals	Clarity, colour, particulate matter, pH, sterility and pyrogen/endotoxins. Powders for injection solutions must include clarity, colour, reconstitution time, water content, pH, sterility, endotoxins and particulate matter. Suspensions would additionally include particle size distribution, redispersibility and rheological features. Emulsions will include phase separation, globule size, viscosity and distribution of dispersed globules.
Large-volume parenterals	Clarity, colour, particulate matter, pH, sterility, pyrogen/endotoxins and volume.

factors (such as product strength, size of container or fill) are tested at all time points as in a full design. Bracketing is based on the assumption that the stability of any intermediate level is represented by the stability of the extremes being evaluated. Bracketing may be applied if tablets of different strengths are compressed from a single powder blend or granulation, or capsules of different strengths are made with different plug sizes using the same powder mix. However, bracketing is not advisable in case of different excipients being used for different strengths. A typical example wherein bracketing is applied would be in the case of a three-batch stability study of a product having three different strengths, namely, 60, 90 and 120 mg packaged in container sizes of 20, 120 and 500 ml, respectively. Using bracketing design, one could avoid stability testing of the intermediate-strength 90-mg product packaged in a 120-ml container. However, all the three batches of the two combinations (60 mg strength packaged in a 20-ml bottle and 120 mg strength packaged in a 500-ml bottle) need to be tested at each time point.

Matrixing is a reduced design for stability testing wherein a selected subset of the total number of possible samples for all the factor combinations are tested at a particular time point. At a successive time point, another subset of samples for all the factor combinations will be tested. The design is based on the assumption that the stability of each subset of samples tested represents the stability of all samples at a given time point. Matrixing designs are applied to closely related formulations of different strengths. Examples would be tablets of different strengths that are compressed from a single powder blend or granulation, or capsules of different strengths made with different plug sizes using the same powder mix. In a matrixing design when time points are matrixed, testing for all selected factor combinations should be done at initial and final time points and few fractions of the designated combinations should be tested at each of the intermediate time points. For example, when applying matrixing design to a tablet having two strengths, all three batches of both tablet strengths would be tested at 0, 12 and 36 months, but one or two of three batches for both strengths would be tested at 3, 6, 9, 18 and 24 months.

7.6.5 ICH Q1E (EVALUATION OF STABILITY DATA)

ICH guideline Q1E gives the requirements for evaluation of stability data and determination of shelf life of pharmaceutical products. It recommends the establishment of

TABLE 7.16
Evaluation Parameters at Different Storage Conditions

Intended Product Storage	Evaluation Parameters
Room temperature	• If there is no change in long-term and accelerated stability data, then statistical analysis of long-term data is not needed. Shelf life can be up to twice as long but should not be more than 12 months beyond period covered by long-term stability. • If there is change in long-term and accelerated stability data, then statistical treatment is needed. Shelf life must not cross the shortest period supported by long-term studies of any batch, that is, 6 months. • If accelerated stability data reveal significant change, then intermediate and long-term stability data will be used to assign shelf life. • If no significant change occurs under intermediate storage conditions, then the proposed shelf life can be one and a half times as long and not exceeding 6 months beyond the period covered by long-term data. • If long-term data are not amenable to statistical analysis and relevant supporting data are available, then the proposed shelf life can be up to 3 months beyond the period covered by long-term data. • If significant change occurs under intermediate conditions, then a shelf life shorter than the period covered by long-term data is suitable.
Refrigerator	• If long-term and accelerated stability data show little change over time and variability, then the shelf life could be one and a half times long but should not exceed 6 months beyond the period covered by long-term data without statistical analysis. • If there is significant change in long-term and accelerated stability data, then the shelf life could be one and a half times long but should not exceed 6 months beyond the time covered by long-term data if statistical analysis is available. The proposed shelf life can only be up to 3 months beyond the period covered by long-term data. • If accelerated stability data show a significant change between 3 and 6 months, then the shelf life should be proposed on long-term data and a shelf life shorter than the period covered by long-term data is suitable. • If accelerated stability data show significant change within the first 3 months, then the shelf life should be proposed on long-term data and a shelf life shorter than the period covered by long-term data is suitable. If long-term data show variability, then the shelf life could be verified through statistical analysis. A stability study may be conducted under accelerated conditions for shorter than 3 months at least for a single batch with frequent sampling and analysis.
Freezer	• The shelf life will be decided on the basis of long-term stability data. • If accelerated storage condition is absent for a drug product intended to be stored in a freezer, then a single batch may be tested at a higher temperature of $5°C \pm 3°C$ or $25°C \pm 2°C$ to study the impact of accelerated condition on the product.
$-20°C$	• The shelf life will be decided on the basis of long-term stability data.

TABLE 7.17
Stability Commitments

Stability Data for Submission	Commitments
Three batches	• To continue long-term studies throughout the proposed shelf life and accelerated studies for 6 months
Less than three batches	• To continue long-term studies through the proposed shelf life and the accelerated studies for 6 months and to place additional production batches to a total of at least three on long-term stability studies throughout the proposed shelf life and on accelerated studies for 6 months
No production batches	• To place the first three production batches on long-term stability studies throughout the proposed shelf life and on accelerated studies for 6 months

shelf lives for drug products intended to be stored at room temperature, in a refrigerator, in a freezer and below −20°C, as shown in Table 7.16 (ICH 2003b; Velagaleti 2008).

7.6.6 STABILITY COMMITMENT

If long-term stability data for primary batches of a drug product do not cover the proposed shelf life granted at the time of product approval, a commitment needs to be made to continue the long-term stability studies after approval to strongly establish the proposed shelf life (ICH 2003a). If data on long-term stability studies on three production batches of the drug product covering the proposed shelf life are available at the time of submission, then a postapproval commitment is not needed. The commitments are shown in Table 7.17.

Stability protocol employed for commitment batches should be the same as the one used for primary batches. If stability testing of primary batches at the accelerated storage conditions shows a significant change and intermediate testing is needed, testing on commitment batches may be conducted at the intermediate or accelerated storage condition. If testing of commitment batches at the accelerated storage condition shows a significant change, then testing at the intermediate condition is conducted.

7.7 PILOT CLINICAL STUDY

Once the product is developed and shows stability under accelerated conditions for a minimum of 1 to 3 months, the next step is to conduct the pilot clinical study. The type of clinical study varies with development approach; that is, in the case of a new drug application product, a proof-of-concept (POC) clinical pilot study (safety, efficacy, dose-ranging study, toxicological study, etc.) will be conducted and analysed against clinical data as specified in QTPP (ICH 2009; Loscalzo 2009). However, in the case of a generic product, a clinical study will be conducted merely to check bio-equivalence against the innovator product. The clinical study should be conducted in compliance with good clinical practices and as specified in ICH and other regulatory guidelines. The pilot clinical studies are conducted in a small number of healthy people, so as to assess the performance of the drug product. The conduct of a pilot clinical study starts with defining several parameters, that is, type of study, type of product and its strength, study design, study subject, sample collection, sampling frequency, sample storage, sample analysis and finally the statistical analysis of data (Table 7.18).

TABLE 7.18
Examples of Considerations of Parameters for Clinical Study

Parameters	Considerations
Type of study	Pharmacodynamic study: specifically for locally acting drug (e.g. rufinamide as colon targeting)
	Pharmacokinetic study: drug shows absorption in blood
	Physical evaluation: for example, resin that shows activity by drug receptor binding
Type of drug product	Metered dose inhaler: pharmacodynamic study
Drug product strength	Linear pharmacokinetics: biowaiver for lower strength based on the in vitro dissolution data
	Nonlinear pharmacokinetics: no biowaiver if nonlinear at the higher side; biowaiver if nonlinear at the lower side
Study design	Replicate design for highly variable drug (e.g. minocycline)
	Crossover design for nonvariable drug
	Parallel design for long half-life drug
Study subject	Normal/patient, medical history, physical/clinical evaluation, demographic profile and so on
Sample collection	Appropriate biological matrix screened
	Ascending phase
	Maximum concentration phase
	Elimination phase
Sampling frequency	Adequate during ascending phase
	Intensive sampling around expected time of maximum concentration (C_{max})
	Sufficient sampling during elimination phase
	Extend the sampling period to at least three longest half-lives of analyte (metabolite)
Sample storage	Appropriate temperature storage
	No analyte loss during storage
Sample analysis	Parent drug and analyte

TABLE 7.19

IVIVC Levels and Their Considerations

Level	Considerations
Level A	It states a point-to-point link between the rate of absorption and dissolution profile. It serves as a surrogate for clinical testing and helps obtain biowaiver.
Level B	Level B correlation is based on statistical moment analysis. In this correlation, the mean in vitro dissolution time is compared with the mean in vivo residence time. This level of correlation is less predictive of bioavailability than level A since various plasma profiles may have the same mean in vivo residence time.
Level C	Level C correlation involves linking one dissolution time point to a single pharmacokinetic parameter such as AUC, C_{max} or T_{max}. The predictability is again lower than that of level A correlation.

If the result of the pilot clinical studies complies with the required clinical data as specified in QTPP, the next step in product development is 'process performance qualification' in a laboratory scale. However, if the pilot studies do not comply with required clinical data as specified in QTPP, then, in such cases, the drug product would be subjected to redevelopment either by modifying the same formula or by adopting a different variant. The redeveloped formulation will be again evaluated in a clinical study to check its compliance with the expected clinical data as specified in QTPP. Multiple failures in pilot clinical study results may lead to increased financial burden of the drug product. Hence, to minimise the risk of failure, the IVIVC technique is often used.

IVIVC involves a quantitative relationship between an in vivo pharmacokinetic parameter such as C_{max} or area under the curve (AUC) and an in vitro property such as dissolution profile to predict the clinical performance of a dosage form and thereby minimise failure in clinical studies (Van Buskirk et al. 2014). IVIVC can be carried out at three levels in descending order of their predictive ability, as shown in Table 7.19.

7.8 PRE–TECHNOLOGY TRANSFER ACTIVITY

Once the POC of developed formulation is satisfactory, the next step before proceeding towards the technology transfer is execution of pre–technology transfer activity, that is, process performance qualification (PPQ) batches at the laboratory scale, specification finalisation (drug substance, in-process materials and finished product) and analytical method validation and method transfer. All these steps run simultaneously to proceed towards the scale-up/technology transfer of the drug product.

7.8.1 PROCESS PERFORMANCE QUALIFICATION (LABORATORY SCALE)

The manufacturing process defined during POC development should be robust enough to demonstrate acceptable quality and performance when transferred from a laboratory scale to a production scale. To ensure robustness in manufacturing process, the PPQ batches are conducted at the laboratory scale. The number of PPQ batches will be selected based on the identified criticality of process parameters of unit operations involved in the overall process. The PPQ batches will be conducted for studying the impact of CPPs using either an empirical or a DoE experimental approach. The data generated from PPQ batches will further help increase process understanding and mitigate some of the high-risk parameters leading to an updated risk assessment. Simultaneously, the control strategy for CPPs will be updated, which ensures robustness in the process to achieve a quality drug product.

The manufacturing process and the critical parameters for the same are different for various dosage forms (Gad 2008; Yu 2008). Considering these aspects, the manufacturing process, the relevant process parameters and the input and output material attributes for the manufacture of different dosage forms are illustrated in Figures 7.3 through 7.8.

7.8.2 Specification Setting for In-Process Materials and Finished Product

After the final formula and manufacturing PPQ are completed at the laboratory scale, the next step is the updating of the in-process material and finished product specifications using the learning gained during this experimentation. The in-process material specifications contain the tests and their limits with which the intermediate product/in-process material must comply to ensure manufacturability of the subsequent process steps and quality of the finished product (Chen et al. 2009; Yu 2008). The finished product specification contains tests and limits with which the product must comply during the initial batch release testing and throughout its shelf life. The specification test methods and limits must be justified and defined considering the design space and control strategy of CMAs and CPPs so that the product quality, safety, performance and stability can be maintained. The specification must also comply with regulatory requirements such as compendial monographs and so on. Many pharmaceutical companies have their own in-house specifications (tighter specifications), so that the product quality will be within CQAs during its entire shelf life. A specimen sample of finished product specification for atorvastatin calcium tablet 10 mg USP is shown in Table 7.20.

7.8.3 Analytical Method Validation

Analytical method validation is the documented evidence that provides a high degree of assurance of the suitability of the test method for its intended use in a drug substance or drug product. Validation of the analytical method depends on whether the test method is a compendial method or a noncompendial method. For compendial methods, truncated validation is carried out, and for noncompendial methods, complete validation as per ICH guideline Q2 (R1) needs to be conducted. The validation of a noncompendial method is done as per ICH or USFDA guidelines (ICH 2005b). The USFDA classifies noncompendial methods in four categories, as shown in Table 7.21.

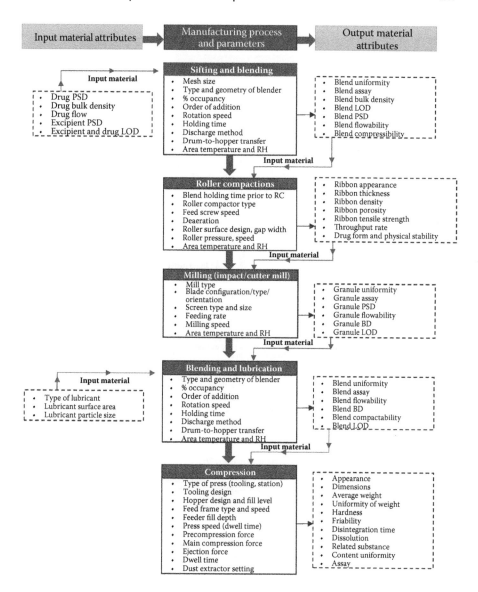

FIGURE 7.3 Process map for tablets (dry granulation).

Several other tests that may be defined as noncompendial include particle size analysis, droplet distribution, optical rotation, differential scanning calorimetry, x-ray diffraction and Raman spectroscopy. All the noncompendial tests should be validated and compared with compendial tests for equivalency.

The analytical method validation in the pharmaceutical industry starts with preparing a validation protocol, executing validation and finally generating the

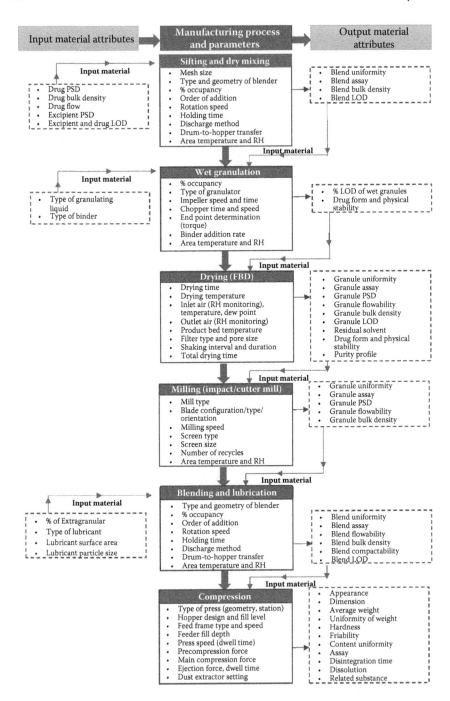

FIGURE 7.4 Process map for tablets (wet granulation).

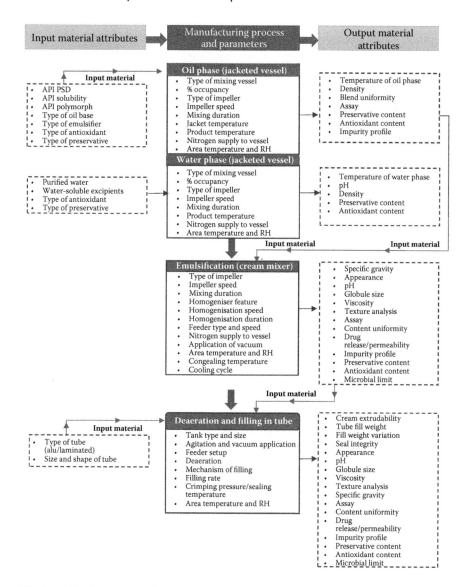

FIGURE 7.5   Process map for creams.

validation report. All these three elements are briefly described in Sections 7.8.3.1 through 7.8.3.3.

7.8.3.1   Validation Protocol

Validation protocol preparation will be based on the regulatory guidelines for product specification. Validation protocol includes the target method to be

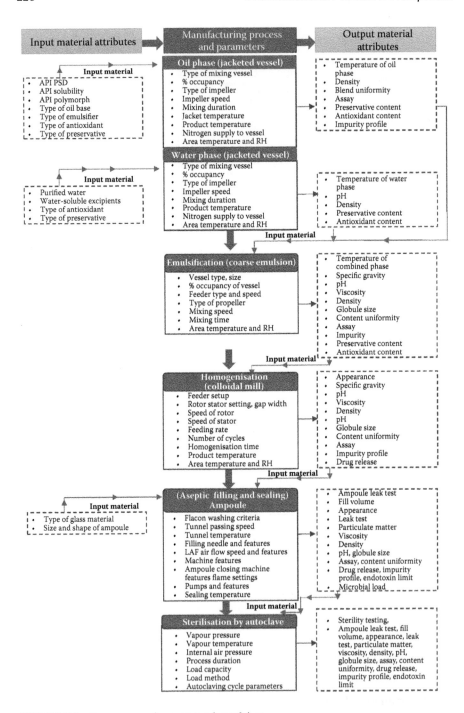

FIGURE 7.6 Process map for parenteral emulsions.

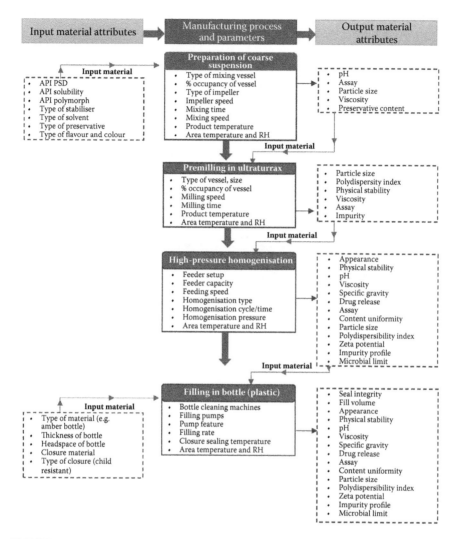

FIGURE 7.7 Process map for oral nanosuspension.

validated, preapproved validation elements and acceptance criteria. Typical validation elements include accuracy, precision (repeatability and intermediate precision), specificity, detection limit, quantitation limit, linearity, range, robustness, solution stability and system suitability. The establishment of the acceptance criteria is based on the type of test method. If any of the test methods does not comply with the acceptance criteria, then the deviation must be documented, signed by respective laboratory management and approved by the quality assurance department.

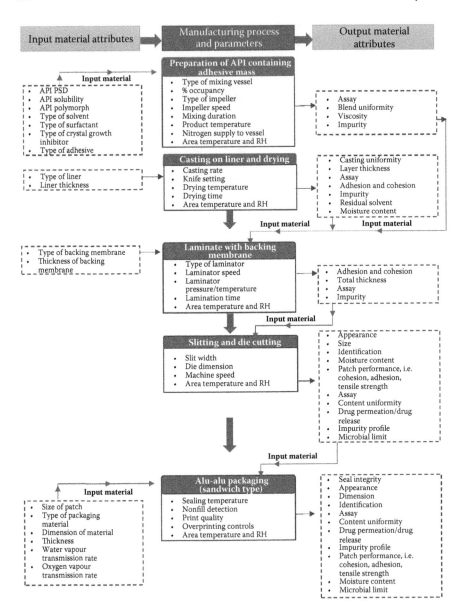

FIGURE 7.8 Process map for transdermal patches (drug in adhesives).

TABLE 7.20
Specimen Sample of Finished Product Specification

Finished Product Specification

Document	Page	**Atorvastatin Calcium Tablets USP 10 mg**
Number	xyz	**Supersedes: None**
Edition 01		Document number: xyz
		Dated: xyz

Label claim: Each film-coated tablet contains atorvastatin calcium USP 10 mg

Sr. No.	Test	Specifications
1.	Appearance	White, elliptical, film-coated tablets, scored on both sides, debossed with PD 155 on one side and 10 on the other side
2.	Average weight	150 mg ± 7.5% w/w
3.	Hardness	Between 60 N and 100 N
4.	Disintegration time	Not more than 15 min
5.	Loss on drying at 105°C up to 5 min	Not more than 5.0% w/w
6.	Identification	
A	By IR	The IR spectrum of the sample is concordant with that of the reference spectrum
B	By HPLC	The retention time of major peak in the sample solution corresponds to that in the standard solution, as obtained in the assay
7.	Assay of atorvastatin calcium (by HPLC)	To meet the requirement as per monograph
8.	Dissolution test	To meet the requirement as per monograph
9.	Uniformity of dosage unit	To meet the requirement as per monograph
10.	Related substances (by HPLC) procedure I	
	Desfluro impurity	Not more than 0.3%
	3S 5R Isomer	Not more than 0.3%
	Desfluro impurity	Not more than 0.3%
	Epoxide impurity	Not more than 0.3%
	Unknown impurity	Not more than 0.1%
	Total impurity	Not more than 1.0%

(Continued)

TABLE 7.20 (CONTINUED)
Specimen Sample of Finished Product Specification

	Finished Product Specification	
11.	Residual solvents	Meets the requirements as per USP general chapter ⟨467⟩
12.	Microbial enumeration test	
	i. Total viable aerobic count	
	Bacteria	Not more than 100 cfu/g
	Fungi	Not more than 10 cfu/g
	ii. Specific organism	
	E. coli	Should be absent in 1 g

TABLE 7.21
Analytical Validation of Noncompendial Methods

Category	Analytical Testing Method
I	Quantification of drug substance in finished product
II	Determination of impurities in drug substance and finished product
III	Performance characteristics (e.g. drug release)
IV	Identification purposes

7.8.3.2 Validation Execution

The method validation is executed as per the validated protocol.

7.8.3.3 Validation Report

The validation report comprises of a validation result including any deviation, if occurred.

7.8.4 ANALYTICAL METHOD TRANSFER

Once the product complies with the POC study, the developed validated analytical method must be transferred to the QC laboratories of the manufacturing site so that the routine QC testing of raw material, in-process controls and drug product can be carried out for scale-up and commercial batches. The successful analytical method transfer can be done in two ways, that is, either both the originating laboratories and receiving laboratories can test the same lot of product and compare the results for closeness or perform mini-validation (specificity, accuracy, precision and linearity) at production QC laboratories (Gibson 2004; Shargel and Kanfer 2014). The documents required for analytical transfer must establish the requirements for the satisfactory transfer of analytical method. The document details are summarised in Sections 7.8.4.1 and 7.8.4.2.

7.8.4.1   Method Transfer Protocol

Method transfer protocol includes test methods and their execution details. The protocol also defines the manner of method transfer, the role of laboratories and the acceptance criteria for successful transfer and reporting items.

7.8.4.2   Method Transfer Report

After the execution of transfer of analytical method, the test results are summarised and statistically compared with the preapproved acceptance criteria to determine whether the receiving laboratory at the manufacturing site is qualified to perform the test method. The transfer report should indicate whether the transfer is successful. All transfer data must be recorded and reviewed. Any deviation from the protocol must be documented and discussed.

7.9   PLANNING FOR EXECUTION OF TECHNOLOGY TRANSFER

During technology transfer at the manufacturing site, the process designed and qualified during development is further evaluated and validated using the equipment train proposed to be used for commercial manufacture (Levin 2001; Shargel and Kanfer 2014; Swarbrick 2006). Sometimes, the performance qualification batches are treated as registration batches, while in other cases, there are two separate sets of batches. This decision is based on the internal norms of each company. For successful execution of PPQ and registration batches during scale-up at the manufacturing site, several elements need to be determined, that is, PPQ/ registration batch sizes, number of PPQ and registration batches, equipment selection and so on (ISPE 2012). These factors are discussed in brief in Sections 7.9.1 and 7.9.2.

7.9.1   PPQ Batch Size and Number of Batches

The determination of PPQ and registration batches, batch size and number of PPQ batches to be conducted during scale-up is an essential element to ensure the robustness in process in large-scale equipment. The selection of PPQ batch size will be based on proposed commercial batch size. As per regulatory requirements, after registration, the batch size can be expanded to a maximum of 10 times for commercial production of the product. Considering this, it is essential that the batch size of the registration batches where the performance is qualified should be a minimum of one-tenth of the proposed commercial batch size or higher. Similarly, the number of PPQ batches will be based on criticality of process parameters. The process that is assessed to be high risk will require more number of PPQ batches, while the process with low risk requires very few PPQ batches. For more details on number of batches to be undertaken during process validation, see Chapter 8.

7.9.2   Equipment Selection

It is a common practice in the industry to use the equipment operating on a similar principle during both development and commercial manufacture, as this helps translate the learning from development to technology transfer. The equipment selection is dependent on the qualified operating range of each equipment and batch sizes. Some processes are scale independent while others are scale dependent, and this has to be taken into consideration while planning for PPQ batches. The equipment geometry used in the production scale is also an important parameter that needs to be assessed for its impact on drug product CQAs. The geometry simulation can be done by dimensional analysis. Dimensional analysis relies on the principle of similitude, where the dimensionless numbers help generate a functional relationship that further facilitates process reproducibility (Levin 2001).

7.9.2.1   Dimensionless Numbers

Dimensionless numbers have no dimensions. Such numbers are used to describe the physical attributes. The following are examples.

Newton (Equation 7.1) is a measure of the power required to overcome friction in fluid flow in a stirred reactor. In mixer granulation applications, this number can be calculated from the power consumption of the impeller.

$$\text{Newton(Ne)} = P/\rho n^3 d^5 \tag{7.1}$$

Froude is described for powder blending and was suggested as a criterion for dynamic similarity as well as scale-up parameter in wet granulation.

$$\text{Froude(Fr)} = n^2 \frac{d}{g} \tag{7.2}$$

Reynolds relates the inertial force to the viscous force and is used to describe mixing processes.

$$\text{Reynolds(Re)} = d^2 \frac{n\rho}{\eta} \tag{7.3}$$

where P is the power consumption, ρ is the specific density of particles, n is the impeller speed, d is the impeller diameter, g is the gravitational constant and η is the dynamic viscosity.

Using the abovementioned equations, several parameters such as mixing of liquid, powder blending and so on can be simulated from a small scale to a large scale.

Once the PPQ batch size, number of PPQ batches and the equipment selection based on geometry are determined, the documents necessary for execution of technology transfer are prepared as per the internal standard operating procedure of the individual company.

7.10 DOCUMENT REQUIREMENTS FOR EXECUTION OF TECHNOLOGY TRANSFER

The documents required for successful execution of technology transfer and responsibility are described briefly in Sections 7.10.1 through 7.10.7 (Gibson 2004; Shargel and Kanfer 2014; Swarbrick 2006).

7.10.1 MASTER FORMULA RECORD

The master formula record (MFR) is a product-specific document that is compiled, checked, approved and authorised by a competent technical person responsible for production and QC. It includes the patent/proprietary name of the product, its strength, generic name, dosage form, product pack and primary packaging materials, MFR number, page number, effective date, ingredient identity, quality and quantity including overages/assay value-based quantities, a brief outline of manufacturing process, equipment details, stepwise manufacturing process, theoretical and practical yield at different stages of manufacture, analytical controls, specification, stability test results and so on.

7.10.2 MASTER PACKAGING RECORD

The master packaging record is a product packaging-specific document that is compiled, checked, approved and authorised by a competent technical person responsible for production and QC. It includes the patent/proprietary name of the product, its strength, generic name, dosage form, packaging type, material used for packaging, pack size, dimensions of pack, fillers such as oxygen buster, moisture absorber, stability profile of packaging, shelf life of packaging and so on.

7.10.3 BATCH MANUFACTURING RECORD

The batch manufacturing record (BMR) gives an insight into the detailed manufacturing process and it is prepared based on the MFR. The BMR must be compiled, checked, approved and authorised by a competent technical person responsible for production and QC. It includes the name of the product, batch number, completion of significant intermediate stage, name of the person responsible for each stage of production, initials of operators who carried out significant processes and initials of persons who check, wherever applicable, quantity, batch number, QC report number of each ingredient actually weighed and amount of any recovered material added, in-process controls carried out and their results, the signature of the person who performed theoretical yield and actual yield at the appropriate stage of production together with an explanation, if variation beyond expectation is observed, and authorisation of any deviation, if made.

7.10.4 SPECIFICATIONS

The specifications of raw materials (API, excipients and packaging materials), in-process controls, finished product release and finished product are required before initiating the scale-up batch.

7.10.5   Standard Testing Procedure

The standard testing procedure for the API, in-process controls and finished product, should be transferred to the production QC unit before initiation of the scale-up batch.

7.10.6   Stability Study Protocol

The stability study protocol includes detailed information about the stability conditions (as per ICH guidelines), their testing frequency and the final pack size in which the sample is required to be charged.

7.10.7   Protocol for Execution of PPQ at the Manufacturing Scale

The protocol for execution of PPQ batches encompasses information on manufacturing conditions, operating parameters, processing limits and raw material inputs. The protocol also defines the data to be collected and how it will be evaluated, the tests to be performed for each significant process step, acceptance criteria for those tests, a sampling plan including sampling points and number of samples, and the frequency of sampling based on statistical rationale. The PPQ protocol aims to transfer knowledge gained during laboratory development to the production scale. The process of transferring knowledge, skill and method of manufacturing from the laboratory scale to the production scale is termed *technology transfer.*

7.11   TECHNOLOGY TRANSFER

Once all the pre–technology transfer activity is performed, the technology transfer will be initiated as per the predefined protocol. The process of technology transfer involves the interplay of several departments, as shown in Table 7.22.

Technology transfer involves assessment of PPQ at the manufacturing scale. The execution of PPQ batches is carried out in a good manufacturing practice (GMP) facility using commercial equipment. The PPQ batch is then executed and evaluated for in-process materials, finished product and other testing as specified in the protocol (ISPE 2012). If the data comply with the acceptance criteria as specified in the protocol, primary packaging is undertaken and submitted for stability testing as per the stability protocol. However, if PPQ batch data do not comply with the acceptance criteria as specified in the protocol, a thorough investigation is needed to identify the root cause and preventive action must be incorporated in future batches as per the quality systems. The action includes a change in specification with suitable justification, or in some cases the formulation batch is subjected to redevelopment (ISPE 2012). The risk associated with CMAs and CPPs is updated based on evaluation of data from PPQ batches. Simultaneously, along with updated risk assessment, the overall control strategy for CMAs and CPPs has to be redefined so that the proposed final formula and process ensures consistent manufacturing of a quality drug product. Finally, the PPQ batch report is prepared, which includes the scale-up BMR, scale-up MFR, in-process and finished product evaluation data, packaging details and stability charging data, which further facilitate the initiation of the registration batch.

TABLE 7.22
Role of Different Departments in Technology Transfer

Department	Responsibility
Research and development department (R&D)	• Successful scale-up by demonstrating that process and analytics are feasible, scalable and controllable • Perform comparability analysis of product quality attributes from scale-up and developmental batches and justify difference if any • Investigate and resolve any process, quality or any other issues attributed to scale-up
Technology transfer department and production department	• Assess feasibility, site capabilities and resource requirements • Review process instructions (with process technologist) to confirm capacity and capability • Check safety implications, for example, solvents, toxicity and sanitising materials • Check impact on local standard operating procedures • Satisfy training requirements of supervisors or operators • Conduct the scale-up exercise (with support from R&D) in line with site-specific equipment and quality manual • Learn the product and process specifics from the product development team • Evaluate concerns with respect to product and process and help resolve them • Assess risk with respect to future product manufacture and give solutions
Engineering department	• Reviews equipment requirement (with production representative) • Initiates required engineering modifications, change or part purchase • Reviews preventative maintenance and calibration impact
Quality control department	• Reviews analytical requirement and availability of instruments • Responsible for analytical method transfer of drug substance and drug product • Absorbs and transfer-in the analytics required for the product keeping site-specific requisites in mind and document learnings • Supports the scale-up exercise (with support from R&D team) in line with the experimental plan • Plans and implements the stability studies; perform comparability analysis of stability trends from scale-up and developmental batches and justify difference if any • Assess risk with respect to future product manufacture and give solutions
Quality assurance department	• Reviews analytical methods with QC to determine its capability of transfer • Absorbs and evaluates the product know-how in line with site-specific equipment and quality manual • Advice on risks involved and mitigation measures • Supports the scale-up process via documents and other controls • Reviews and ensures that the process and documentation are as per the GMP requirements as well as regulatory requirements

7.12 PLANNING FOR REGISTRATION BATCHES

Before initiation of the registration batch, the registration batch execution protocol and several other documents are also prepared, namely, revised specification (based on scale-up batch data), scale-up report, registration batch MFR, BMR, stability protocol, packaging protocol and so on.

7.13 REGISTRATION BATCHES AND DEFINITIVE STABILITY STUDIES

The registration batches will be manufactured as per the prepared protocol and evaluated as per the required acceptance criteria specified in the protocol. If the data comply with acceptance criteria as specified in the protocol, they will be packed in finalised primary pack and submitted for stability testing as per the stability protocol. If the stability data of the registration batch shows compliance under accelerated conditions for a minimum of 1 to 3 months, the next step is to conduct a pivotal clinical study. However, if PPQ batch data do not comply with acceptance criteria as specified in the protocol, a thorough investigation is needed to identify the root cause and preventive action must be incorporated in future batches as per the quality systems. The action includes a change in specification with suitable justification, or in some cases the formulation batch is subjected to redevelopment.

Finally, the risk is updated on the basis of the registration batch data so as to ensure that all criticality is moved down to the lower side. Simultaneously, along with updated risk assessment, the overall control strategy for CMAs and CPPs will be redefined so that the proposed final formula ensures consistent development of the quality product.

7.14 PIVOTAL CLINICAL STUDY

A pivotal clinical study of the registration batch is to be conducted in a larger population as compared to the pilot study. The selection of pivotal study design will be based on clinical data obtained in the pilot clinical study. If the pivotal clinical study complies with expected clinical data as specified in QTPP, the next step will be preparation of dossier for regulatory submission. If the pivotal clinical study does not comply, the formulation will undergo redevelopment.

7.15 DOSSIER COMPILATION FOR REGULATORY SUBMISSION

Once the registration batches are prepared and their pivotal clinical data comply with the assigned QTPP, the next and essential step in product development is compilation of all the documents as per the Common Technical Document (CTD) for their regulatory submission. The CTD has five modules, and for detailed information, see Chapter 3. Module 3 of CTD is composed of product development-related dossier submission requirement, that is, 3.2.S for a drug substance (Table 7.23) and 3.2.P (Table 7.24) for a drug product (ICH 2002a).

TABLE 7.23

CTD Module 3.2.S (Drug Substance)

3.2.S		Drug Substance
3.2.S.1	General information	• Nomenclature • Structure • Physicochemical properties
3.2.S.2	Manufacture	• Details of manufacturing process and process controls: control of materials • Control of critical steps and intermediates • Process validation • Development of manufacturing process
3.2.S.3	Characterisation	• Structural elucidation • Impurities
3.2.S.4	Control of drug substance	• Specification • Analytical procedures • Validation of analytical procedure • Analysis of batches • Justification of specifications
3.2.S.5	Reference standards or materials	• Reference standards used for testing drug substance
3.2.S.6	Container-closure system	• Description of primary package, secondary pack and suitability with regard to light protection, moisture protection and compatibility
3.2.S.7	Stability	• Stability summary and conclusion • Postapproval stability protocol and stability commitment • Stability data

After the successful development of a product and its dossier submission to the respective regulatory bodies, the next step towards commercialisation of a product is process validation (for details, see Chapter 8). During the regulatory approval process, there may be queries from a regulatory agency and a robust and complete development would help enhance the ability of the team to respond quickly and completely to the regulatory queries. Adopting the QbD and quality risk management approach to the development of a drug substance or drug product therefore improves the efficiency and effectiveness of the research and development team. It also ensures speedy regulatory approval and thus helps in augmenting business potential for the company. In addition, the in-depth product and process understanding gained is documented as knowledge management, which helps the commercial

TABLE 7.24
CTD Module 3.2.P (Drug Product)

3.2.P		Drug Product
3.2.P.1	Description and composition of the drug product	• Description about dosage form • Composition • Type of container closure for dosage form
3.2.P.2	Pharmaceutical development	• Drug product: name and type of dosage form • Components of drug product • Formulation development • Overages • Physicochemical and biological properties • Manufacturing process development • Container-closure system • Microbiological attributes • Compatibility
3.2.P.3	Manufacture	• Manufacturer name and address of manufacturing facility • Batch formula • Description of manufacturing process and controls • Control of critical steps and intermediates • Process validation and evaluation
3.2.P.4	Control of excipients	• Specifications used with their justification • Analytical procedures • Validation, justification of specifications • Excipients of human or animal origin: provide information on source, specifications, testing and viral safety data • Novel excipients: complete details of manufacture, characterisation and controls, with cross referencing to supporting safety data (nonclinical or clinical), should be included according to the drug substance format
3.2.P.5	Control of drug product	• Specifications used with their justification • Analytical procedures • Validation of analytical procedures • Batch analyses • Characterisation of impurities
3.2.P.6	Reference standards or materials	• Reference standards for testing of the drug product
3.2.P.7	Container-closure system	• Description of primary package, secondary pack and suitability with regard to light protection, moisture protection and compatibility

(Continued)

TABLE 7.24 (CONTINUED)

Specimen Sample of Finished Product Specification

	Finished Product Specification
3.2.P.8 Stability	• Stability summary and conclusion
	• Postapproval stability protocol and stability commitment
3.2.R Regional information	
3.3 Literature references	

manufacturing team to consistently and reliably manufacture the product and, in case of any unanticipated problem, helps in the quick resolution of the same so that the supply chain is not interrupted.

REFERENCES

Bajaj, S., Singla, D., Sakhuja, N. 2012. Stability Testing of Pharmaceutical Products. *J Appl Pharm Sci* 2:129–138.

Bandi, N. 2012. Topic 2: Development of Clinically Relevant Dissolution Specifications. CMC Focus Group Meeting. https://www.aaps.org/uploadedFiles/Content/Sections _and_Groups/Focus_Groups/Chemistry,_Manufacturing,_and_Controls/CMCFGmtg Topic2May2012.pdf (accessed July 11, 2015).

Chen, W., Qui, Y., Zheng, J. 2009. *Developing Solid Oral Dosage Forms: Pharmaceutical Theory & Practice*. New York: Elsevier.

Dickinson, P.A., Lee, W.W., Stott, P.W. 2008. Clinical Relevance of Dissolution Testing in Quality by Design. *AAPS J* 10(2):280–290.

DiFeo, T.J. 2003. Drug Product Development: A Technical Review of Chemistry, Manufacturing, and Controls Information for the Support of Pharmaceutical Compound Licensing Activities. *Drug Dev Ind Pharm* 29(9):939–958.

Gad, S.C. 2008. *Pharmaceutical Manufacturing Handbook: Production and Processes*, 1st edition. New Jersey: John Wiley & Sons, Inc.

Gao, Q., Sanvordeker, D.R. 2014. Analytical Methods Development and Methods Validation for Oral Solid Dosage Forms. In: *Generic Drug Product Development, Solid Oral Dosage Forms*, eds. L. Shargel and I. Kanfer, 32–50. Boca Raton, FL: CRC Press, Taylor & Francis Group.

Gibson, M. 2004. *Pharmaceutical Preformulation and Formulation: A Practical Guide from Candidate Drug Selection to Commercial Dosage Form*. Boca Raton, FL: Interpharm/ CRC Press.

ICH. 1996a. International Conference on Harmonisation of Technical Requirements for Registration of Pharmaceuticals for Human Use, ICH Harmonised Tripartite Guideline, Stability Testing: Photostability Testing of New Drug Substances and Products Q1B. http://www.ich.org/fileadmin/Public_Web_Site/ICH_Products/Guidelines/Quality /Q1B/Step4/Q1B_Guideline.pdf (accessed July 11, 2015).

ICH. 1996b. International Conference on Harmonisation of Technical Requirements for Registration of Pharmaceuticals for Human Use, ICH Harmonised Tripartite Guideline, Stability Testing for New Dosage Forms, Annex to the ICH Harmonised Tripartite

Guideline on Stability Testing for New Drugs and Products Q1C. http://www.ich
.org/fileadmin/Public_Web_Site/ICH_Products/Guidelines/Quality/Q1C/Step4/Q1C
_Guideline.pdf (accessed July 11, 2015).

ICH. 1999. International Conference on Harmonisation of Technical Requirements for
Registration of Pharmaceuticals for Human Use ICH Harmonised Tripartite Guideline
Specifications: Test Procedures and Acceptance Criteria for New Drug Substances and
New Drug Products: Chemical Substances. http://www.ich.org/fileadmin/Public_Web
_Site/ICH.../Q6A/.../Q6Astep4.pdf (assessed May 15, 2015).

ICH. 2002a. International Conference on Harmonisation of Technical Requirements for
Registration of Pharmaceuticals for Human Use, ICH Harmonised Tripartite Guideline,
The Common Technical Document for the Registration of Pharmaceuticals for Human
Use: Quality – M4Q(R1), Quality Overall Summary of Module 2, Module 3: Quality.
http://www.ich.org/fileadmin/Public_Web_Site/ICH_Products/CTD/M4_R1_Quality
/M4Q__R1_.pdf (accessed July 11, 2015).

ICH. 2002b. International Conference on Harmonisation of Technical Requirements for
Registration of Pharmaceuticals for Human Use, ICH Harmonised Tripartite Guideline,
Bracketing and Matrixing Designs for Stability Testing of New Drug Substances
and Products Q1D. http://www.ich.org/fileadmin/Public_Web_Site/ICH_Products
/Guidelines/Quality/Q1D/Step4/Q1D_Guideline.pdf (accessed July 11, 2015).

ICH. 2003a. International Conference on Harmonisation of Technical Requirements for
Registration of Pharmaceuticals for Human Use, ICH Harmonised Tripartite Guideline,
Stability Testing of New Drug Substances and Products Q1 A(R2). http://www.ich.org
/fileadmin/Public_Web_Site/ICH_Products/Guidelines/Quality/Q1A_R2/Step4/Q1A
_R2__Guideline.pdf (accessed July 11, 2015).

ICH. 2003b. International Conference on Harmonisation of Technical Requirements for
Registration of Pharmaceuticals for Human Use, ICH Harmonised Tripartite Guideline,
Evaluation for Stability Data Q1E. http://www.ich.org/fileadmin/Public_Web_Site
/ICH_Products/Guidelines/Quality/Q1E/Step4/Q1E_Guideline.pdf (accessed July 11,
2015).

ICH. 2005a. International Conference on Harmonisation of Technical Requirements
for Registration of Pharmaceuticals for Human Use, ICH Harmonised Tripartite
Guideline, Quality Risk Management Q9. http://www.ich.org/fileadmin/Public_Web
_Site/ICH_Products/Guidelines/Quality/Q9/Step4/Q9_Guideline.pdf (accessed June
15, 2015).

ICH. 2005b. International Conference on Harmonisation of Technical Requirements for
Registration of Pharmaceuticals for Human Use, ICH Harmonised Tripartite Guideline,
Validation of Analytical Procedures: Text and Methodology Q2 (R1). http://www.ich
.org/fileadmin/Public.../ICH.../Q2_R1/.../Q2_R1__Guideline.pdf (assessed June 15,
2015).

ICH. 2008. International Conference on Harmonisation of Technical Requirements for
Registration of Pharmaceuticals for Human Use, ICH Harmonised Tripartite Guideline,
Pharmaceutical Quality System Q10. http://www.ich.org/fileadmin/Public_Web_Site
/ICH_Products/Guidelines/Quality/Q10/Step4/Q10_Guideline.pdf (accessed June 15,
2015).

ICH. 2009. International Conference on Harmonisation of Technical Requirements for
Registration of Pharmaceuticals for Human Use, ICH Harmonised Tripartite Guideline,
Pharmaceutical Development Q8 (R2). http://www.ich.org/fileadmin/Public_Web_Site
/ICH_Products/Guidelines/Quality/Q8_R1/Step4/Q8_R2_Guideline.pdf (accessed June
15, 2015).

ISPE. 2012. Topic 1 – Stage 2 Process Validation: Determining and Justifying the Number
of Process Performance Qualification Batches. http://www.ispe.org/discussion-papers
/stage-2-process-validation.pdf (assessed June 15, 2015).

Kuwana, R. 2007. Dissolution Testing, Evaluation of Quality and Interchangeability of Medicinal Products, Training Workshop for Evaluators from National Medicines Regulatory Authorities in East Africa Community. World Health Organisation. http://apps.who.int/prequal/trainingresources/pq_pres/DarEsSalam-Sept07/DissolTesting.PPT (accessed July 11, 2015).

Levin, M. 2001. *Pharmaceutical Process Scale-up*, 1st edition. New York: Marcel Dekker.

Loscalzo, J. 2009. Pilot Trials in Clinical Research: Of What Value Are They? *Circulation* 119:1694–1696.

Shargel, L., Kanfer, I. 2014. *Generic Drug Product Development—Solid Oral Dosage Form*, 2nd edition. Boca Raton, FL: CRC Press, Taylor & Francis Group.

Swarbrick, J. 2006. *Encyclopedia of Pharmaceutical Technology*, 3rd edition. Boca Raton, FL: CRC Press, Taylor & Francis Group.

Van Buskirk, G.A., Asotra, S., Balducci, C. 2014. Best Practices for the Development, Scale-Up, and Post-Approval Change Control of IR and MR Dosage Forms in the Current Quality-by-Design Paradigm. *AAPS PharmSciTech* 15(3):665–693.

Velagaleti, R. 2008. Stability and Shelf Life of Pharmaceutical Products. In: *Pharmaceutical Manufacturing Handbook: Regulations and Quality*, ed. S.C. Gad, 559–582. New York: John Wiley & Sons, Inc.

Vogt, F.G., Kord, A. 2011. Development of Quality by Design Analytical Methods. *J Pharm Sci* 100(3):797–812.

WHO. 2009. Annex 2, Stability Testing of Active Pharmaceutical Ingredients and Finished Pharmaceutical Products. WHO Technical Report Series. http://www.ich.org/fileadmin/Public_Web_Site/ICH_Products/Guidelines/Quality/Q1F/Stability_Guideline_WHO.pdf (accessed July 11, 2015).

Yu, L.X. 2008. Pharmaceutical Quality by Design: Product and Process Development, Understanding, and Control. *Pharm Res* 25(4):781–791.

Yu, L., Amidon, G., Khan, M. 2014. Understanding Pharmaceutical Quality by Design. *AAPS J* 16(4):771–783.

8 Process Validation and Postapproval Changes

Amit G. Mirani and Maharukh T. Rustomjee

CONTENTS

8.1 PROCESS VALIDATION

After the successful development of a product and its dossier submission to the respective regulatory bodies, the next step towards commercialisation of the product is process validation. Process validation is establishing documented evidence, which provides a high degree of assurance that a specific process will consistently produce a product meeting its predetermined specifications and quality characteristics. The concept of process validation was introduced 50 years ago by Byers and Loftus (Byers 1974; Loftus 1978), which was later drafted as a specific guideline by the US Food and Drug Administration (USFDA) in 1987 entitled 'General Principles of Process Validation'. There have been paradigm shifts in terms of concept of process validation (Figure 8.1). In the past, the concept of process validation was conceived to represent consistency and reproducibility in manufacturing process conditions that were used for the development of a drug product without necessarily defining, understanding or controlling critical process parameters or critical product quality attributes. Thus, the inherent variability associated with the process remained unidentified. To overcome this issue, USFDA

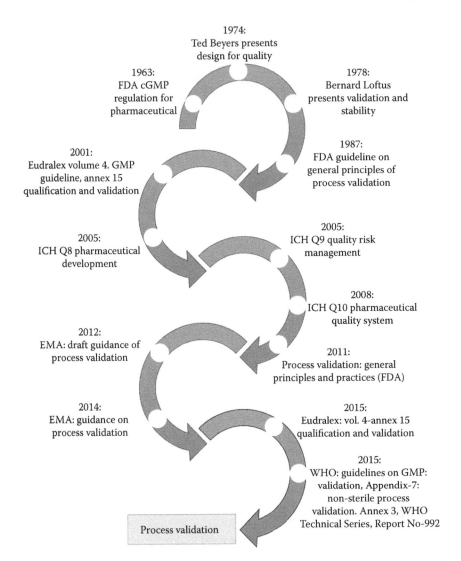

FIGURE 8.1   Evolution of process development.

(2011), EMEA (2014) and ICH (2005, 2008, 2009, 2012) came up with a new paradigm for process development, which involves a risk-based approach to identify the critical process parameters and hence meet the defined critical quality attributes. The process development, risk management and quality system strategies and tactics outlined in ICH Q8 Pharmaceutical Development (ICH 2009), Q9 Quality Risk Management (ICH 2005) and Q10 Pharmaceutical Quality System (ICH 2008), respectively, have been discussed in Chapter 2. The life cycle approach for process validation is described in three different stages, namely, process design, process qualification and continued process verification. The three stages and their execution plan are illustrated in Figure 8.2.

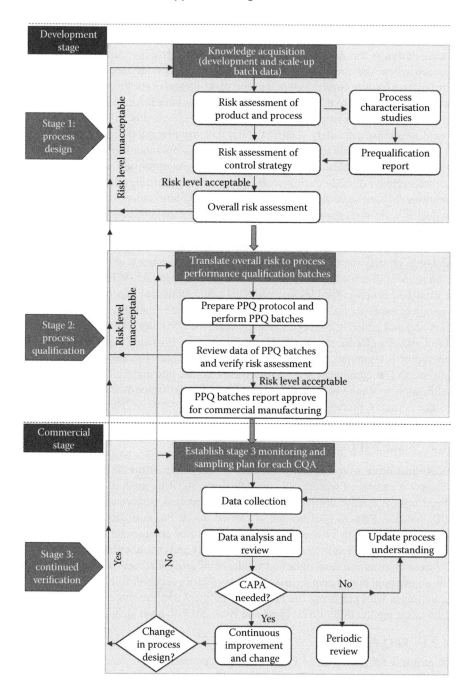

FIGURE 8.2 Process validation stage and execution plan.

8.1.1 Stage 1 (Process Design)

Process design is the first step in the execution of the process validation life cycle approach where the commercial manufacturing process is defined on the basis of knowledge gained through development and scale-up activities. Process design overcomes the limitations of the old paradigm as it allows identification and understanding of how the process responds to input variation and how this will have an impact on quality and patient safety (USFDA 2011). During process design, scalability of the unit operation needs to be considered. Some unit operations are scale independent and so the variables associated with them can be optimised at the laboratory scale and applied directly during scale-up (e.g. drying, milling). However, the unit operations that are scale dependent (e.g. granulation, lubrication) require optimisation during scale-up at the manufacturing site. These should be conducted in a structured manner through the use of various risk assessment techniques (USFDA 2011; Hyde and Hyde 2013). The risk assessment techniques as discussed in Chapters 1 and 4 help identify the failure modes in terms of critical process attributes to ensure the attainment of critical quality attributes of the drug product (ICH 2005). Further, the identified critical variables are evaluated using statistical experimentation, that is, design of experiments (DoE), to efficiently screen the variables using a mathematical model and generate a design space to define the optimum responses in terms of control strategy (for thorough details, see Chapter 5). Finally, the process development report is compiled, which includes the process control strategy and qualification report. The equipment and facility qualification, raw material and component qualification and analytical qualification should be confirmed before the execution of Stage 2, that is, process qualification (Ryan et al. 2013; Soto 2014).

8.1.2 Stage 2 (Process Qualification)

Process qualification incorporating risk management, ensures that the process designed in Stage 1 is capable of reproducibly manufacturing the drug product, meeting its predefined critical quality attributes in the form of process performance qualification (PPQ) batches. Process qualification is carried out at the manufacturing site using equipment proposed to be used for the manufacture of commercial batches. Several components of Stage 1 such as qualification of facility and critical utilities, qualification of process systems and equipment and validation of analytical methods are considered to be an essential component for the execution of process qualification. Process qualification starts with preparing the PPQ batch protocol and execution of the PPQ and qualification report (ISPE 2012a; Hyde and Hyde 2013; Ryan et al. 2013; Soto 2014).

8.1.2.1 PPQ Protocol

The protocol for execution of PPQ batches encompasses information on the manufacturing conditions, operating parameters, processing limits and raw material inputs. The protocol also defines the data to be collected and how they will be evaluated, the tests to be performed for each significant process step, acceptance criteria for those tests, a sampling plan including sampling points and number of samples and the frequency of sampling based upon statistical rationale.

8.1.2.2 PPQ Batches

The decision on the number of batches to be undertaken for the PPQ is an important stage as it ensures successful validation of the process. The process that exhibits high risk will require a greater number of PPQ batches, while the process with low risk requires very few PPQ batches. The number of PPQ batches could be defined on the basis of the following two suggested approaches (any other suitable alternative approach can also be used for PPQ batches selection) (ISPE 2012a):

Approach 1 (based on rationale and experience): In this approach, the number of PPQ batches will be based on the residual risk level (Table 8.1). The residual risk (Table 8.2) is arrived at following risk assessment of the product understanding, process knowledge and control strategy after process design (Stage 1).

TABLE 8.1
Number of PPQ Batches Based on Residual Risk Assessment

Risk Level	No. of PPQ Batches	Rationale
Severe	Not ready for PPQ	Redevelopment of process to reduce the risk
High	10	Large number of PPQ batches to ensure consistency in process. If not achieved, then redevelopment of process to reduce the risk
Moderate	5	Five PPQ batches to demonstrate process consistency
Low	3	Three PPQ batches appropriate for demonstrating process consistency
Minimal	1–2	Strong knowledge and high degree of control to minimise risk, for example, using the PAT approach

TABLE 8.2
Residual Risk of the Manufacturing Process

Risk Level	Description
Severe	Multiple factors have high-risk rating
High	Few factors have high-risk rating
Moderate	Medium-risk level for multiple factors
Low	Medium-risk level for few factors
Minimal	Low-risk level for all factors

Approach 2 (based on target process confidence and target process capability): The assurance in performance of manufacturing process can be elaborated using two terminologies, that is, *target process capability* and *target process confidence*.

Target process capability: Statistically, process capability (Cpk) is a measure of process robustness that can be used to assess the ability/capability of the process to meet the required CQAs (ISPE 2012a).

Cpk of ≥1 is defined as a capable process, therefore demonstration of a Cpk of at least 1 as a starting point for assessing capability of process seems reasonable. However, the level of confidence in this assessment coupled with the residual risk of the manufacturing process (Table 8.2) gives an understanding of an overall confidence of the process. A Cpk of ≥1 with 90% confidence for a process with low residual risk is deemed to be capable based on the limited experience and data available at commercial product launch. Hence, it would not be expected to require a larger number of PPQ batches. However, as the residual product risk increases, the need to provide higher level of confidence that the process is actually performing at acceptable level of capability is required (ISPE 2012a).

Target process confidence: The level of confidence needed in the Cpk calculated using the stage 1 and stage 2 data is an important factor for reaching a high level of confidence of quality consistency between batches. It is expected that each individual batch manufactured during stage 2 will meet quality requirements before being released for commercial distribution. However, determining the robustness of the process actually may take a number of batches to experience the full range of variability into the process and the resultant impact on product CQAs (ISPE 2012a).

TABLE 8.3

Correlation of Residual Risk with Target Confidence Levels

Risk Level	Target Confidence Required	Comment
Severe	N/A	High risk indicates major gaps in knowledge and understanding.
High	97%	High confidence interval is needed to provide high assurance.
Moderate	95%	Target confidence designed to provide reasonable assurance of process capability, supporting commercial distribution.
Low	90%	
Minimal	N/A	Minimal risk indicating high level of confidence of existing understanding and the capability of the control strategy. It is not necessary to base the number of PPQ on target variability.

TABLE 8.4
Number of PPQ Batches as per the Target Confidence

Risk Level	Number of Batches	Target Confidence
Severe	Not ready for PPQ	N/A
High	14	97%
Moderate	11	95%
Low	7	90%
Minimal	1–3	N/A

Table 8.3 provides an example of how target confidence level can be determined based on the residual risk.

The number of PPQ batches can be determined statistically using the *target process capability* (Cpk ≥1) and *target process confidence* as mentioned in following Table 8.4.

8.1.2.3 PPQ Report

Once the desired number of PPQ batches is prepared, the data can be evaluated using statistical approaches such as the Box–Cox plot, a multivariate chart and other charts that would provide a visual summary of intra- and interbatch variability. Finally, the data are documented in a report summarising the testing done along with the results and their conformance with expectations that confirm consistency of the manufacturing operations.

8.1.3 STAGE 3 (CONTINUED PROCESS VERIFICATION)

The goal of the Stage 3 validation is continued assurance that the process remains in a state of control (the validated state) during the commercial manufacture. The assurance of in-process control parameters over the entire product lifetime will be conducted by risk analysis. This can be achieved by collecting and analysing process data so as to verify that the quality attributes are being controlled throughout the process (ISPE 2012b; Ryan et al. 2013; Soto 2014).

The sequential steps for the execution of continued process verification (CPV) are described in Sections 8.1.3.1 through 8.1.3.3.

8.1.3.1 Establish CPV Monitoring Plan

The CPV monitoring plan is prepared to ensure a high level of confidence in the process throughout the commercial phase of the product life cycle. The monitoring plan should be established to evaluate process capability and variability so as to increase the process understanding and generate variability estimates using statistical tools. The monitoring plan should address the attributes to be monitored for each product, along with the method of collection and analysis of the data.

8.1.3.2 Selection of Attributes to Be Monitored

The selection of attributes to be monitored will be mainly based on the current understanding of the manufacturing process in relation to product quality attributes. The number of process attributes may increase or decrease as further knowledge is gained from monitoring experiences and specific risks are reduced or new risks are identified.

8.1.3.3 Data Analysis

The data generated during the manufacturing process are analysed to assure that the process is in a controlled state. The data should be analysed using several statistical principles by trained personnel and further evaluated for intrabatch variation, out of specification and so on. The statistical tools used for analysis include descriptive statistical analysis (time series plot, Box–Cox plot, etc.), statistical process control (SPC) charts (I-MR, X-R, etc.) and process capability (C_p, C_{pk}, etc.). The statistically analysed data are further evaluated for the presence of outliers within the results when compared with the set SPC limits of output attributes. The SPC limit differs from the specification or proven acceptable range or in-process control, as depicted in Figure 8.3. The SPC limit is an indicator of inherent unavoidable variability in the process represented as the 'voice of process'. Preferably, if the inherent variability of the process allows, the SPC limit should be established as a tighter limit compared to the specification limit. The tighter SPC limit acts as an alert level that can help prevent occurrence of out of specification owing to the potential variability. In case the data lie beyond the SPC limit but are within the specification limit, an investigation is needed to identify whether the variability that occurred is critical or not. If data lie beyond the specifications, a thorough investigation is needed to identify the root cause and preventive action needs to be taken. The preventive action should be confirmed through the monitoring plan and documented in a corrective and preventive action system. The agreed action may require a change in Stage 1 (process design) or Stage 2 (process qualification). The collected information is used to update the process understanding and the process design via knowledge management so as to assure consistency throughout the life cycle of the product. In this way, the continuous monitoring of process parameters and quality attributes helps minimise variability in the process.

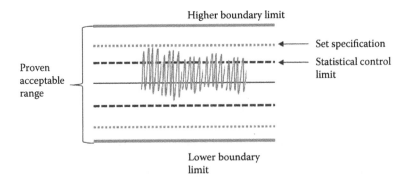

FIGURE 8.3 Statistical control limit.

Process validation is completed when all the stages are fully executed and data are evaluated. Once process validation and regulatory approval are obtained, the next step is to release the batch for commercial distribution.

8.2 POSTAPPROVAL CHANGES

Changes in manufacturing process and drug product may occur during the scale-up and postapproval stage of a drug product life cycle. Such changes, that is, in raw materials, equipment, manufacturing site, manufacturing process, batch size and so on, significantly affect the product quality attributes, which are directly linked to safety and efficacy. Considering this, the regulatory authorities of different countries such as the FDA; the European Commission Agency (Europe); the Ministry of Health, Labour and Welfare (Japan); Canada Ministry of Health (Canada); and so on have enforced the guidelines on postapproval changes so as to provide continuous improvement of the manufacturability and the commercial viability of the product. This section describes various guidelines on postapproval changes levied by different regulatory bodies and the measures taken to harmonise the regulation so as to assure the quality and performance of pharmaceuticals. This section also briefly discusses the integration of the quality by design (QbD) tool in the control of scale-up and postapproval changes.

8.2.1 REGULATORY REQUIREMENTS FOR POSTAPPROVAL CHANGES

The regulatory requirements for postapproval changes from different regulatory agencies are briefly described in Sections 8.2.1.1 and 8.2.1.2.

8.2.1.1 FDA Regulation on SUPAC

The FDA Modernization Act (FDAMA) (November 1997), an amendment to the current Food, Drug and Cosmetics Act, has added Section 506A (21U.S.C. 356a), which includes level of manufacturing changes recommended to an approved application. In accordance with the act, the FDA issued Guidance for Industry: Changes to an Approved NDA or ANDA (finalised in 2004). This guidance is the current standard document used in the pharmaceutical industry to assess and report manufacturing changes.

Before FDAMA, the Scale-Up and Post-Approval Changes (SUPAC) Guidance for Industry, that is, SUPAC-Immediate release solid dosage form, was the first attempt to provide information to the industry on submission requirements for postapproval changes (USFDA 1997a). Several types of postapproval changes considered under this guidance are changes in the components and composition, manufacturing sites, manufacturing process, specifications, container closure system, labelling and so on. The changes can be further classified in terms of 'levels' (i.e. Level 1, Level 2 and Level 3) based on intensity of their effect on formulation. Level 1 signifies the changes that have a minimal impact on the formulation quality and hence less documentation is required in the annual review report. Changes in accordance with Level 2 could have a significant impact on the formulation quality and performance of the dosage form. The supplement for such changes should be submitted to the FDA 30 days before the product to which the change is made to be distributed. Level 3 changes are most likely to have a significant impact on the formulation quality and performance and hence extensive documentation justifying the changes should be submitted to the FDA before implementation of changes. On the basis of the abovementioned levels, several recommendations with respect to Chemistry and Manufacturing Controls documentation, in vitro dissolution and in vivo bioequivalence test, are required to be submitted to the FDA. The in vitro and in vivo test data submission is a mandatory regulatory requirement as it is directly linked to the patient needs, that is, safety and efficacy. However, the in vitro–in vivo correlation (IVIVC) study can obviate the need for an in vivo study and thereby reduce the regulatory submission requirements for an approved drug product. Similarly, IVIVC of higher strength can be utilised to waive lower strength if both exhibit qualitative similarity. The IVIVC study cannot be applicable if the said changes are related to dosage strength, new formulation approach, excipient with high impact on absorption and so on. The different types of changes and their considerations under SUPAC Guideline for Industry: Changes to Approved NDA or ANDA (USFDA 2004) are summarised in Table 8.5.

The conditions for dissolution testing of different dosage forms are shown in Table 8.6. Similarly, SUPAC guidelines specific for immediate release (USFDA 1995), extended release (USFDA 1997a), delayed release (USFDA 1997a) and nonsterile

TABLE 8.5
Types of Changes under 'Changes to Approved NDA or ANDA'

Types of Changes	Consideration
Change in component and composition	Changes in composition and component are not covered in detail because of the complexity involved in the recommendations. Hence, specific SUPAC guidelines should be considered, for example, SUPAC-IR and SUPAC-MR.
Change in manufacturing site	Movement to a different manufacturing site for processing, testing, packaging or labelling of a drug product.
Change in manufacturing process	Change in process parameters, replacement of equipment, change in production scale involving use of different equipment and so on.
Change in specification	Changes in specifications for analytical method, in-process control and finished product.
Change in container closure system	Changes include size and shape of container, change in packaging component and change in closure.
Change in labelling	Changes in the package insert and package container label are included in the labelling changes.
Miscellaneous change	Apart from categories mentioned above, changes like stability protocol, expiration period and addition of stability protocol or comparability protocol have been included in the miscellaneous category.

TABLE 8.6

Dissolution Condition Requirement for Different Dosage Forms

Case	Dosage Form	Requirements
Case 1	Immediate release	**Multipoint dissolution:** Dissolution in compendial medium with sampling points 15, 30, 45, 60 and 120 min or until an asymptote is reached in compendial medium.
Case 2	Modified release	**Multipoint dissolution:** In addition to dissolution in compendial medium, other media, for example, in water, 0.1 N HCl and pH 4.5 and 6.8 buffer with adequate sampling until either 80% of the drug from the drug product is released or an asymptote is reached.
Case 3	Delayed release	**Multipoint dissolution:** In addition to dissolution in compendial medium, other media, for example, 0.1 N HCl for 2 h (acid stage) followed by testing in pH 4.5–6.8 buffer (buffer stage) under standard testing conditions. Adequate sampling should be performed during buffer stage until either 80% of the drug from the drug product is released or an asymptote is reached.
Case 4	Nonsterile semisolid	In vitro release study of prechange lot and postchange lot using Franz diffusion cell. The test passes if the 90% confidence interval falls within 75%–133.33%. If not passed, four additional runs of six cells should be carried out and total run should comply within 75%–133.33% confidence interval.

semisolid (USFDA 1997b) drug products are issued by the FDA. The same principles of assessment of changes, that is, level 1, 2 or 3, as used in these guidelines, are summarised in Figures 8.4 through 8.8.

8.2.1.2 European Regulation on SUPAC

Europe is one of the largest pharmaceutical sectors all over the globe. EUDRALEX pharmaceutical unit in Europe is responsible for making pharmaceutical legislation, guidelines and notices for applicants to ensure the safety and efficacy of a pharmaceutical product. In 2003, EUDRALEX implemented a regulation for granting postapproval variations; the intent was to simplify reporting procedures without negotiating any quality attribute of the drug product (European Parliament and European Council 2009; European Commission 2012). In the European Union, the changes to an approved application are called variations. Mainly, there are two types of postapproval variations, that is, Type I (Type IA, Type IB) and Type II (European Commission 2012; EMA 2015).

Type I variation: These variations can also be defined as minor variations and are simply administrative in nature. This type of variation does not affect the quality, safety or efficacy of the product. Type I is further subdivided into two types: Type IA and Type IB.

		Documentation				
Level	Change consideration	Immediate release (IR)	Modified release (MR)	Delayed release (DR)	Nonsterile semisolid (SS)	Filing documentation
Level 1 Minor changes	1. No impact on formulation quality and performance 2. Changes in excipient (%w/w) of total formulation <5%	1. Notification of change in component/composition 2. Updated batch record 3. Compendial requirements for dissolution and chemistry documentation 4. No biostudy				Annual report
Level 2 Moderate changes	1. Significant impact on formulation quality and performance 2. Changes in grade of an excipient 3. Changes in excipient >5% but <10%	1. Notification of component and updated batch record 2. Stability: One batch long-term stability in the annual report and one batch 3-month accelerated stability in supplement				1. Supplement – changes being effected 2. Accelerated stability data for ER, DR and SS in prior supplement 3. Long-term stability data in the annual report
		3. Dissolution: Case 1* 4. No biostudy	Dissolution: Case 2* Biostudy for drug with narrow therapeutic index	Dissolution: Case 3* No biostudy	Dissolution: Case 4* No biostudy	
Level 3 Major changes	1. Significant impact on formulation quality and performance 2. Qualitative and quantitative change in excipient >10%	1. Notification of component and updated batch record 2. Stability: One batch long-term stability in the annual report and one batch 3-month accelerated stability in prior supplement				1. Prior approval supplement 2. Accelerated stability data for ER, DR and SS in prior supplement 3. Long-term stability data in the annual report
		3. Dissolution: Case 1* 4. Biostudy or IVIVC	Dissolution: Case 2* Biostudy or IVIVC	Dissolution: Case 3* Biostudy or IVIVC	Dissolution: Case 4* Biostudy or IVIVC	

* The dissolution conditions are as specified in Table 8.6.

FIGURE 8.4   Change in component and composition.

Level	Change consideration	Documentation				
		Immediate release (IR)	Modified release (MR)	Delayed release (DR)	Nonsterile semisolid (SS)	Filing documentation
Level 1 Minor changes	1. Change in process such as mixing time, speeds within application/validation ranges	1. Notification of change in the manufacturing process 2. Updated batch record 3. Compendial requirements for dissolution and chemistry documentation 4. No biostudy				Annual report
Level 2 Moderate changes	1. Change outside of application/validation range (mixing time, operating speed, etc.)	1. Notification of process change and updated batch record				1. Supplement – changes being effected 2. Accelerated stability data for ER, DR and SS in prior supplement 3. Long-term stability data in the annual report
		2. Stability: One batch long-term stability	2. One batch long-term stability in the annual report and one batch 3-month accelerated stability in supplement			
		3. Dissolution: Case 1* 4. No biostudy	3. Dissolution: Case 2* No biostudy	3. Dissolution: Case 3* No biostudy	Dissolution: Case 4* No biostudy	
Level 3 Major changes	1. Change in type of process (e.g. wet granulation to direct compression)	1. Notification of component and updated batch record			No level 3 changes anticipated	1. Prior approval supplement 2. Accelerated stability data for ER, DR and SS in prior supplement 3. Long-term stability data in the annual report
		2. Stability: one batch long-term stability data	2. One batch long-term stability in the annual report and one batch 3-month accelerated stability in prior supplement			
		3. Dissolution: Case 1* 4. No biostudy	3. Dissolution: Case 2* Biostudy or IVIVC	3. Dissolution: Case 3* Biostudy or IVIVC		

* The dissolution conditions are as specified in Table 8.6.

FIGURE 8.5 Change in manufacturing process.

Level	Change consideration	Documentation				Filing documentation
		Immediate release (IR)	Modified release (MR)	Delayed release (DR)	Nonsterile semisolid (SS)	
Level 1 Minor changes	1. From nonautomated to automated equipment 2. Alternative equipment of same design, principle and capacity	1. Notification of change in manufacturing equipment 2. Updated batch record 3. Compendial requirements for dissolution and chemistry documentation 4. No biostudy				Annual report
Level 2 Moderate changes	1. Changes in equipment with a different design and operating principle	1. Notification of process change and updated batch record 2. Stability: One batch long-term stability 3. Dissolution: Case 1* 4. No biostudy	One batch long-term stability in the annual report and one batch 3-month accelerated stability in supplement Dissolution: Case 2* No biostudy	Dissolution: Case 3* No biostudy	Dissolution: Case 4* No biostudy	1. Supplement – changes being effected 2. Accelerated stability data for ER, DR and SS in prior supplement 3. Long-term stability data in the annual report
Level 3 Major changes		No Level 3 change anticipated in this category				

* The dissolution conditions are as specified in Table 8.6.

FIGURE 8.6   Change in manufacturing equipment.

		Documentation				
Level	Change consideration	Immediate release (IR)	Modified release (MR)	Delayed release (DR)	Nonsterile semisolid (SS)	Filing documentation
Level 1 Minor changes	1. Site change within the facility using the same equipment SOP 2. No changes made to manufacturing batch records	1. Notification of change in the manufacturing site 2. Updated batch record 3. Compendial requirements for dissolution and chemistry documentation 4. No biostudy				Annual report
Level 2 Moderate changes	1. Site change between the facility but within the campus using the same equipment SOP 2. No changes made to manufacturing batch records	1. Location of manufacturing site change and updated batch record 2. Stability: One batch long-term stability	One batch long-term stability in the annual report and one batch 3-month accelerated stability in supplement			1. Supplement – changes being effected 2. Accelerated stability data for ER, DR and SS in prior supplement 3. Long-term stability data in the annual report
		3. Dissolution: Case 1* 4. No biostudy	Dissolution: Case 2* No biostudy	Dissolution: Case 3* No biostudy	Dissolution: Case 4* No biostudy	
Level 3 Major changes	1. Change in type of process (e.g. wet granulation to direct compression)	1. Location of manufacturing site change and updated batch record 2. Stability: One batch long-term stability	One batch long-term stability in annual report and one batch 3-month accelerated stability in prior supplement			1. Prior approval supplement 2. Accelerated stability data for ER, DR and SS in prior supplement 3. Long-term stability data in the annual report
		3. Dissolution: Case 1* 4. No biostudy	Dissolution: Case 2* Biostudy or IVIVC	Dissolution: Case 3* Biostudy or IVIVC	Dissolution: Case 4* No biostudy	

* The dissolution conditions are as specified in Table 8.6.

FIGURE 8.7 Change in manufacturing site.

Level	Change consideration	Documentation: Immediate release (IR)	Documentation: Modified release (MR)	Documentation: Delayed release (DR)	Documentation: Nonsterile semisolid (SS)	Filing documentation
Level 1 Minor changes	1. Change in batch size less than 10 times the pilot/bio-batch using the same equipment having the same design and operating principle as that used to produce the test batch	1. Notification of change in batch size 2. Updated batch record 3. Compendial requirements for dissolution and chemistry documentation 4. One batch long-term stability study in the annual report 5. No biostudy				Annual report
Level 2 Moderate changes	1. Change in batch size beyond 10 times the pilot/bio-batch using the same equipment having the same design and operating principle as that used to produce the test batch	1. Notification of change in batch size and updated batch record 2. Stability: one batch long-term stability 3. Dissolution: Case 1* 4. No biostudy	One batch long-term stability in the annual report and one batch 3-month accelerated stability in supplement Dissolution: Case 2* No biostudy	Dissolution: Case 3* No biostudy	Dissolution: Case 4* No biostudy	1. Supplement – changes being effected 2. Accelerated stability data for ER, DR and SS in prior supplement 3. Long-term stability data in the annual report
Level 3 Major changes		No Level 3 change anticipated in this category				

* The dissolution conditions are as specified in Table 8.6.

FIGURE 8.8 Change in batch size: scale-up/scale-down.

Type IA variation: The variations under Type IA do not require any prior approval but must be notified by the holder within 12 months after implementation ('do and tell' procedure). However, certain minor variations of Type IA require immediate notification after implementation, in order to ensure the continuous supervision of the medicinal product. For example, changes in batch size of the finished product up to 10-fold compared to the original batch size approved at the time of marketing authorisation are considered as type IA variation, and it requires data analysis of at least one production batch of both the currently approved and the proposed batch sizes.

Type IB variation: The variation under Type IB must be notified before implementation ('tell, wait and do' procedure). The regulatory approval is expected within 30 days of submission. For example, changes in batch size of the finished product more than 10-fold compared to the original batch size approved at the time of marketing authorisation are considered as Type IB variation, and it requires data analysis as mentioned above, a copy of release and shelf life specification, and the batch number used in the validation study should be indicated.

Type II variation: This variation can also be defined as a major variation. Type II variations are the most severe category (changes in product summary characteristics such as indication, posology, contraindications, warnings, etc.), which may cause significant safety concerns about the medicinal product. The approval timing for a Type II variation is typically 3–6 months.

Information on the complete list of all changes linked to Type I and Type II can be obtained in the 'Guideline on Dossier Requirement for Type I and Type II Notifications'.

8.2.2 INTEGRATION OF THE QbD TOOL FOR POSTAPPROVAL CHANGES

QbD is an essential tool to ensure quality in the finished product. Implication of the QbD approach starts with defining the quality target product profile and critical quality attributes (CQAs), prioritisation of the critical process and product parameters (based on risk assessment studies), statistical optimisation of critical parameters using DoE, generation of a design space and finally assigning a control strategy to ensure the development of quality products. For details, see Chapter 4.

The control strategy defined from generated design space ensures the continued development of quality products. Hence, working within the control strategy/design space may not be considered as a change and therefore regulatory relief may be expected. The changes that are out of the design space necessitate a regulatory postapproval change process as described in Section 8.2.1. The role of the QbD approach for postapproval changes can be further explained using several examples such as batch size change, manufacturing site change and so on (Buskirk et al. 2014), as shown in Figure 8.9.

Type of change	Current SUPAC requirements	Proposed QbD approach	Rationale for proposed QbD approach
IR Product (Level 1) Batch size change <10× biobatch (final blend size)	Annual notification with updated batch record, one lot on long-term stability	No notification if within approved blending design space or uniformity is controlled with in-line NIR	Additional stability data may not be required since no change in stability indicating CQA (assay) while monitoring using NIR
	Release as per compendial specification	Documentation managed through documented risk assessment and justification for stability continuous process verification plan	Additional dissolution testing may not be required based on risk assessment data
IR Product (Level 2) Batch size change >10× biobatch (final blend size)	Supplement with updated batch records One batch with 3-month accelerated stability data and one batch with long-term stability data	No notification if within approved blending design space or uniformity is controlled with in-line NIR	Additional stability data may not be required since no change in stability indicating CQA (assay) while monitoring using NIR
	Multipoint dissolution profile vs. compendial medium	Documentation managed through documented risk assessment and justification for stability continuous process verification plan	Additional dissolution testing may not be required based on risk assessment data
MR Product Change in manufacturing site	Prior approval supplement and updated executed batch records	Notify via changes being effected supplement with no changes to approved design space and control strategy compared to the original site	The postchange design space confirmed with prechange batch and hence no change in control strategy
	Three batches with 3-month accelerated stability data in prior approval supplement	Documentation managed through documented risk assessment and justification for stability continuous process verification plan	The risk assessment does not show any process change; the bioequivalence study may not be required
	Long-term data in the annual report Multipoint dissolution vs. compendial or prechange or reference lot Bioequivalence study	Emphasis on demonstrating equivalent quality and performance in supplement through direct confirmation of design space at the new site	Additional dissolution testing may not be required based on risk assessment data

FIGURE 8.9   Application of the QbD tool for postapproval changes.

8.2.3   ICH Q12

Considering the lack of harmonised control in technical and regulatory considerations for postapproval changes, the ICH has issued a concept paper proposing a guideline, Q12, entitled 'Technical and Regulatory Considerations for Pharmaceutical Product Lifecycle Management' (ICH 2014). This guideline will introduce regulatory tools for prospective change management, that is, postapproval change management plans and protocols, which aims to investigate any change in quality attributes and their implementation based on a control strategy derived from design space and supports a continual improvement process thereby resulting in decreased variability and increased efficiency.

In conclusion, the newer paradigm of process validation assures design and development of manufacturing processes with comprehensive understanding and monitoring using a science- and risk-based approach. Thus, implementation of this new paradigm results in fewer process deviations, excursions and failures, thereby improving the consistency and reliability of manufacturing processes. However, some changes could be inevitable (such as vendors, manufacturing composition, manufacturing process, manufacturing site, etc.) during scale-up or commercial manufacturing stage. The science- and risk-based approach needs to be adopted for this change management such that efficacy, safety and quality of the product are not compromised. This change management is facilitated by guidelines from different regulatory bodies that provide an insight on the postregistration filing requirements based on severity of change and its risk assessment.

REFERENCES

Buskirk, V., Asotra, S., Balducci, C. 2014. Best practices for the development, scale-up, and post-approval change control of IR and MR dosage forms in the current quality-by-design paradigm. *AAPS PharmSciTech* 15(3):665–693.

Byers, T. E. 1974. GMP's and design for quality. *PDA J Pharm Sci Technol* 32(1):22–25.

EMA. 2015. European Medicines Agency post-authorisation procedural advice for users of the centralised procedure. http://www.ema.europa.eu/ema/pages/includes/document/open_document.jsp?webContentId=WC500003981 (accessed June 15, 2015).

EMEA. 2014. Guideline on process validation for finished products-information and data to be provided in regulatory submissions. http://www.ema.europa.eu/docs/en_GB/document_library/Scientific_guideline/2014/02/WC500162136.pdf (accessed June 15, 2015).

European Commission. 2012. Commission Regulation (EU) No 712/2012 amendment to Regulation (EC) No 1234/2008, as regards to Communication from the Commission-Guideline on the details of the various categories of variations to the terms of marketing authorizations for medicinal products for human use and veterinary medicinal products. *Official Journal of the European Union.* http://ec.europa.eu/health/files/eudralex/vol-1/reg_2012_712/reg_2012_712_en.pdf (assessed June 15, 2015).

European Parliament and European Council. 2009. Directive 2009/53/EC amendment to Directive 2001/82/EC and Directive 2001/83/EC, as regards variations to the terms of marketing authorisations for medicinal products. http://ec.europa.eu/health/files/eudralex/vol-1/dir2009 53/dir200953en.pdf.

Hyde, J., Hyde, A. 2013. FDA's 2011 process validation guidance: A blueprint for modern pharmaceutical manufacturing. http://www.ivtnetwork.com/article/fda%E2%80%99s-2011-process-validation-guidanceblueprint-modern-pharmaceutical-manufacturing.

ICH. 2005. International Conference on Harmonisation of technical requirements for registration of pharmaceuticals for human use, ICH harmonised tripartite guideline, quality risk management Q9. http://www.ich.org/fileadmin/Public_Web_Site/ICH_Products /Guidelines/Quality/Q9/Step4/Q9_Guideline.pdf (accessed June 15, 2015).

ICH. 2008. International Conference on Harmonisation of technical requirements for registration of pharmaceuticals for human use, ICH harmonised tripartite guideline, pharmaceutical quality system Q10. http://www.ich.org/fileadmin/Public_Web_Site/ICH_Products /Guidelines/Quality/Q10/Step4/Q10_Guideline.pdf (accessed June 15, 2015).

ICH. 2009. International Conference on Harmonisation of technical requirements for registration of pharmaceuticals for human use, ICH harmonised tripartite guideline, pharmaceutical development Q8 (R2). http://www.ich.org/fileadmin/Public_Web_Site/ICH_Products /Guidelines/Quality/Q8_R1/Step4/Q8_R2_Guideline.pdf (accessed June 15, 2015).

ICH. 2012. International Conference on Harmonisation of technical requirements for registration of pharmaceuticals for human use, ICH harmonised tripartite guideline, development and manufacture of drug substances (chemical entities and biotechnological/ biological entities) Q11. http://www.ich.org/fileadmin/Public_Web_Site/ICH_Products /Guidelines/Quality/Q11/Q11_Step_4.pdf (accessed June 15, 2015).

ICH. 2014. International Conference on Harmonisation, final concept paper Q12: Technical and regulatory considerations for pharmaceutical product lifecycle management. http:// www.ich.org/fileadmin/Public_Web_Site/ICH_Products/Guidelines/Quality/Q12/Q12 _Final_Concept_Paper_July_2014.pdf (accessed June 15, 2015).

ISPE. 2012a. Topic 1 – Stage 2 process validation: Determining and justifying the number of process performance qualification batches. http://www.ispe.org/discussion-papers /stage-2-process-validation.pdf (accessed June 15, 2015).

ISPE. 2012b. Topic 2 – Stage 3 process validation: Applying continued process verification expectations to new and existing products. http://www.ispe.org/discussion-papers/stage -3-process-validation.pdf (accessed June 15, 2015).

Loftus, B. T. 1978. Validation and stability. *PDA J Pharm Sci Technol* 32(6):268–272.

Ryan, D., Greene, A., Calnan, N. 2013. Process validation: Begin with the end in mind – An industry survey on continued process verification. *J Valid Technol* 19(4). http://www .ivtnetwork.com/article/process-validation-begin-end-mind%E2%80%94-industry -survey-continued-process-verification.

Soto, I. 2014. Process validation FDA guidance – Implementation approach. http://www .ivtnetwork.com/article/process-validation-fda-guidance-implementation-approach.

USFDA. 1995. FDA guidance for industry: SUPAC–IR: Immediate release solid oral dosage forms: Scale-up and post-approval changes: Chemistry, manufacturing and controls, in vitro dissolution testing, and in vivo bioequivalence documentation. http://www.fda .gov/downloads/Drugs/Guidances/UCM070636.pdf (accessed June 15, 2015).

USFDA. 1997a. FDA guidance for industry: SUPAC–MR: Modified release solid oral dosage forms: Scale-up and post-approval changes: Chemistry, manufacturing, and controls; in vitro dissolution testing and in vivo bioequivalence documentation. http://www.fda.gov /downloads/Drugs/Guidances/ucm070640.pdf (accessed June 15, 2015).

USFDA. 1997b. FDA guidance for industry: SUPAC–SS: Nonsterile semisolid dosage forms: Scale-up and post-approval changes: Chemistry, manufacturing and controls; in vitro release testing and in vivo bioequivalence documentation. http://www.fda.gov /downloads/Drugs/Guidances/UCM070930.pdf (accessed June 15, 2015).

USFDA. 2004. FDA guidance for industry: Changes to an approved NDA or ANDA. http:// www.fda.gov/downloads/drugs/guidancecomplianceregulatoryinformation/guidances /ucm077097.pdf (accessed June 15, 2015).

USFDA. 2011. FDA guidance for industry, process validation: General principles and practices. http://www.fda.gov/downloads/Drugs/Guidances/UCM070336.pdf (accessed June 15, 2015).

9 Case Studies on Pharmaceutical Product Development

Ankit A. Agrawal, Manasi M. Chogale,
Preshita P. Desai, Priyanka S. Prabhu,
Sandip M. Gite, Vandana B. Patravale,
Maharukh T. Rustomjee and John I. Disouza

CONTENTS

9.1   INTRODUCTION

The key aspects of pharmaceutical product development with respect to design and strategy, formulation and its manufacturing applying quality by design (QbD) and risk management principles have been discussed in Chapters 4 and 7, respectively. The aim of this chapter is to further elucidate the process of pharmaceutical product development for different dosage forms intended to be administered through different routes of administration. The chapter illustrates hypothetical case studies on oral, topical/transdermal, rectal, vaginal, parenteral, inhalation (pulmonary), ophthalmic and nasal routes of administration.

9.2   ORAL

Oral drug delivery is the most patient-compliant route of administration and captures more than 50% of the global drug delivery market. The range of dosage forms that can be administered via this route includes liquids (solutions, suspensions, emulsions), semisolids (sip gels, etc.), solids (tablets, capsules, lozenges, etc.) and the advanced drug delivery systems thereof. Upon oral administration, the drug undergoes dissolution in gastrointestinal (GI) milieu followed by its permeation across the GI membrane to reach systemic circulation. Thus, the resultant oral bioavailability depends upon the extent of dissolution and permeation. In view of this, oral drug delivery systems are very effective for the drugs with high solubility and permeability (Biopharmaceutics Classification System [BCS] class I drugs) but offer a challenge for those with limited solubility/permeability/stability (BCS class II, III and IV) (see Chapter 4 for details) (Desai et al. 2011). This section deals with a detailed case study of a new pharmaceutical product development of an immediate-release tablet of acetriptan (20 mg) prescribed for migraine treatment. Herein, a systematic QbD approach was employed as a tool for product development and was performed in compliance with Food and Drug Administration (FDA) guidelines (CMC-IM 2008).

9.2.1   CASE STUDY – ACETRIPTAN TABLET

The rationale towards product development considered the two most critical product attributes, namely, faster onset of action towards alleviation of migraine symptoms and a patient-compliant oral dosage form. The product development process is summarised as follows:

- With the critical points in vision, first, the desired quality target product profile (QTPP) with acceptance criteria was determined. This was followed by meticulous identification of critical quality attributes (CQAs) as described in Figure 9.1. From the profile, the quality attributes considered to be critical for satisfactory quality were assay, impurities, content uniformity, hardness and in vitro dissolution of tablet.
- The formulation type chosen was an oral standard-release tablet, in consideration of the need for a rapid onset of action and known pharmacokinetic characteristics of the molecule.
- A roller compaction granulation process was chosen on the basis of prior scientific knowledge of this molecule and of other similar products and technologies and equipment available. The decision was also driven by the known heat-labile nature of the drug and its poor flow properties.

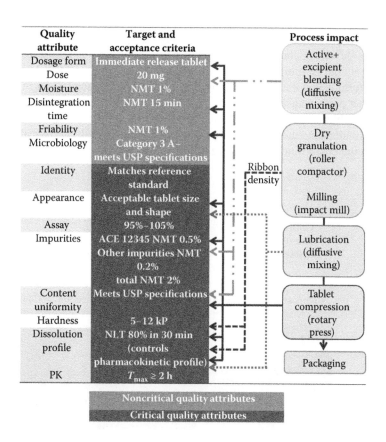

FIGURE 9.1 Target product profile and impact of process on quality attributes.

- A risk analysis, in accordance with ICH Q9, was used to establish the formulation variables and unit operations that were likely to have the greatest impact on product quality. Prior knowledge was used to arrive at the variables having a low potential for causing risk to product quality. In Figure 9.2, the potential risk factors for the product are shown and were studied further during development.
- The material attributes of the active pharmaceutical ingredient (API) and excipients were evaluated for their potential to affect the CQAs. For the unit operations with the potential to affect quality, a further risk assessment was used to identify process parameters and material attributes that could affect product quality.
- The formulation and process variables that were identified as high risk were studied extensively by experimentation designed to help understand their impact on the CQAs and arrive at an appropriate control strategy to be developed, thereby mitigating the risk to quality.

The key point to consider here is the poor aqueous solubility of the BCS class II drug acetriptan that possesses the greatest rate-limiting step in achieving the desired faster onset of action.

In order to achieve acetriptan tablets meeting the specifications of CQA, it was essential to develop a compliant manufacturing process that was robust and reproducible. From the preliminary assessment as described in Figures 9.1 and 9.2, it was clear that formulation composition as well as each unit operation has an impact on final product performance.

Thus, to achieve a product fitting into the acceptance criteria, the foremost necessity was to understand the variation in product performance resulting from alteration in formulation composition as well as each unit operation.

| | Variables, unit operations | | | | |
| | Critical material attributes | | Critical process attributes | | |
CQA	Composition	Blending	Roller compaction and milling	Lubrication	Compression
Identity	Low	Low	Low	Low	Low
Impurity	High	Low	Low	Low	Low
Assay	Low	Low	Low	Low	High
Content uniformity	High	High	High	Low	High
Appearance	Low	Low	Low	High	High
Hardness	Low	Low	High	Low	High
Dissolution	High	Low	High	High	High

FIGURE 9.2 Risk analysis for the acetriptan tablet.

As faster dissolution (not less than [NLT] 80% in 30 min) was the most critical acceptance parameter and directly linked to efficacy of the product, an in vitro discriminatory dissolution method was essential for this development and was developed as per ICH Q6A.

9.2.2 Development of Discriminatory Analysis Method

For this, a systematic dissolution study in aqueous buffers (pH 1.2 to 6.8) was performed, but it failed to establish a routine discriminatory control test owing to low solubility of acetriptan (BCS class II). Thus, as per the guidelines and preliminary surfactant solubility studies, Tween 80 and sodium lauryl sulphate (SLS) were screened, amongst which 1% SLS was confirmed to be the optimum concentration that enabled controlled dissolution with possible discrimination in formulation and process variables. Further, it also ensured the desired dissolution (>80% in 30 min), which helped in final product specification selection.

Once the discriminatory method was confirmed, the next step was to select the formulation key composition that involved API specifications and excipients, namely, diluent, disintegrant, lubricant and glidant (critical material attributes [CMAs]) so as to ensure unit tablet dosage form with faster disintegration and dissolution. The selection process is as described in Section 9.2.3.

9.2.3 Product Composition Selection

9.2.3.1 API Selection

Acetriptan offers very low solubility (BCS class II) and thus the salt form of the drug was examined to identify if it has higher solubility over the base form. The results revealed that the salt form of acetriptan did not show significant solubility enhancement over the free base form. Thus, the free base form of acetriptan was selected as it additionally precluded the interference owing to a varied crystalline form and polymorphs. After confirming the API base form, the particle size of acetriptan (d_{90} value) was restricted between 10 and 40 μm based on prior knowledge. Further, the particle size and impurities of acetriptan were identified as the most critical risk variables affecting dissolution and content uniformity, respectively.

9.2.3.2 Excipient Selection

Considering the tablet dosage form and drug granulation process, the excipients selected were microcrystalline cellulose (diluent), lactose monohydrate (diluent), croscarmellose sodium (disintegrant), magnesium stearate (lubricant) and talc (glidant), achieving a tablet of 200 mg containing 20 mg (single dose) of drug. All the inactive excipients were observed to be compliant with United States Pharmacopoeia (USP), European Pharmacopoeia and Japanese Pharmacopoeia. Further, drug–excipient compatibility was assessed for formulation feasibility.

Additionally, the selected excipients needed further concentration optimisation. In view of this, a design space was identified and optimised to control the CMAs.

For this, a QbD module was adapted in the form of a central composite response surface design involving three variables, namely, concentration of acetriptan, croscarmellose sodium, and magnesium stearate in the tablet. The impact of these variables was assessed over the critical response parameters and design space was decided subject to the compliance with acceptance criteria as described in Figure 9.3. The important outcome of the study was confirming the acetriptan loading to 10% w/w, which was continued in further product development.

Interestingly, an interaction was observed between drug load (concentration and particle size) and magnesium stearate that affected the resulting tablet hardness. To preclude the interference of these parameters, systematic response surface design was implemented over acetriptan particle size and magnesium stearate concentration (Figure 9.4). Figure 9.5 depicts the impact of both these parameters on tablet hardness and resulting dissolution profiles. From the figure, it was corroborated that the desired hardness of 5–12 kP was possible within the given design space at all the possible combinations. However, decrease in hardness was observed with an increase in concentration of both components. As a conclusion, it was revealed that the hardness could be controlled by compensating the large particle size of the API by reducing the lubricant concentration. Based on this, magnesium stearate concentration was proposed to be 1%–2% w/w for small to medium API particle size (d_{90} ~ 10–35 μm) and 1%–1.75% for large API particle size (d_{90} ~ 35–40 μm), as depicted in Figure 9.4.

Factor	Range investigated	Design space
Acetriptan (% w/w of tablet)	5–15	10
Croscarmellose sodium (% w/w of tablet)	1–4	3–4
Magnesium stearate (% w/w of tablet)	0.75–2.25	0.75–2.25

Response	Goal	Acceptance
Tablet hardness at fixed compression pressure (kP)	5–12	5–12
In vitro dissolution (% in 30 min)	NLT 80	NLT 80
Tablet weight uniformity (mg)	200 ± <3%	200 ± <3%

FIGURE 9.3 Impact of API and key excipient concentration.

Factor	Range investigated	Design space	
Acetriptan (10% w/w of tablet) particle size (μm)	10–40	10–35	35–40
Magnesium stearate (% w/w of tablet)	1–2	1–2	1–1.75

Response	Goal	Acceptance
Tablet hardness at fixed compression pressure (kP)	5–12	5–12
In vitro dissolution (% in 30 min)	NLT 80	NLT 80

FIGURE 9.4 Impact of API particle size and magnesium stearate concentration.

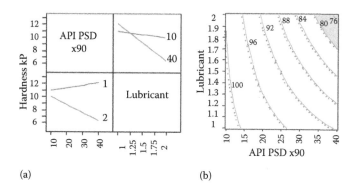

(a)   (b)

FIGURE 9.5 Impact of API particle size and lubricant concentration on hardness (a) at a fixed compression pressure and (b) at constant hardness of 12 kP.

The formulation composition space was thus successfully optimised to comply with desired product profile. This was then followed by optimisation of each of the unit operations (flow chart as depicted in Figure 9.1) leading to development of tablet unit dosage form.

9.2.4 PROCESS OPTIMISATION

From the thorough final product quality risk assessment, the most critical unit operations identified were blending, roller compaction, milling and compression, which were optimised in a step-by-step manner as described below.

9.2.4.1 Blending Optimisation

Blending was observed to be the most crucial unit operation controlling content uniformity of the final dosage form. Diffusive blenders were opted for the said process and the end point was determined using the near-infrared (NIR)-based spectrophotometric technique indicating the uniformity of blend.

The critical parameters affecting this unit operation were identified to be humidity (20%–70%), acetriptan particle size (10–40 µm), microcrystalline cellulose (30–90 µm) particle size, type of blender and number of revolutions.

Studies concluded that the particle size of microcrystalline cellulose should be controlled between 40 and 80 µm to achieve uniform blending, whereas all other parameters can be retained within the above proposed range and any diffusive blender can be utilised for the said unit operation.

Most importantly, the developers observed that the time required to achieve uniform blending varied dramatically and they could not establish any direct relationship between blender volume and mixing revolutions (in turn time). This demanded identification of a control measure to confirm the uniform blending end point, and the NIR-based spectrophotometric technique was then considered as a rescue tool. As an outcome, an in-line NIR (as an in-process quality control tool) was introduced as an end point determiner that assured uniform blending at all stages.

9.2.4.2　Roller Compaction and Milling Optimisation

Having assured blend content uniformity, roller compaction and milling were analysed for control strategy. Understanding the operation, six variables were identified as listed in Figure 9.6.

Here, it must be noted that roller compaction pressure predominantly affects the ribbon density and mill screen size, and speed determines the granule surface area, which in turn affects the dissolution rate, and thus both were chosen as in-process response parameters.

For roller compaction, a two-level design of experiment (DoE) study was executed (Figure 9.7), which revealed that particle size had a predominant effect on dissolution rate followed by roller pressure and magnesium stearate level. From the figure, it was well identified that large particle size (>35 μm) with a magnesium stearate concentration of 2% at high roller pressure resulted in an unacceptable dissolution profile. Here, one must note that the combination of acetriptan particle size and magnesium stearate concentration is critical for product performance. Also, the roller pressure is a critical process attribute. Thus, in the context of this unit operation, this alteration in dissolution profile was inversely correlated to roller pressure and in turn the ribbon density. On the basis of these critical observations, the design space and control strategy were confirmed, as depicted in Figure 9.6.

Further, milling studies (screen size and speed) exhibited a direct relation with granule surface area, which is directly proportional to dissolution rate. This unit operation was concluded to be relatively robust in the predicted design space as the resultant broad range of granule surface area (12,000 to 41,000 cm^2/100 g) gave a desired dissolution profile.

The next step in the process was lubrication with extragranular magnesium stearate (0.25%) and talc (5%). This was concluded to be a noncritical step as it was observed to be very robust and a conventional mixing approach was employed.

Factor	Range investigated	Design space	
Acetriptan (10% w/w of tablet) particle size (μm)	10–40	10–35	35–40
Magnesium stearate (% w/w of tablet)	1–2	1–2	1–1.75
Croscarmellose sodium (% w/w of tablet)	3–4	3–4	
Roller pressure (bars)	50–150	50–150	
Mill screen size (inch)	0.039–0.062	0.039–0.062	
Mill speed (rpm)	600–1200	600–1200	

Response	Goal	Acceptance
In vitro dissolution (% in 30 min)	NLT 80	NLT 80
Ribbon density (g/cm^3)	0.68–0.81	0.68–0.81
Granule surface area (cm^2/100 g)	12,000–41,000	12,000–41,000

FIGURE 9.6　Impact of key variables on roller compaction and milling.

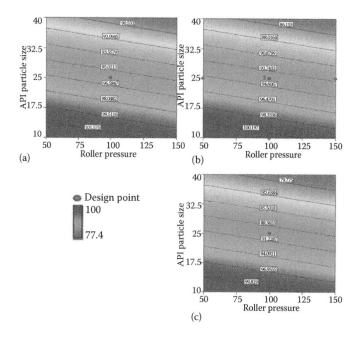

FIGURE 9.7 **(See colour insert.)** Contour plot response curve for API particle size and roller pressure against % dissolution at 30 min. (a) Magnesium stearate concentration 1% w/w, (b) magnesium stearate concentration 1.5% w/w and (c) magnesium stearate concentration 2% w/w.

9.2.4.3 Tablet Compression Optimisation

This unit operation was performed using a rotary tablet press. Having established the design space with respect to excipient composition and uniform blending and granulation, the critical tablet press parameters affecting this unit operation were identified to be precompression force (0.3–2.9 kN), compression force (7.4–12.9 kN) and press speed (26,000–94,000 tablets/h). The control response parameters assessed in this study were hardness, friability, disintegration time, weight variation and in vitro dissolution.

The DoE-based analysis over these critical parameters revealed that hardness increases proportionally with an increase in compression force and precompression force affects the hardness only in conjunction with compression force. The rotary machine speed was observed to have a nonsignificant effect on the final product quality within the defined design space. Understanding this, the control strategy here was to set the compression force limit within the design space limits and tablets witnessing the compression force beyond this limit were automatically discarded by the rotary tablet press.

9.2.5 IN VIVO PRODUCT PERFORMANCE ASSESSMENT

From the product composition and manufacturing process optimisation, design space was identified and the most critical factors affecting the product target profile were identified to be acetriptan particle size, ribbon density and magnesium stearate concentration. To analyse the impact of these parameters on product clinical performance, a crossover pharmacokinetic study (12 volunteers) was undertaken with five formulations representing variants of aforementioned factors (within the design space). The formulations exhibiting more than 80% dissolution within 30 min achieved T_{max} in less than 2 h (average ~ 1.33 h) with acceptable C_{max} and area under the curve (AUC). Thus, dissolution control (NLT 80% in 30 min) was confirmed as the critical discriminating test to select suitable formulation.

9.2.6 MICROBIAL GROWTH ASSESSMENT

Excipients as well as three primary stability batches were assessed for microbial growth test and water activity test as per USP and were observed to possess negligible tendency towards microbial growth support, which was thus considered to be a low-risk parameter.

9.2.7 CONTAINER AND CLOSURE SYSTEM

The developed tablets were packed in two ways, namely, 10 tablets per 30-cc high-density polyethylene bottle having cotton wadding with heat induction seal and 6 tablets per Aclar blister with push-through foil lid further enclosed in a secondary cardboard carton.

9.2.8 DESIGN SPACE

Based on the systematic study as described in Sections 9.2.3 through 9.2.7, design space and product specifications (both in-process and final) were confirmed and are summarised in Table 9.1.

9.2.9 CONTROL STRATEGY

All the high-risk variables and unit operations identified before product development (Figure 9.2) were investigated and optimised systematically and a design space and control strategy was proposed. After successful product optimisation, all the variables and unit operations were reassessed towards the risk and it was observed that the risk was minimised at each stage with the proposed control strategy, as depicted in Table 9.2.

The finalised product was then scaled up at a batch size of 100 kg in compliance with design space and in-line control parameters along with manufacturing control strategy. The successful scale-up was achieved with a product well within the target profile. This confirmed process optimisation with minimum risk. The stability studies were carried out as per ICH guidelines and the standard dossier filing process was initiated.

TABLE 9.1

Design Space and In-Process Specifications for Acetriptan Tablets

Parameter		Design Space		Control Criteria
		Range 1	Range 2	
Acetriptan	Particle size, d_{90} (μm)	10–35	35–40	Critical material control
	Concentration (% w/w)	10		–
Microcrystalline cellulose	Particle size, d_{90} (μm)	40–80		Critical material control
	Concentration (% w/w)	40		–
Lactose monohydrate	Particle size, d_{90} (μm)	70–100		Critical material control
	Concentration (% w/w)	38.75–40.75	39–40.75	–
Magnesium stearate	Intragranular (% w/w)	1–2	1–1.75	–
	Extragranular (% w/w)	0.25		–
Croscarmellose sodium (% w/w)		3–4		–
Talc (% w/w)		5		–
Humidity (% RH)		20–70		Process control
Unit Operation: Blending				
Blender (diffusive blender)		% CV NMT 5%		Critical process control
Unit Operation: Roller Compaction and Milling				
Roller pressure (bar)		50–150		Critical process control
Mill screen size (inch)		0.039–0.062		Critical process control
Mill speed (rpm)		600–1200		Critical process control
Ribbon density (g/cm³)		0.68–0.81		Intermediate material control
Granule surface area (cm²/100 g)		12,000–41,000		Intermediate material control
Unit Operation: Compression				
Precompression force (kN)		0.3–2.9		Critical process control
Press speed (tablets/h)		26,000–94,000		Critical process control
Compression force (kN)		7.4–12.9		Critical process control

(Continued)

TABLE 9.1 (CONTINUED)
Design Space and In-Process Specifications for Acetriptan Tablets

Parameter	Design Space		Control Criteria
	Range 1	Range 2	
Critical Product Attributes			
Hardness (kN)	5–12		Final product control
Dissolution profile	NLT 80% in 30 min		Final product control
Pharmacokinetics	$T_{max} \leq 2$ h		Confirmed from pharmacokinetic studies
Tablet weight (mg)	Mean of 20 tablets, 194–206		Final product control

9.3 TOPICAL AND TRANSDERMAL

Transdermal delivery provides an alternative to the oral and parenteral delivery of drugs. It offers a myriad of advantages including avoidance of first-pass metabolism, release over prolonged periods and consistent drug plasma levels, resulting in patient compliance for chronic conditions (Prausnitz and Langer 2009). It offers a relatively cheaper option to expensive hospitalisation costs (Paudel et al. 2011).

A transdermal patch is a medicated adhesive patch that is applied on the skin to deliver drug continuously through the skin to treat a systemic disease. Transdermal patches are broadly classified as reservoir systems and matrix (drug-in-adhesive) systems (Dhiman et al. 2011).

Reservoir systems are composed of a drug reservoir flanked by a backing membrane on one side and the rate-controlling membrane and adhesive on the other side. The monolithic drug-in-adhesive type of transdermal patch is one in which the adhesive layer not only serves as an adherent layer to the skin but also is responsible for the release of the drug.

9.3.1 CASE STUDY – RIVASTIGMINE DRUG-IN-ADHESIVE TRANSDERMAL PATCH

The case study discussed herein deals with the fabrication of a monolithic drug-in-adhesive type of transdermal patch of rivastigmine, an acetylcholinesterase inhibitor for the therapy of Alzheimer's disease (AD) (see Figure 9.8). Currently, no curative treatment is available for AD. Symptomatic drug treatments need to be taken lifelong and their efficacy depends on dose. With oral dosage forms, particularly at high doses, GI side effects are common and often result in suboptimal doses or discontinuation of treatment. Therefore, the development of alternative

TABLE 9.2
Control Strategy for Acetriptan Tablets

	Control Strategy				
	Critical Material Attributes	Variables, Unit Operations			
		Critical Process Attributes			
CQA	Composition	Blending	Roller Compaction and Milling	Lubrication	Compression
Identity	Low	Low	Low	Low	Low
Impurity	Control on purity and compatibility	Low	Low	Low	Low
Assay	Low	Low	Low	Low	
Content uniformity	Control on concentration and particle size	In-line NIR control (% CV NMT 5%)	Control on ribbon density In-line NIR control (ribbon density 0.68–0.81 g/cm³)	Low	In-line control on tablet weight (mean of 20 tablets, 194–206 mg)
Appearance	Low	Low	Low	No significant impact observed	Control on compression force
Hardness	Low	Low	Control on ribbon density	Low	Automatically discard tablets witnessing force beyond the limit (tablet hardness 5–12 kN)
Dissolution	Control on concentration and particle size	Low	In-line NIR control (ribbon density 0.68–0.81 g/cm³)	No significant impact observed	

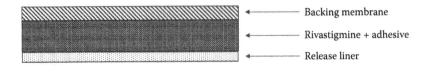

FIGURE 9.8 Schematic representation of the rivastigmine drug-in-adhesive transdermal patch.

formulations to improve tolerability and adherence is of key importance. Hence, rivastigmine has been formulated as a transdermal patch for its uninterrupted use in the therapy of AD.

Rivastigmine is a small (<500 Da) lipophilic molecule and passes rapidly through the skin and into the bloodstream and is therefore considered as a viable patch medication. Moreover, rivastigmine is a potent cholinesterase inhibitor, requiring only small doses of the drug for effective treatment (Kurz et al. 2009). This enables the patches to be small and discrete, thereby improving adhesion and reducing the risk of frequent adverse skin reactions.

9.3.2 Characterisation of the Marketed Product

Rivastigmine is marketed as a drug-in-adhesive transdermal patch, and this marketed product was used as a reference for the development of a generic product in this case study. The marketed product is available as a '9.5 mg/24 h' transdermal patch (i.e. the patch releases 9.5 mg of rivastigmine over a period of 24 h once applied to the skin), wherein each transdermal patch of 10 cm^2 contains 18 mg of rivastigmine.

To provide the necessary concentration gradient to drive the diffusion process through the skin, all rivastigmine transdermal patches are loaded with a greater amount of rivastigmine than will be absorbed into the bloodstream.

The marketed product consists of a lacquered polyethylene terephthalate backing film and fluoropolymer-coated polyester release liner. The medicinal product matrix is composed of alpha-tocopherol and acrylic copolymer along with the drug, and the adhesive matrix is composed of alpha-tocopherol, silicone oil and dimethicone.

This marketed product was evaluated for its physical and release properties. The results of this evaluation are listed in Table 9.3.

9.3.3 Quality Target Product Profile

On the basis of the evaluation results of the marketed product, the QTPP for the rivastigmine transdermal patch formulation is presented in Table 9.4 incorporating the summary of quality characteristics of the formulation from a patient point of view.

TABLE 9.3

Evaluation of the Marketed Rivastigmine Transdermal Patch

Parameters	Results ($n = 6$)
Thickness	0.11 ± 0.04 mm
Moisture content	$3.52\% \pm 0.21\%$
Peel adhesion test	10.5 N/25 mm
Shear strength	24 ± 1.5 min
Tack	87 ± 4 mm
Residual solvent	<0.2 ppm
Transdermal flux	7.5 ± 0.07 µg/cm²/h
In vitro drug release	4.6 mg in 24 h

9.3.4 CRITICAL QUALITY ATTRIBUTES

The CQAs of the formulation identified on the basis of the QTPP defined are shown in Table 9.5.

9.3.5 FORMULATION PROCESS OF THE RIVASTIGMINE TRANSDERMAL PATCH

The monolithic drug-in-adhesive patch was prepared by the solvent casting method. The excipients used were selected on the basis of compatibility studies carried out with the API. Since peel adhesion is an important parameter of transdermal patches, adhesion of placebo patches of different grades of the BIO-PSA polymer was evaluated as mentioned in Table 9.6, and on the basis of this study, BIO-PSA 4302 was selected as the pressure-sensitive adhesive (PSA).

2-Propanol was selected as the penetration enhancer (PE) from a range of compounds screened since it ensured high transdermal flux, was found to be compatible with rivastigmine and showed lower potential of toxicity. Polyethylene glycol (PEG) 400 was screened to be used as a plasticiser. A hydrophilic polymer (HPMC K100M) was also incorporated in the formulation to further facilitate the release of the active ingredient.

For fabrication of the patch, rivastigmine was weighed accurately and mixed with ethyl acetate followed by addition of the PSA, PE, plasticiser and the polymer and mixed well to obtain a homogeneous mixture of formulation. The mixture was kept for 10 to 15 min to remove entrapped air bubbles. A Mathis lab dryer was used to

TABLE 9.4
QTPP of the Rivastigmine Transdermal Patch

Quality Target Product Profile	Target	Justification
Dosage form	Drug in adhesive transdermal patch	Transdermal delivery helps achieve the systemic blood levels faster and reduces side effects associated with oral administration
Route of administration	Topical (skin)	Rapid delivery
Dosage strength	18 mg/10 cm^2 of transdermal patch	To ensure efficacy of the product
Patch Characteristics		
Physical attributes (appearance, size, shape)	Equivalent to the marketed product	To ensure patient acceptability and compliance
Thickness	Equivalent to the marketed product	To ensure convenience of use by the patient
Moisture content	Equivalent to the marketed product	To ensure efficacy of the formulation
Peel adhesion test	Equivalent to the marketed product	To ensure convenience of use by the patient
Shear strength	Equivalent to the marketed product	To ensure efficacy of the formulation
Tack	Equivalent to the marketed product	To ensure convenience of use by the patient
Assay	95%–105% of the label claim	To ensure safety and efficacy of the formulation
Content uniformity	As per pharmacopoeial requirement	To ensure consistency in the efficacy of the patches
Residual solvents	<0.2 ppm	To ensure safety of the formulation
Impurities	In compliance with ICH requirements	To ensure safety of the formulation
In vitro drug release	Equivalent to the marketed product	To ensure safety and efficacy of the formulation

(*Continued*)

TABLE 9.4 (CONTINUED)
QTPP of the Rivastigmine Transdermal Patch

Quality Target Product Profile	Target	Justification
Transdermal flux	Equivalent to the marketed product	To ensure safety and efficacy of the formulation
Container closure system	Internally lined aluminium pouch	Packaging in a lined aluminium pouch will protect the patch over its shelf life and prevent moisture ingress or any other instability
Stability	At least 24 months shelf-life at room temperature	To maintain therapeutic potential of the drug during storage period

cast the patches. The resulting liquid formulation was layered onto a fluoropolymer-treated polyester release liner with a thickness-adjusting knife. The film was dried at room temperature for 10 min and then placed in an oven for drying at 40°C–50°C for 40–50 min to remove the organic solvent. Once the film was dried, it was laminated with the polyester backing membrane. The backing membrane was fixed by moving 2 kg of roller on casted film. The patches were then cut into the desired size and shape. A schematic description of the formulation flow chart of the transdermal patch is illustrated in Figure 9.9.

The developed patches were evaluated for their appearance and adhesion properties, in vitro release of rivastigmine and flux.

9.3.6 RISK ESTIMATION MATRIX

The prioritisation for selecting the high-risk factors was carried out by constructing the risk estimation matrix (REM), showing the potential risks associated with each of the material attributes and process parameter(s) of the rivastigmine transdermal patch on its potential CQAs by assigning low, medium and high values to each of the studied factors. Figure 9.10 illustrates the REM for the rivastigmine transdermal patch for discerning high-risk parameters, if any.

- Assessment of low-risk parameters
  The composition and processing variables were found to be of low risk with respect to the adhesive properties of the patch.
- Assessment of medium-risk parameters
  The composition variables were found to be of medium risk with respect to the moisture content and level of impurities. This is because these attributes of the final product can be regulated by minimising the level of these

TABLE 9.5

CQAs for the Rivastigmine Transdermal Patch

Quality Attribute of Drug Product	Target	Is This a CQA?	Justification
Physical attributes (appearance, size, shape)	Patient acceptable and compliant	No	Physical attributes of the formulation were not considered as critical, as these are not directly allied to the efficacy and safety.
Thickness	0.10–0.15 mm	No	The thickness of the patch is crucial from the perspective of handling the dosage form by the patient, but it is not linked with the safety and efficacy of the formulation.
Moisture content	2%–5%	Yes	Moisture content and moisture uptake of the patch is important from the stability viewpoint of the formulation.
Peel adhesion test	10–25 N/ 25 mm	Yes	Peel adhesion is an important test to ensure prolonged adherence of the patch with the skin.
Shear strength	20–30 min	Yes	Shear strength is the measure of time required for the removal of the patch from its site of application.
Tack	70–90 mm	Yes	Tack is the ability of the patch to adhere to the surface under light pressure.
Assay	95%–105% of the label claim	Yes	Assay is a crucial parameter to ensure the efficacy of the formulation.
Content uniformity	As per pharmacopoeial specifications	Yes	Content uniformity is critical to ensure consistency of performance within the patches.
Residual solvents	Less than 0.2 ppm	Yes	Presence of residual solvents may hamper the safety and efficacy of the patch.
Impurities	In compliance with ICH specifications	Yes	Presence of impurities or degradants may deter the quality of the product.
In vitro drug release	More than 90% in 24 h	Yes	Performance of a delivery system is dependent on the amount of drug releasing from the system. Hence, this parameter is considered to be highly critical.
Transdermal flux	7–10 µg/cm²/h	Yes	Performance of a transdermal patch is primarily dependent on the amount of drug permeating through the skin and into the systemic circulation.

TABLE 9.6

Screening of PSA

Grades of PSA	Peel Adhesion Values (N/25 mm)
Bio-PSA AC 7-4102	3.6–4.5
Bio-PSA AC 7-4202	4–5.2
Bio-PSA AC 7-4302	6.5–8

impurities in the initial products. Similarly, the processing variables were of medium risk with respect to these attributes, since the temperature and time of processing moderately influence the moisture content and level of impurities. These variables were found to be of medium risk with respect to the residual solvents, since the drying operation should be satisfactory to lower the level of residual solvents. The medium-risk processing parameters were optimised by classical experimentation, the results of which are depicted in Table 9.7.

- Assessment of high-risk parameters

    Transdermal flux and peel adhesion were denoted as high-risk attributes, since they play a crucial role in the efficacy and performance of the patch and hence were to be optimised by DoE.

9.3.7 APPLICATION OF OPTIMISATION DESIGN AND DETERMINATION OF DESIGN SPACE

The three independent variables selected on the basis of the REM include concentration of permeation enhancer (X_1), concentration of PSA (X_2) and concentration of plasticiser (X_3). The critical response CQAs to be studied were peel adhesion (Y_1) and transdermal flux (Y_2). See Table 9.8 for the independent variables and response CQAs.

Central composite design was used for the optimisation of the three variables at two levels with five centre points. The impact of the independent variables on the critical responses studied is displayed in the form of three-dimensional (3D) plots. Figure 9.11a and b shows the impact of concentration of PSA and permeation enhancer on the peel adhesion and transdermal flux, respectively, of the patch at a constant plasticiser concentration of 15%.

Figure 9.11c displays the overlay plot depicting the design space (shaded in yellow) at a constant plasticiser concentration of 15%. The design space represents that area of the independent variables that yields a product possessing CQAs within the desired range.

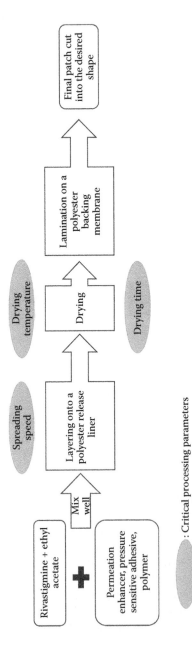

FIGURE 9.9 Procedure for manufacturing the rivastigmine transdermal patch.

Drug product CQAs	Critical composition variables			Critical processing parameters		
	Concentration of PSA	Concentration of PE	Plasticiser	Spreading speed	Drying temperature	Drying time
Moisture content	Medium	Medium	Medium	Medium	Medium	Medium
Peel adhesion test	High	Low	Low	Low	Low	Low
Shear strength	Low	Low	Low	Low	Low	Low
Tack	Low	Low	Low	Low	Low	Low
Assay	Medium	Medium	Medium	Medium	Low	Low
Content uniformity	Medium	Medium	Medium	Medium	Low	Low
Residual solvents	Low	Low	Low	Low	Medium	Medium
Impurities and related substances	Medium	Medium	Medium	Medium	Medium	Medium
In vitro drug release	Medium	Medium	Medium	Low	Low	Low
Transdermal flux	High	High	High	Low	Low	Low

FIGURE 9.10 REM for the manufacture of the rivastigmine transdermal patch.

TABLE 9.7

Classical Optimisation of the Processing Parameters

Factor	Range	Optimum Range
Spreading speed (m/min)	0.2–0.7	0.45–0.52
Drying temperature (°C)	40–50	44–47
Drying time (min)	40–50	43–47

TABLE 9.8

Independent Variables and Their Levels

Independent Variables	Material	Concentration Range Studied
X_1	Concentration of PE	1%–10%
X_2	Concentration of PSA	10%–50%
X_3	Concentration of plasticiser	10%–20%

9.3.8 UPDATED RISK ASSESSMENT AND CONTROL STRATEGY

The updated risk assessment is depicted in Table 9.9, which shows that all the high-risk parameters have been converted to low-risk parameters by maintaining the concentrations of the composition within the obtained design space (see Figure 9.11c) and keeping the processing parameters within the optimised ranges (see Table 9.7) and hence the QTPP can be achieved for the product.

On the basis of the optimisation studies carried out and the design space obtained, a control strategy is established indicating the levels of each of the independent variables, which will ensure a product within the range specified in the CQAs. Table 9.10 shows the overall control strategy to obtain the desired product.

9.3.9 CONTAINER AND CLOSURE SYSTEM

The final transdermal patch would be packed in a high-moisture and high-oxygen barrier, internally lacquered aluminium pouch.

Rivastigmine was successfully formulated as a transdermal patch with biopharmaceutical efficacy equivalent to that of the marketed formulation using QbD and risk management principles.

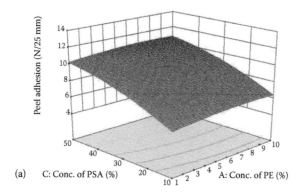

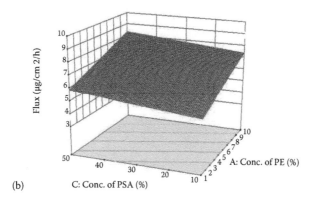

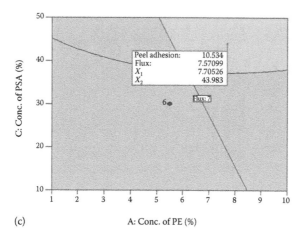

FIGURE 9.11 (See colour insert.) 3D plots for (a) peel adhesion, (b) transdermal flux and (c) overlay plot.

TABLE 9.9
Updated Risk Assessment for Formulation of the Rivastigmine Transdermal Patch

Drug Product CQAs	Critical Composition Variables			Critical Processing Parameters		
	Concentration of PSA	Concentration of PE	Concentration of Plasticiser	Spreading Speed	Drying Temperature	Drying Time
Moisture content	Low	Low	Low	Low	Low	Low
Peel adhesion test	Low (concentration in the range of 39.08–49.88)	Low	Low	Low	Low	Low
Shear strength	Low	Low	Low	Low	Low	Low
Tack	Low	Low	Low	Low	Low	Low
Assay	Low	Low	Low	Low	Low	Low
Content uniformity	Low	Low	Low	Low	Low	Low
Residual solvents	Low	Low	Low	Low	Low	Low
Impurities and related substances	Low	Low	Low	Low	Low	Low
In vitro drug release	Low	Low	Low	Low	Low	Low
Transdermal flux	Low (concentration in the range of 39.08%–49.88%)	Low (concentration in the range of 5.13%–9.98%)	Low (concentration in the range of 15.54%–19.94%)	Low	Low	Low

TABLE 9.10

Control Strategy for the Manufacture of the Rivastigmine Transdermal Patch

Factor	Units	Range	Design Space	Fixed Value	Purpose
Critical Composition Variables					
Concentration of permeation enhancer (X_1)	%	1–10	5.13–9.98	7.70	To ensure development of a stable and efficacious transdermal patch
Concentration of PSA (X_2)	%	10–50	15.54–19.94	18.00	
Concentration of plasticiser (X_3)	%	10–20	38.08–49.88	43.98	
Critical Processing Variables					
Spreading speed	m/min	0.2–0.7	0.45–0.52	–	
Drying temperature	°C	40–50	44–47	–	
Drying time	min	40–50	43–47	–	

9.4 RECTAL

The rectal route of administration is preferred for treatment of local as well as systemic conditions. This can be looked upon as an alternative route when oral or parenteral administration is not feasible considering the patient condition (age, unconsciousness, emesis, etc.) or drug properties (e.g. first-pass metabolism can be circumvented using rectal formulations, etc.). Various dosage forms that can be administered via this route include suppositories (~1 g for children and ~2 g for adults), gels, enemas and so on. Some considerations specific to this route include size and shape of dosage form to ensure patient compliance, formulation pH (between 5 and 7) to avoid local irritation, dissolution consideration in view of low rectal fluid content (~30 mL), softening of suppository at rectal temperature to ensure drug release (liquefaction point), use of special catheters for rectal administration and so on (Laxmi et al. 2012).

9.4.1 Case Study – Self-Microemulsifying Suppositories of β-Artemether

The present hypothetical case study aims at novel pharmaceutical product development of self-microemulsifying suppositories (SMES) of β-artemether (40 mg single dose) to be administered via rectal route for malaria therapy. Here, a scientific and risk analysis-based optimisation approach was followed to develop a product meeting the desired quality and specifications.

β-Artemether is one of the listed essential drugs by the World Health Organization (WHO) for management of severe multidrug-resistant malaria and cerebral malaria. The drug is currently available in the market as oral tablets and intramuscular oily injection. In the current clinical setting, the oral administration of β-artemether offers low bioavailability (~40%) because of low aqueous solubility (BCS class II), instability in gastric fluid and first-pass metabolism whereas the intramuscular route is associated with painful administration and often results in noncompliance (Joshi et al. 2008). Furthermore, vomiting, convulsions and unconsciousness are common symptoms in patients (especially children) with severe malaria, which further restrict the effective use of oral therapy. Thus, it presented a perfect rationale to develop rectally administrable β-artemether suppositories to overcome first-pass metabolism and the aforementioned disadvantages associated with oral/intramuscular route.

9.4.2 TARGET PRODUCT PROFILE IDENTIFICATION

Considering the novel dosage form development, the target product profile (TPP) on the basis of prior knowledge of β-artemether and general considerations for rectal dosage form was proposed and is depicted in Table 9.11.

9.4.3 PROPOSED FORMULATION AND PROCESS

While designing a suppository system for rectal administration, some points need critical consideration and are discussed below. Rectal tissue is lipoidal in nature and shows presence of lower, middle and upper hemorrhoidal veins. Amongst these, the lower and middle vein carry blood to the inferior vena cava, and thus, if the rectal dosage form is retained in the anterior region of the rectum, first-pass metabolism can be prevented (Laxmi et al. 2012).

β-Artemether belongs to BCS class II and thus presents dissolution-controlled drug absorption. To solubilise the drug and enhance its bioavailability, SMES formulation was proposed. SMES can be defined as a suppository consisting of lipid(s) along with surfactant(s), cosurfactant(s) and stabilisers, which, upon contact with aqueous media, forms a spontaneous submicron emulsion. The proposed formulation strategy was anticipated to offer the following advantages:

- Conventionally preferred PEG/glycerogelatin-based suppositories often lead to rectal irritation and soften in the posterior rectum, thus restricting the drug absorption owing to first-pass metabolism. This can be overruled by the SMES as they spontaneously form oil-in-water-type microemulsion in the presence of aqueous media in the anterior rectum. This not only ensures early softening but also will enhance drug absorption owing to higher interfacial surface area and can bypass first-pass metabolism.
- The hydrophobic drug β-artemether can be retained in the lipid matrix of a submicron oil-in-water microemulsion that enhances the rate and extent of absorption.

TABLE 9.11

TPP for β-Artemether Suppositories

TPP Element	Target	Justification
Dosage form	Self-microemulsifying suppository	To solubilise the drug and to achieve enhanced drug absorption via rectal route
Route of administration	Rectal	Route of choice when oral and intramuscular delivery is not feasible for severe/cerebral malaria treatment
Physical attributes and appearance	~2 g suppository weight, torpedo-shaped unit entity, nongreasy, nonappearance of fissuring, pitting, fat blooming and sedimentation	Patient compliance for ease of administration and marketing need
Dosage strength	40 mg/suppository	Dose selected for preliminary studies is the same as the oral dose Dose or dosing frequency can be reduced if enhanced bioavailability is observed
Identity	Meets reference standard of β-artemether	Clinical safety and efficacy
Assay of β-artemether	95%–105% of label claim	Pharmacopoeial specification (Indian Pharmacopoeia), clinical safety and efficacy
Uniformity of dosage	Should meet the requirements as per Indian Pharmacopoeia	Targeted for consistent clinical effectiveness
Related substances	Should meet the requirements as per Indian Pharmacopoeia	To ensure patient safety, and needed for clinical effectiveness Individual known impurity limits will be finalised on the basis of stability data of drug product
Liquefaction time	NMT 30 min	Pharmacopoeial limit (Indian Pharmacopoeia)
Mechanical strength/crushing test	1.8–2 kg	Pharmacopoeial limit to ensure intactness of unit dosage form
Drug dissolution	NLT 80% in 60 min in pH 6.8 phosphate buffer, paddle, 50 rpm	To ensure batch-to-batch uniformity and a tool to ensure better pharmacokinetic profile

(*Continued*)

TABLE 9.11 (CONTINUED)
TPP for β-Artemether Suppositories

TPP Element	Target	Justification
Pharmacokinetics	C_{max} and AUC should show bioequivalence with the reference product as per guidelines	To ensure therapeutic efficacy and safety equivalent to reference product
Microbiological evaluation	Meets pharmacopoeial standard (Indian Pharmacopoeia)	To ensure patient safety
Storage condition	Store at a temperature of 15°C–25°C	To ensure stability and maintenance of unit dosage form
Container and closure system	Alu-Alu blister pack	To ensure protection against moisture, stability and maintenance of unit dosage form
Stability	At least 24 months shelf-life at specified storage condition	Stability condition as per ICH Q1 (A) and to maintain therapeutic efficacy

Thus, it was decided to develop SMES of β-artemether for malaria treatment. The drug source was selected on the basis of its compliance with Indian pharmacopoeial specifications. The excipient selection for the proposed SMES formulation was done based on prior knowledge and literature information coupled with preliminary screening experiments.

9.4.3.1 Lipid Selection

Use of hard fat (EP/NF/JPE) is recommended for the rectal and vaginal route. Based on this, various synthetic grades of Suppocire (Gattefosse Pvt. Ltd.) with functional and nonfunctional alterations were considered. To select a suitable lipid, two sets of studies were conducted, namely, the saturation solubility study of β-artemether in various grades of Suppocire and dispersibility of Suppocire grades in water (desired character as the aim was to develop a self-microemulsifying system) (Table 9.12). From the studies, it was observed that the drug shows maximum solubility in Suppocire D but Suppocire AP was considered to be optimum for the formulation as it exhibited sufficient solubility and was the only grade that was dispersible in water. Chemically, Suppocire AP is a saturated polyglycolised glyceride base with PEG and glycerol (Gattefosse 2015). This makes it amphiphilic in nature, which may further enhance solubilisation and dissolution of BCS class II molecule β-artemether when formulated as SMES.

9.4.3.2 Surfactant and Stabiliser Selection

For surfactant selection, the rectally permitted surfactants were screened based on two parameters: the solubility of drug in surfactant and miscibility of surfactant with the lipid base (Table 9.13). From the results, Gelucire 44/14 was selected as it

TABLE 9.12

Screening of Various Grades of Suppocire

Suppocire Grade	β-Artemether Solubility (mg/g)	Miscibility with Water
Suppocire A	57 ± 3.6	Insoluble
Suppocire C	64.7 ± 3.6	Insoluble
Suppocire AP	68 ± 4.3	Dispersible
Suppocire D	76 ± 6.8	Insoluble
Suppocire BM	49.5 ± 2.8	Insoluble
Suppocire SNB	61 ± 4.8	Insoluble
Suppocire NCX	53 ± 6.2	Insoluble

TABLE 9.13

Screening of Various Rectally Approved Surfactants

Surfactant	β-Artemether Solubility (mg/g)	Miscibility with Suppocire AP
Tween 20	50 ± 2.4	Phase separation observed on storage
Tween 80	12 ± 9.4	Phase separation observed on storage
Labrasol	21 ± 2.1	Miscible
Gelucire 44/14	44 ± 1.3	Miscible
Transcutol P	30 ± 3.4	Miscible
PEG 400	2 ± 4.6	Miscible

exhibited acceptable solubility for the drug and very good miscibility with Suppocire AP. Additionally, Gelucire 44/14 is a well-documented microemulsifier as well as PE. To further stabilise the microemulsion, Transcutol P was selected as a cosurfactant as it exhibited better microemulsification and miscibility with lipid base when used in combination with Gelucire 44/14.

On the basis of preliminary investigations, the tentative proposed formula for SMES is as mentioned in Table 9.14.

The product is expected to be hygroscopic because of its contents. Therefore, a moisture protective pack is required for the product to ensure stability during its shelf life. Thus, the Alu-Alu blister pack was proposed as a primary pack.

TABLE 9.14

Tentative Proposed Formula for β-Artemether SMES

Ingredient	Function	Quantity
β-Artemether[a]	Active	40 mg/suppository
	Suppository Base	
Suppocire AP	Lipid base	5%–15% w/w
Gelucire 44/14	Surfactant (microemulsifying agent)	78%–90% w/w
Transcutol P	Stabiliser	2%–10% w/w

[a] The final formula will be calculated considering the displacement value of the drug for the given base as 0.5.

9.4.3.2.1   Selection of Formulation Process and Equipments

After deciding the formulation, the next step was to identify the formulation process along with the equipments to be used and the in-process quality control tests to be performed to manufacture the final product with desired quality attributes. The same is listed in Figure 9.12.

The continuous manufacturing process enabled with the process analytical technique (PAT) (to ensure in-process quality control) was proposed for the said formulation. Briefly, the process involves melt mixing of drug and excipients to achieve uniform microemulsion preconcentrate. The preconcentrate is then moulded to a suppository by controlled cooling. From the understanding of the process, it was identified that assurance of a uniform preconcentrate during melt mixing is the most critical step and thus an in-line NIR tool was introduced in the mixing line to continuously check and ensure blend uniformity. Further, the rate of cooling during suppository moulding is the other critical parameter as it will affect the self-microemulsification time, globule size after microemulsification and cumulative drug release. Thus, this process parameter was studied in greater detail to understand its impact on product quality attributes.

9.4.4   Determination of CQAs

After proposing the TPP, formulation and manufacturing process, CQAs were identified. For this, various quality attributes based on TPP were screened for criticality and are described in Figure 9.13.

Based on evaluation (Figure 9.13), assay, content uniformity, self-microemulsification time, globule size and drug release were identified to be potential CQAs and were subjected to further risk analysis.

9.4.5   Risk Analysis

From the understanding of SMES composition and the manufacturing process, it was concluded that prior meticulous selection of the API manufacturer and other

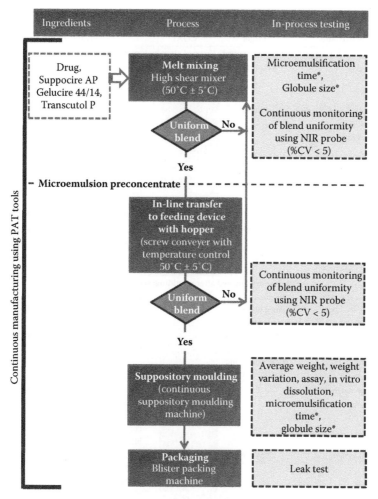

| Ingredients | Process | In-process testing |

Drug, Suppocire AP Gelucire 44/14, Transcutol P

Melt mixing High shear mixer (50°C ± 5°C)

Microemulsification time*, Globule size*

Uniform blend — No

Continuous monitoring of blend uniformity using NIR probe (%CV < 5)

Yes

– Microemulsion preconcentrate – – – – – – – – – – – –

In-line transfer to feeding device with hopper (screw conveyer with temperature control 50°C ± 5°C)

Uniform blend — No

Continuous monitoring of blend uniformity using NIR probe (%CV < 5)

Yes

Suppository moulding (continuous suppository moulding machine)

Average weight, weight variation, assay, in vitro dissolution, microemulsification time*, globule size*

Packaging Blister packing machine

Leak test

Continuous manufacturing using PAT tools

* In-process parameters analysed only during development phase

FIGURE 9.12   β-Artemether SMES process flow diagram.

excipients reduces the criticality of these material attributes. However, the composition of formulation (concentration of excipients) plays a very critical role. Understanding of the process revealed that the melt mixing unit operation and the rate of cooling of suppository molten mass during moulding are the most critical steps as they will affect the final unit dosage form characteristics. The REM (Figure 9.14) was then constructed to understand the impact of critical formulation composition and process attributes on the predetermined CQAs. The risk criticality was represented as high, medium and low. The attributes having high-risk impact on CQAs were further subjected to empirical or classical optimisation to understand their effect on CQAs and to arrive at an appropriate design space and overall control strategy to mitigate the risk and to achieve a quality product well within the specifications.

Quality attribute	Target	Is this a CQA?	Justification
Microemulsion preconcentrate (formed after melt mixing; see Figure 9.12)			
Blend assay	95%–105%	YES (High)	This will affect the safety and efficacy. Process variables can affect the assay and thus will be evaluated at each step.
Self-microemulsification time at rectal temperature	45–75 s	YES (High)	This will ensure maintenance of unit dosage form during rectal insertion followed by faster self-microemulsification. It varies significantly with change in composition; hence, it is critical.
Droplet size upon emulsification with aqueous media	<150 nm	YES (High)	This will enhance absorption via rectal route and varies significantly with change in composition, hence, it is critical.
Polydispersity index	≤0.2	YES (Medium)	This will ensure uniform droplet size distribution after microemulsification and can be controlled, thus medium critical.
Suppository (formed after moulding; see Figure 9.12)			
Physical attributes	Weight, torpedo-shaped unit entity, nongreasy, nonappearance of fissuring, pitting fat blooming and sedimentation	NO (Low)	These were not considered to be critical as they do not affect safety and efficacy and can be achieved in an acceptable range in the development under consideration.
Average weight	2 g	YES (Medium)	Patient compliance and to incorporate the drug dose and dose uniformity but herein can be controlled by controlling the uniformity of molten blend.
Weight variation	NMT 5%	YES (Medium)	Affects the uniformity of dosage and controlled by controlling the uniformity of molten blend and thus it is critical but graded as medium.
Liquefaction time	NMT 10 min	YES (Medium)	In-house specification (less than the pharmacopoeial limit of 30 min) as it affects self-microemulsification and the drug dissolution. Medium critical as the softening temperature of the SMES blend is ~37°C–39°C ensuring immediate softening.
Droplet size after liquefaction and microemulsification	<150 nm	YES (High)	This will enhance absorption via rectal route and varies significantly with change in composition; hence, it is critical.

FIGURE 9.13   CQA identification for β-artemether SMES.   (*Continued*)

Quality attribute	Target	Is this a CQA?	Justification
Polydispersity index after liquefaction and microemulsification	≤0.2	YES (Medium)	This will ensure uniform droplet size distribution and can be moderately controlled, thus considered to be medium-critical.
Identity	Meets reference standard of β-artemether	NO (Low)	This will affect the clinical safety and efficacy but can be controlled with appropriate API source selection.
Assay of β-artemether	95%–105% of label claim	YES (High)	This will affect the safety and efficacy. Process variables can affect the assay and thus will be evaluated at each step, which makes it critical.
Uniformity of dosage	Should meet the requirements as per Indian pharmacopoeia	YES (High)	This is necessary to ensure consistent clinical effectiveness and any alteration in the process may affect the dosage uniformity; thus, it is critical.
Related substances	Should meet the requirements as per Indian pharmacopoeia	YES (Medium)	This will affect the safety and efficacy. Process variables can cause drug degradation but the drug is highly stable to processing parameters under consideration; thus, the criticality is medium.
Drug release	NLT 80% in 60 min in pH 7.4 phosphate buffer, paddle, 50 rpm	YES (High)	Any alteration in this will result in batch-to-batch variation and will affect the pharmocokinetic profile; thus, it is highly critical.
Microbiological evaluation	Should meet the requirements as per Indian pharmacopoeia	NO (Low)	This will affect the safety but can be controlled in the current scenario by controlling the manufacturing area and by following good manufacturing practice; thus, it is not very critical.

FIGURE 9.13 (CONTINUED) CQA identification for β-artemether SMES.

9.4.6 Design Space Determination

9.4.6.1 Optimisation of Formulation Composition: SMES Base

The suppository base is composed of Suppocire AP, Gelucire 44/14 and Transcutol P. Optimisation of this blend concentration is the most critical aspect of this formulation to ensure faster self-microemulsification, optimum droplet size and blend uniformity. Considering a very low drug loading per suppository (~2% w/w), the formulation composition was optimised for placebo suppository base (as a very small part of the suppository base will be replaced by drug in the final formulation based on the displacement value of the drug).

As the composition of blend should account for unity, the D-optimal design was implemented for DoE (see Chapter 5 for a detailed description). The impact of independent composition variables was assessed over the critical response parameters of self-microemulsification time and droplet size (Figure 9.15).

Drug product CQAs	Formulation composition (quantitative composition)	CPPs	
		Melt mixing	Suppository moulding (rate of cooling is critical)
Assay	Low	Medium	Medium
Content uniformity	High	High	Medium
Self-microemulsification time	High	High	High
Globule size	High	Medium	High
Cumulative drug release	Medium	Medium	High

FIGURE 9.14 Initial risk assessment for β-artemether SMES.

Factor	Range investigated (% w/w) such that A + B + C = 100%	Design space
Suppocire AP (A)	5%–15% w/w	8%–12% w/w
Gelucire 44/14 (B)	78%–90% w/w	78%–85% w/w
Transcutol P (C)	2%–10% w/w	2%–10% w/w
Response	Goal	Acceptance
Self microemulsification time	45–75 s	Less than 60 s
Droplet size	Less than 150 nm	Less than 150 nm

FIGURE 9.15 Impact of formulation composition.

The surface response plot generated as a result of this D-optimal design (Figure 9.16) experiment revealed that an increase in lipid Suppocire concentration leads to higher self-microemulsification time and microemulsion droplet size and thus needs critical consideration. Further, as the concentration of the surfactant Gelucire 44/14 increases, both self-microemulsification time and microemulsion droplet size decreases. Variation in the stabiliser Transcutol P concentration did not show any significant impact on response parameters. On the basis of the target of achieving a droplet size of less than 150 nm and a self-microemulsification time between 45 and 75 s (time selected based on optimum balance between insertion time and rapid microemulsification), the overlay plot was generated and design space was defined (Figure 9.15).

On the basis of design space, a final composition formula with β-artemether (40 mg/suppository) is depicted in Table 9.15.

The formulation composition space was thus successfully optimised to comply with the desired product profile. This was then followed by optimisation of each of the unit operations (flow chart as depicted in Figure 9.12).

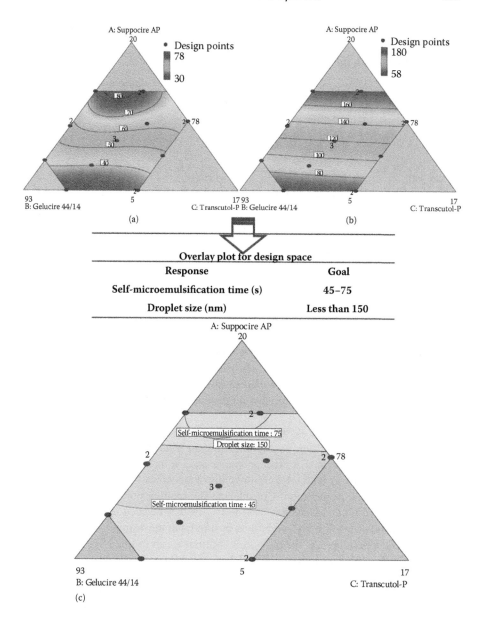

FIGURE 9.16 (**See colour insert.**) Impact of formulation composition on (a) self-microemulsification time (s), (b) microemulsion droplet size (nm) and (c) overlay plot.

TABLE 9.15

Optimised Formula for β-Artemether SMES

Ingredient		Blank Suppository(mg)	Quantity per Medicated Suppository(mg)
β-Artemether[a]	Active		40[a]
Suppocire AP	Lipid base	200	200
Gelucire 44/14	Surfactant (microemulsifying agent)	1680	1600
Transcutol P	Stabiliser	120	120
Total weight		2000	1960

[a] Displacement value of the drug for the given base is 0.5; thus, 40 mg of the drug will replace 80 mg of the base and the same is calculated for final dose consideration.

9.4.6.2 Optimisation of Critical Process Parameters

The continuous manufacturing process was selected for the suppository manufacture and in-line NIR PAT tools were installed to investigate and control the in-process quality parameters that make the process more simple and confident to control.

- Melt mixing (see step 1 of Figure 9.12) was observed to be the most crucial unit operation controlling content uniformity, self-microemulsification time and globule size of SMES preconcentrate and, in turn, final dosage form. A high-shear mixer with temperature control jacket was selected as equipment for this unit operation. Considering the melting point of Suppocire AP (35°C–36°C) and Gelucire 44/14 (44°C), it was decided to perform the melt mixing at 50°C ± 5°C, and it was ensured that the drug is stable at this process temperature. As the process is continuous, an in-line NIR spectrophotometric tool was introduced to determine and ensure the uniform melt mixing via continuous analysis (end point: % CV not more than [NMT] 5%). From in-line continuous analysis, the optimum mixing time for this unit operation was revealed to be 15 ± 5 min. The so-formed uniform melt was considered to be optimum as it ensured the melting of excipients and uniform mixing and exhibited acceptable flow properties for introduction in suppository moulds of the rotary suppository moulding machine. The molten mass was then transferred in-line to the feeding device with hopper (an in-line NIR tool was introduced at this stage also to reensure the blend uniformity before its introduction in the rotary suppository moulding machine).

- Suppository moulding of microemulsion preconcentrate (see step 2 of Figure 9.12) to form a torpedo-shaped suppository dosage form is achieved via controlled cooling of the melt mix (from 50°C ± 5°C to 20°C ± 5°C). Herein, the rate of cooling is very critical, and based on prior process knowledge, the controlled cooling rate of 1–5°C/min was studied at three levels (1, 3 and 5°C/min), and its self-microemulsification time, globule size after microemulsification and % cumulative drug release at the end of 30 min were studied as the response variables. The results are indicated in Table 9.16. From the results, it was understood that the moulding process has a measurable impact only on self-microemulsification time, which was found to be increasing with an increase in rate of cooling. This can be attributed to sudden cooling of the molten matrix, which may lead to crystalline matrix formation and delay in the liquefaction and in turn the microemulsification. Thus, the optimum rate of cooling was decided to be between 1 and 3°C/min.

9.4.6.3 Design Space

Based on the systematic study as described in Sections 9.4.6.1 and 9.4.6.2, design space and product specifications (both in-process and final) were confirmed and are summarised in Table 9.17.

9.4.7 UPDATED RISK ASSESSMENT AND CONTROL STRATEGY FOR β-ARTEMETHER SMES

After the optimisation of critical formulation attributes (CFAs) and critical process parameters (CPPs), an updated risk analysis was performed. The results indicated that all the attributes under consideration now present a low risk and a quality

---

TABLE 9.16

Impact of Suppository Mould Cooling Rate on CQAs

Rate of Cooling (°C/min)	Self-Microemulsification Time (X s) Specification: $45 \leq X \leq 75$	Globule Size after Microemulsification (nm) Specification: >150 nm	Cumulative Drug Release at the End of 30 min Specification: ≥80%
1	61 ± 4	122.4 ± 5.3	96.3 ± 3.4
3	67 ± 3	128.4 ± 6.1	93.6 ± 2.6
5	74 ± 4[a]	126.3 ± 4.3	91.38 ± 1.7

[a] Self-microemulsification time was exceeded beyond specifications at 5°C/min cooling rate, and thus the optimum rate of cooling was decided to be 1–3°C/min.

TABLE 9.17

Design Space for β-Artemether SMES

Parameter	Design Space	Control Criteria
β-Artemether	40 mg/suppository	Critical formulation attribute
Suppocire AP (A)	8%–12% w/w	
Gelucire 44/14 (B)	78%–85% w/w	
Transcutol P (C)	2%–10% w/w	
Unit Operation: Melt Mixing		
High-shear mixer with temperature control jacket	% CV NMT 5%	Critical process control
Unit Operation: Suppository Moulding and Unit Dosage Form Generation		
Rate of cooling	1–3°C/min	Critical process control

TABLE 9.18

Updated Risk Assessment with Control Strategy for β-Artemether SMES

Drug Product CQAs	Formulation Composition (Quantitative Composition)	CPPs	
		Melt Mixing	Suppository Moulding (Rate of Cooling Is Critical)
Assay	Low	Low	Low
Content uniformity	Low Control on excipient concentration	Low In-line NIR control (% CV NMT 5%)	Low In-line control on suppository weight
Self-microemulsification time	Low Control on excipient concentration	Low In-line NIR control (% CV NMT 5%)	Low In-line NIR control (% CV NMT 5% and controlled rate of cooling)
Globule size	Low Control on excipient concentration	Low No significant impact observed	Low No significant impact observed
Cumulative drug release	Low Control on excipient concentration	Low	Low In-line NIR control (% CV NMT 5% and controlled rate of cooling)

product within the desired specifications can be achieved within the framework of the proposed control strategy. The amalgamated table of updated risk assessment along with the control strategy is described in Table 9.18.

The finalised product was thus successfully developed and optimised using in-line control parameters along with a formulation and manufacturing control strategy.

9.5 VAGINAL

The presence of a dense network of blood vessels has made the vagina an excellent route of drug delivery for both systemic and local effects. The principal advantages of vaginal drug delivery over conventional drug delivery routes with respect to systemic delivery are its ability to bypass first-pass metabolism, ease of administration and high permeability for low-molecular-weight drugs. Vaginal absorption of the drugs is dependent on the physiological factors of the vagina including but not limited to cyclic changes in thickness of vaginal epithelium, fluid volume and composition, pH and sexual arousal. Physicochemical properties of the drug candidates such as molecular weight, lipophilicity, ionisation, surface charge and chemical nature also have an impact on the vaginal absorption of the drug candidates. It is generally accepted that low-molecular-weight hydrophilic and lipophilic drugs are likely to be absorbed more as compared to their higher-molecular-weight counterparts (Hussain and Ahsan 2005).

Formulation systems for vaginal delivery traditionally included solutions, pessaries, gels, foams and tablets. Vaginal rings have been recently introduced for the delivery of contraceptives and hormone therapy. Novel strategies for development of vaginal delivery systems include the formulation of bioadhesive formulations, in situ gels and vaginal immunisation therapy (Bash 2000; Roumen 2008).

The case study described in Section 9.5.1 deals with the development of a vaginal gel. The critical formulation parameters to be considered with respect to a vaginal gel include the pH and the viscosity of the final gel formulation. The pH of the formulation should be compatible with the pH of the vaginal fluids to prevent any irritation. Moreover, the vagina is home to various microorganisms such as lactobacilli, which produce bacteriocins and lactic acid that lower the pH of the vagina. This low pH of the vagina is crucial to ward off any unnecessary infections and hence the pH of the formulation should be such that it does not alter the physiological pH of the vagina (Sharma et al. 2012). The consistency of gel is responsible for the residence time and adhesion of the gel and thereby has an impact on its duration of action. Hence, the viscosity should be sufficient to retain the gel at the site of administration for a satisfactory period.

9.5.1 CASE STUDY – TENOFOVIR-LOADED MICROEMULSION-BASED VAGINAL GEL

The case study illustrated here is the development of an antiretroviral tenofovir (TNF)-loaded microemulsion-based gel. TNF belongs to a class of antiretroviral drugs known as nucleotide analogue reverse transcriptase inhibitors, which block reverse transcriptase, an enzyme crucial to viral production in an HIV-infected population. The approach employed in the study was to incorporate TNF in a micro-emulsion. This microemulsion formulation was further viscosised as a hydrophilic gel for final application.

TNF is a water-insoluble drug with a log P value of 1.25 (Drug Bank: Tenofovir [DB00300] 2015). The hydrophobic nature of TNF poses issues for the development of a suitable topical dosage form for vaginal delivery. Hence, for solubilisation of TNF, development of a 'phospholipid-based microemulsion system' appeared to be a viable approach. Microemulsions have gained great attention for delivery of hydrophobic agents for systemic and local therapy. The solubilisation of TNF in microemulsions would improve its vaginal availability. However, it is also essential to have a dosage form that adheres to the vaginal mucosa and increases the residence time of TNF in vagina. This functionality can be imparted by gelling of the TNF microemulsion using a bioadhesive agent. Thus, in the present investigation, the potential of a microemulsion-based bioadhesive gel of TNF was explored for vaginal delivery. The strategy for the development of this formulation is classified into two sections:

- Formulation of clear and stable microemulsion of TNF
- Final formulation of this microemulsion into a bioadhesive gel system by incorporating a suitable gelling polymer

Formulation optimisation of each of the aforementioned two steps has been discussed separately.

9.5.2 QUALITY TARGET PRODUCT PROFILE

As per the QbD paradigm, product development is initiated with the establishment of the QTPP. The QTPP for this product was defined encompassing the summary of quality characteristics for the development of a microemulsion-based gel of TNF for accomplishing maximal therapeutic efficacy. The QTPP for this formulation is mentioned in Table 9.19.

9.5.3 CRITICAL QUALITY ATTRIBUTES

On the basis of the various elements of QTPP, the CQAs of the product components and process affecting the performance of TNF microemulsion-based gel were identified. The CQAs are listed in Table 9.20.

9.5.4 SELECTION OF THE EXCIPIENTS AND FORMULATION PROCESS

For the formulation of a microemulsion, the selection of the oil that acts as a solvent for the drug and the surfactant and cosurfactant are critical to the development. The screening of oil, surfactants and cosurfactants required for the formulation of the microemulsion was performed by determining the solubility of TNF in these ingredients. Solubility studies were carried out by placing an excess amount of TNF in a

TABLE 9.19

QTPP for the TNF-Loaded Microemulsion-Based Vaginal Gel

QTPP Element	Target	Justification
Dosage form	Microemulsion-based gel	Selection of phospholipid-based microemulsion helps in the enhanced topical delivery and aids in solubilisation of the hydrophobic API
Physical Attributes		
Colour	No addition of artificial colours	Patient acceptability and patient compliance
Odour	No objectionable odour	
Appearance	Clear transparent gel	
Route of administration	Topical (vaginal)	Direct targeting to the site of action
Dosage strength	4% w/w	To ensure efficacy of the formulation
Globule size of microemulsion	Z-average: 50–100 nm	Critical for stability and efficacy of the formulation
Polydispersity index	0.1–0.3	Critical for stability and efficacy of the formulation
Viscosity	40,000–60,000 cps	Optimum viscosity of the product is essential to ensure vaginal retention of the formulation for an optimal duration
Assay	95%–105% of the label claim	To ensure activity and efficacy of the formulation
Content uniformity	As per pharmacopoeial specifications	To ensure activity and efficacy of the formulation
pH	3.0–4.5	The pH of the formulation should be compatible with the pH of the vaginal environment to prevent irritation on application
In vitro release	Sustained release of more than 75% over 12 h	To ensure efficacy of the product
In vitro bioadhesion	>20 min	To ensure performance and efficacy of the formulation

(Continued)

TABLE 9.19 (CONTINUED)
QTPP for the TNF-Loaded Microemulsion-Based Vaginal Gel

QTPP Element	Target	Justification
Impurities	In compliance with ICH requirement	To ensure safety and efficacy of the formulation
Microbial content	<100 cfu	To ensure safety of the formulation
Preservative efficacy testing	Meets the pharmacopoeial requirement	To ensure safety and stability of the formulation
Stability	At least 24 months shelf-life at room temperature	To maintain therapeutic potential of the drug during storage period
Container closure system	Internally lacquered aluminium tubes with an enclosed applicator	The TNF-loaded microemulsion gel can be easily incorporated in aluminium tubes for improved patient compliance, portability and manufacturing ease. The applicator will aid in ease of application of the formulation

vial containing 1 mL of the respective component. After shaking for 72 h, the excess undissolved drug was separated by centrifugation and estimated photometrically. The components showing maximum solubility for TNF were selected. On the basis of these solubility studies, isopropyl myristate (IPM) was selected as the oil phase. A combination of nonionic surfactant (Tween 80) and saturated lipids (Labrasol) in the ratio of 1:1 was employed as the surfactant mixture, while ethanol was selected as the cosurfactant. Pseudoternary phase diagrams of oil, surfactant, cosurfactant and water were constructed using the water titration method to obtain their corresponding concentration ranges that result in large areas of microemulsion formation. The surfactant was blended with the cosurfactant in fixed weight ratios (1:1, 2:1 and 3:1). This blend is referred to as S_{mix}. Each of these blends was mixed with the selected oil at room temperature followed by drop-wise addition of water under continuous stirring. The resulting system was visually observed for transparency. The samples were marked as points in the phase diagram. The cumulative area covered by these points was considered as the 'microemulsion region of existence'. For each phase diagram, the ratios of oil to S_{mix} were varied from 1:9 to 9:1. On the basis of the microemulsion region in the phase diagram, the microemulsion formulations were selected at the desired component ratios (see Figure 9.17 for the phase diagrams).

As observed in Figure 9.17, the phase diagram with an S_{mix} ratio of 3:1 was found to provide the maximum microemulsion region and hence this ratio was used for further formulation. A separate preservative was not added to the formulation, since the high surfactant concentration was sufficient to ensure freedom from microbial contamination. Figure 9.18 illustrates a detailed schematic of the process flow of formulation development of TNF-loaded microemulsion.

TABLE 9.20

CQAs for the TNF-Loaded Microemulsion-Based Vaginal Gel

Quality Attributes of the Drug Product	Target	Is This a CQA?	Justification
Colour	Clear	No	Physical attributes of the formulation were not considered as critical, as these are not directly linked to the efficacy and safety
Odour	No unpleasant odour	No	
Appearance	Clear and transparent gel	No	
Globule size	Z-average: 50–100 nm	Yes	Smaller globule size of microemulsion will facilitate in better permeation and retention of the drugs within the vaginal layers
Polydispersity index	0.1–0.3	Yes	A narrow range of PDI helps ensure stability of the formulation
Viscosity	40,000–60,000 cps	Yes	The viscosity of a vaginal gel determines its residence time at the site of application and consequently its duration of action
Assay	95%–105%	Yes	Assay of the formulation is critical to ensure availability of the drug in the right dose for absorption
Content uniformity	As per pharmacopoeial requirement	Yes	Content uniformity is a critical parameter to assure consistency in the performance of the delivery system
pH	3.0–4.5	Yes	pH of the formulation is essential for the maintenance of the vaginal pH, which is crucial for the preservation of the vaginal microflora
Impurities	In compliance with ICH specifications	Yes	The level of impurities should be maintained below the set limits to ensure safety and efficacy of the formulation
Microbial content	<100 cfu	Yes	Microbial content of the formulation must be maintained below the specified limit to ensure safety of the formulation

(*Continued*)

TABLE 9.20 (CONTINUED)
CQAs for the TNF-Loaded Microemulsion-Based Vaginal Gel

Quality Attributes of the Drug Product	Target	Is This a CQA?	Justification
In vitro drug release	0–4 h: Not less than 35% 4–8 h: 35%–65% 8–12 h: Not less than 75%	Yes	In vitro drug release is an essential parameter to assess the topical delivery potential of the formulation for enhanced therapeutic efficacy
In vitro bioadhesion	>20 min	Yes	In vitro bioadhesion is an essential tool to ensure adherence of the formulation to the site of application

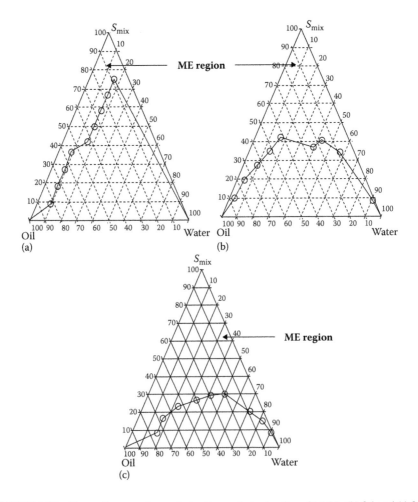

FIGURE 9.17 Phase diagrams for surfactant/cosurfactant ratios of (a) 1:1, (b) 2:1 and (c) 3:1.

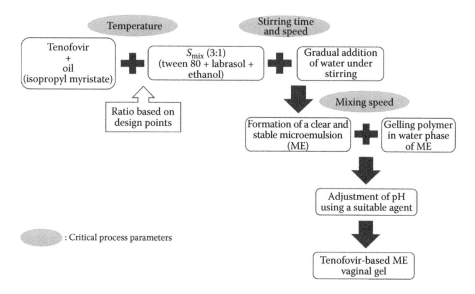

FIGURE 9.18 Schematic representing the flow chart of formulation of the TNF-loaded microemulsion-based vaginal gel.

9.5.5 REM for the Formulation of TNF-Loaded Microemulsion

The prioritisation for selecting the high-risk factors was carried out by constructing the REM, showing the potential risks associated with each of the material attributes and process parameter(s) of TNF-loaded microemulsions on its potential CQAs by assigning low, medium and high values to each of the studied factors. Figure 9.19 illustrates the REM for TNF-loaded microemulsions for discerning high-risk parameters, if any.

- Assessment of low-risk parameters

    The processing parameters were found to be of low risk with respect to the in vitro drug release since it was not influenced by the processing parameters. The composition and processing variables were found to be below risk with respect to pH, since the microemulsion preparation step did not have any impact on the pH of the formulation.

- Assessment of medium-risk parameters

    The composition and processing variables were found to be of medium risk with respect to assay, content uniformity and in vitro bioadhesion since they could be converted to low risk by empirical optimisation of the said parameters. Also, these variables were reckoned as medium risk with respect to impurities and related substances (RS) as the impurities in the final product could be minimised by prescreening the API and the excipients to ensure that their impurity level is below the specified limit.

Drug product CQAs	Critical composition variables			Critical processing parameters		
	Concentration of IPM	Concentration of S_{mix}	Water	Microemulsion stirring time	Microemulsion stirring speed	Temperature
Globule size	High	High	High	Medium	Low	Low
PDI	Medium	Medium	Medium	Medium	Medium	Medium
Viscosity	Medium	Medium	Medium	Low	Low	Low
Assay	Medium	Medium	Medium	Medium	Medium	Medium
Content uniformity	Medium	Medium	Medium	Medium	Medium	Medium
pH	Low	Low	Low	Low	Low	Low
Impurities	Medium	Medium	Medium	Medium	Low	Medium
Microbial content	Medium	Medium	Medium	Low	Low	Low
In vitro drug release	High	High	High	Low	Low	Low
In vitro bioadhesion	Medium	Medium	Medium	Low	Low	Low

FIGURE 9.19   REM for formulation of TNF-loaded microemulsion.

TABLE 9.21

Optimisation of the Critical Processing Variables of Microemulsion Formulation

Factor	Range Studied	Optimised Range
Stirring speed (rpm)	50–200	120–180
Stirring time (min)	5–30	15–20
Temperature (°C)	30–40	33–37

The processing variables were converted from medium risk to low risk by empirical optimisation. The optimised processing variables are depicted in Table 9.21.

- Assessment of high-risk parameters
   The high-risk parameters (i.e. concentration of oil, S_{mix} and water) were to be optimised by DoE, thereby reducing their criticality and turning them to low-risk parameters.

9.5.5.1 Application of Optimisation Design and Determination of Design Space

The levels of three independent composition variables, namely, oil (X_1), S_{mix} (X_2) and water (X_3) were chosen on the basis of pseudoternary phase diagrams. These three factors were varied with respect to each other such that the total sum was 100%. The variables were optimised using D-optimal mixture design since the sum of concentrations of all the independent variables was always equivalent to 100%. The globule size (Y_1) and in vitro drug release (Y_2) were selected as the response CQAs based on the REM (see Table 9.22).

The necessary experiments for optimisation were carried out and the obtained results are represented in the form of 3D plots. Figure 9.20a depicts the 3D plots

TABLE 9.22

Independent Variables and Their Levels

Independent Variables	Material	Concentration Range Studied
X_1	Concentration of oil	$9\% \leq X_1 \leq 10\%$
X_2	Concentration of S_{mix}	$30\% \leq X_2 \leq 41\%$
X_3	Concentration of water	$50\% \leq X_3 \leq 61\%$

obtained for globule size; while Figure 9.20b through d represents the 3D plots for in vitro drug release at 4, 8 and 12 h, respectively. From the 3D plots, it can be observed that the globule size reduces as a direct function of the S_{mix} concentration. The in vitro release rate was found to rise with a corresponding increase in the concentration of the oil.

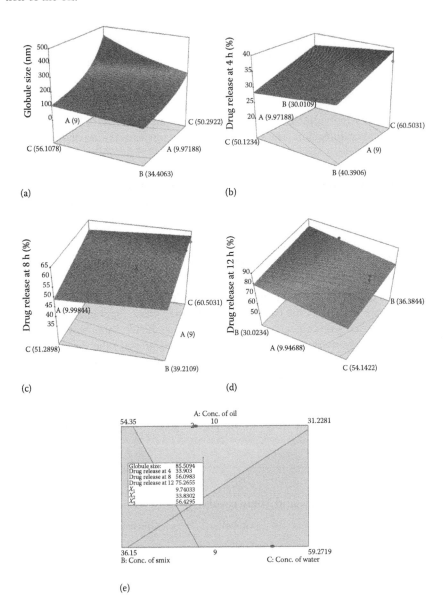

FIGURE 9.20 (See colour insert.) Effect of product variables on the tenofovir microemulsion-based vaginal gel: (a) globule size, (b) in vitro drug release at 4 h, (c) in vitro drug release at 8 h, (d) in vitro drug release at 12 h and (e) overlay plot.

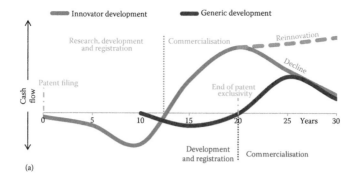

(a)

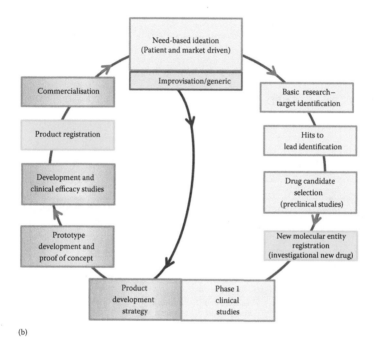

(b)

FIGURE 1.2 Pharmaceutical product life cycle. (a) Financial aspect. (b) Specific activities.

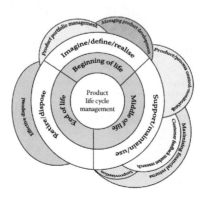

FIGURE 1.3 Overview of PLM.

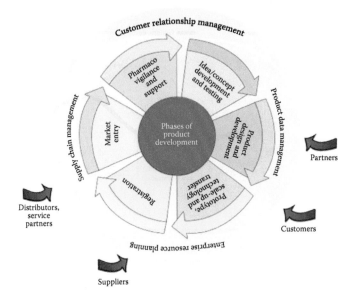

FIGURE 1.5 Phases of NPD.

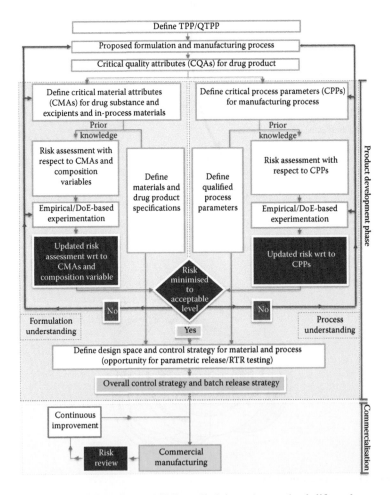

FIGURE 4.3 The iterative nature of QbD applied through a product's life cycle.

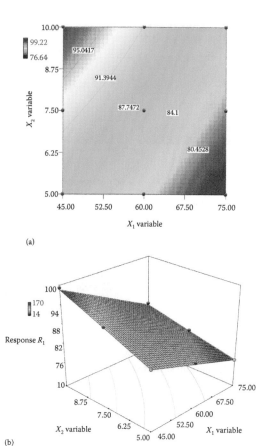

(a)

(b)

FIGURE 5.13 (a) Contour plot. (b) 3-D plot.

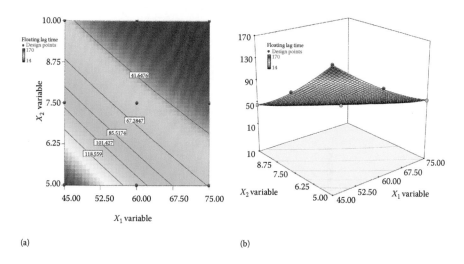

(a)

(b)

FIGURE 5.17 Response surface plot. (a) 2-D contour plot of floating lag time. (b) 3-D response surface plot of floating lag time.

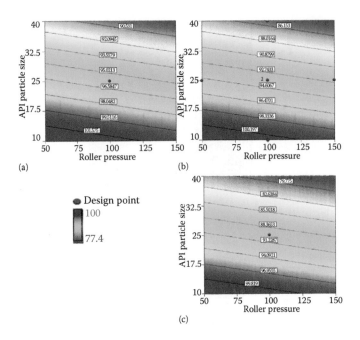

FIGURE 9.7 Contour plot response curve for API particle size and roller pressure against % dissolution at 30 min. (a) Magnesium stearate concentration 1% w/w, (b) magnesium stearate concentration 1.5% w/w and (c) magnesium stearate concentration 2% w/w.

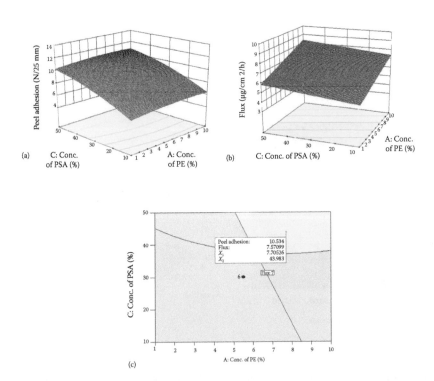

FIGURE 9.11 3D plots for (a) peel adhesion, (b) transdermal flux and (c) overlay plot.

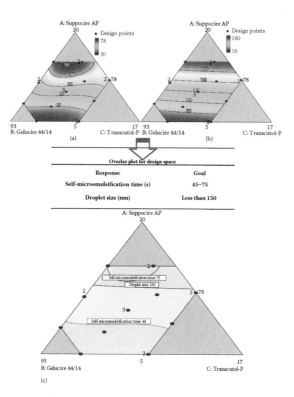

FIGURE 9.16 Impact of formulation composition on (a) self-microemulsification time (s), (b) microemulsion droplet size (nm) and (c) overlay plot.

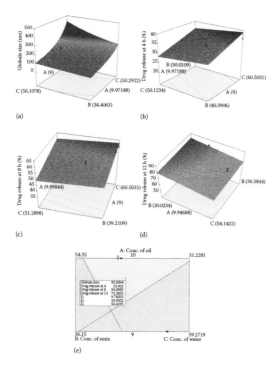

FIGURE 9.20 Effect of product variables on the tenofovir microemulsion-based vaginal gel: (a) globule size, (b) in vitro drug release at 4 h, (c) in vitro drug release at 8 h, (d) in vitro drug release at 12 h and (e) overlay plot.

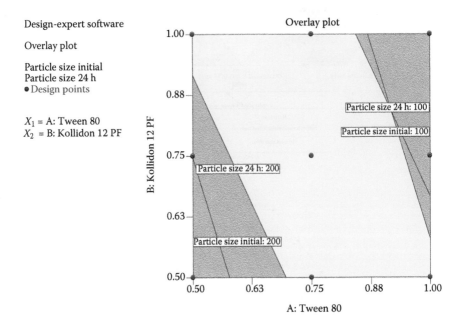

FIGURE 9.26 Impact of stabiliser and wetting agent concentration on the particle size of the nanosuspension and the particle size of the nanosuspension after 24 h at a constant homogenisation pressure (1200 bars) and number of HPH cycles (25 cycles).

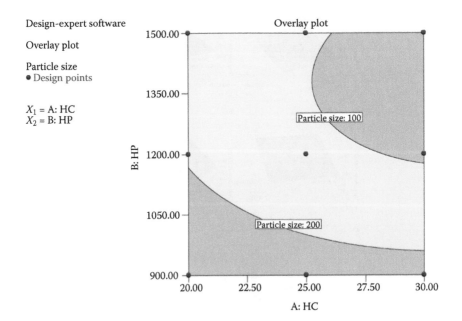

FIGURE 9.28 Impact of homogenisation pressure (HP) and number of HPH cycles (HC) on the particle size of the nanosuspension at a constant ratio of drug/stabiliser/wetting agent.

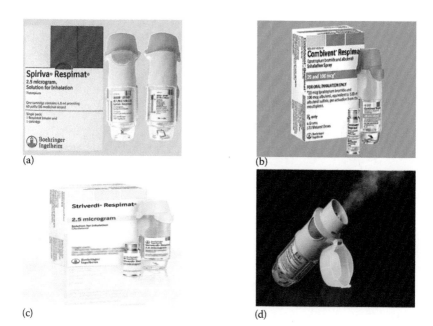

(a)

(b)

(c)

(d)

FIGURE 9.35 Marketed products that use Respimat as a delivery device for COPD treatment: (a) SPIRIVA – tiotropium bromide, (b) COMBIVENT – ipratropium bromide and albuterol sulfate spray, (c) STRIVERDI – olodaterol and (d) STIOLTO – tiotropium bromide and olodaterol. (From Combivent Respimat Inhalation 2015; SPIRIVA Respimat Inhaler 2015; STIOLTO Respimat 2015; Striverdi Respimat 2015.)

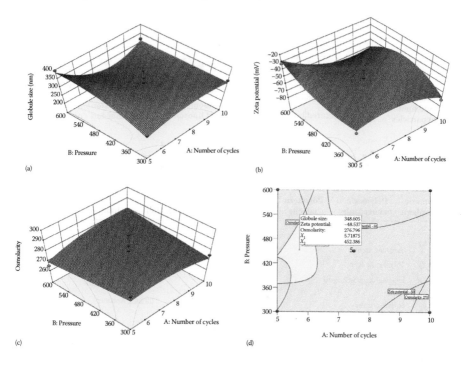

FIGURE 9.39 Effect of product and process variables on nanoemulsion: (a) globule size, (b) zeta potential, (c) osmolarity and (d) overlay plot depicting controlled design space.

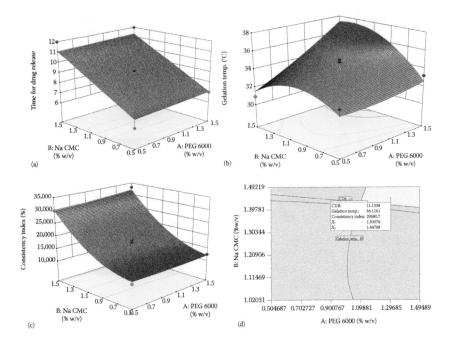

FIGURE 9.41 Effect of product variables on the apomorphine in situ gel: (a) time for 100% drug release, (b) gelation temperature, (c) consistency index and (d) overlay plot.

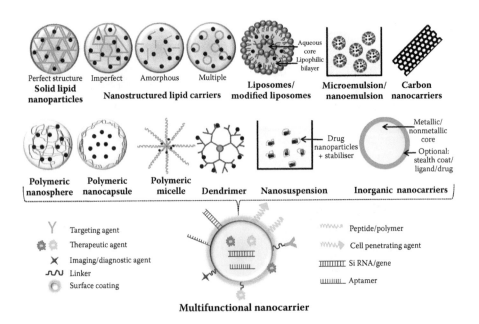

FIGURE 12.1 Nanopharmaceuticals: spectrum and advances.

On the basis of the results obtained from the experimental runs (as shown in Figure 9.20a through d) and the range of responses desired, the software provided the design space giving the ranges of values of the independent variables to be maintained in order to obtain the desired responses. Figure 9.20e displays the overlay plot depicting the controlled design space.

The yellow region in the overlay plot depicted in Figure 9.20e shows the concentration of the composition variables sufficient to obtain the globule size and % drug release within the desired range. On the basis of this overlay plot, the updated risk assessment and control strategy was established.

9.5.5.2 Updated Risk Assessment for the Formulation of TNF-Loaded Microemulsion

The updated risk assessment represented in Table 9.23 shows that all the high-risk parameters (see Figure 9.19) can be converted to low risk by maintaining the concentrations of the variables within the obtained design space (see Figure 9.20e) and hence the QTPP can be achieved. The medium-risk parameters (i.e. the critical processing parameters) were converted to low risk by empirical optimisation.

9.5.6 Initial Risk Assessment of the Gel Formation Process

The final step in the formulation involves conversion of the formulated microemulsion into a hydrophilic gel by incorporating a hydrophilic gelling polymer in the external water phase of the microemulsion under constant stirring. A number of gelling agents at various concentrations were screened for this purpose. Table 9.24 enlists the various gelling agents screened along with the observations associated with each of them.

On the basis of this initial screening, 'Carbopol 940' was selected as the gelling polymer. Figure 9.21 depicts the initial risk assessment for the formulation of microemulsion-based gel showing the potential risks associated with each of the material attributes and process parameter(s) of the gel on its potential CQAs.

Mixing speed and time of microemulsion with the gelling polymer were designated as low-risk parameters. The concentration of the gelling polymer was not found to have a substantial effect on globule size, assay or the release rate of the drug. However, the final viscosity of the microemulsion-based gel was found to be primarily dependent on the concentration of the gelling polymer and hence was designated as a high-risk parameter. Concentration of carbopol was considered as a medium-risk parameter with respect to pH and in vitro bioadhesion since formulation of a carbopol gel requires a final pH adjustment step. Moreover, carbopol being a bioadhesive polymer will determine the bioadhesion of the final formulation.

9.5.6.1 Formulation of Microemulsion-Based Gel of TNF

The microemulsion was prepared as per the quantities in the design space depicted in Figure 9.20e. To this, 1% w/w carbopol was dispersed in the external water phase of the microemulsion under constant stirring till the entire quantity of carbopol was dissolved.

TABLE 9.23

Updated Risk Assessment for Formulation of TNF-Loaded Microemulsion

Drug Product CQAs	Critical Composition Variables			Critical Processing Parameters		
	Concentration of IPM	Concentration of S_{mix}	Water	Microemulsion Stirring Time	Microemulsion Stirring Speed	Temperature
Globule size	Low Concentration of IPM: 9.00%–9.24%	Low Concentration of S_{mix}: 31.84%–32.79%	Low Concentration of water: 58.20%–59.50%	Low	Low	Low
PDI	Low	Low	Low	Low	Low	Low
Viscosity	Low	Low	Low	Low	Low	Low
Assay	Low	Low	Low	Low	Low	Low
Content uniformity	Low	Low	Low	Low	Low	Low
pH	Low	Low	Low	Low	Low	Low
Impurities	Low	Low	Low	Low	Low	Low
Microbial content	Low	Low	Low	Low	Low	Low
In vitro drug release	Low Concentration of IPM: 9.00%–9.24%	Low Concentration of S_{mix}: 31.84%–32.79%	Low Concentration of water: 58.20%–59.50%	Low	Low	Low

TABLE 9.24

Screening of Gelling Polymers

Gelling Polymer	Concentration (% w/w)	Observations
Klucel HXF	1	No gel formation
(hydroxypropyl	2	Formation of a smooth translucent gel
cellulose)	3	Formation of a smooth translucent gel
Natrosol HHX	1	No gel formation
(hydroxyethylcellulose)	2	No gel formation
	3	Formation of clear viscous gel
Carbopol 940	0.2	No gel formation
	0.5	Formation of a clear gel
	1.0	Formation of a clear and viscous gel
Chitosan (medium	1	No gel formation
molecular weight)	2	Formation of a clear and viscous gel
	3	Formation of a clear and viscous gel
Sodium alginate	4	No gel formation
	6	Formation of a turbid gel
	8	Formation of a turbid gel

Drug product CQAs	Critical composition variable Concentration of carbopol	CCP Mixing speed and time
Globule size	Low	Low
PDI	Low	Low
Viscosity	High	Low
Assay	Low	Low
Content uniformity	Low	Low
pH	Medium	Low
Impurities	Low	Low
Microbial content	Low	Low
In vitro drug release	Low	Low
In vitro bioadhesion	Medium	Low

FIGURE 9.21   REM for formulation of the TNF-loaded microemulsion-based gel.

The pH of the system was adjusted to its desired pH by gradual addition of a 1% w/v solution of triethanolamine followed by stirring. The stirring speed was studied between a range of 150–400 rpm of which a range of 250–300 rpm was found to be suitable for intimate blending of the preformed microemulsion with the gelling polymer. The stirring time was deliberated within a range of 15–45 min, wherein the stirring time in a range of 25–35 min was found to be optimum. The obtained gel was evaluated for its clarity and consistency. The final optimised gel was a clear and smooth preparation and showed pH in the range of 4.1 ± 0.03, and the viscosity was approximately 45,000 cps. The in vitro bioadhesion time was found to be 38 ± 0.87 min.

9.5.6.2   Updated Risk Assessment for the Gelling Process

The updated risk assessment depicted in Table 9.25 shows that the high-risk parameters have been converted to low risk by maintaining the mentioned parameters within the optimised ranges.

On the basis of the optimisation studies, carbopol was selected as the gelling polymer at the concentration of 1% w/w. The obtained gel was found to be clear and viscous. The viscosity and pH of the formulated gel were within the limits and hence could meet the QTPP of the product.

9.5.7   Control Strategy

On the basis of the optimisation studies carried out, an overall control strategy is established indicating the concentration of each of the independent variables and the ranges of the optimised processing conditions to be followed to obtain a product as per the QTPP. Table 9.26 shows the overall control strategy to obtain the desired product.

TABLE 9.25
Updated REM for Formulation of the TNF-Loaded Microemulsion-Based Gel

	Critical Composition Variable	CCP
Drug Product CQAs	**Concentration of Carbopol**	**Mixing Speed and Time**
Globule size	Low	Low
PDI	Low	Low
Viscosity	Low Concentration of carbopol: 1% w/w	Low
Assay and content uniformity	Low	Low
Content uniformity	Low	Low
pH	Low	Low
Impurities	Low	Low
Microbial content	Low	Low
In vitro drug release	Low	Low
In vitro bioadhesion	Low	Low

TABLE 9.26

Control Strategy for Formulation of the TNF-Loaded Microemulsion-Based Vaginal Gel

Factor	Unit	Range	Design Space	Fixed Point	Purpose
Critical Composition Variables					
Concentration of oil (X_1)	%	9–10	9.00–9.24	9.09	To ensure development of
Concentration of S_{mix} (X_2)	%	30–60	31.84–32.79	32.23	a stable and efficacious vaginal gel
Concentration of water (X_3)	%	50–70	58.20–59.50	58.68	formulation
Concentration of carbopol	%	0.2–1.0	–	1.0	
Critical Processing Variables					
Microemulsion Preparation Step					
Stirring time	min	50–200	120–180	–	
Stirring speed	rpm	5–30	15–20	–	
Temperature	°C	30–40	33–37	–	
Gel Preparation Step					
Mixing time	min	150–400	250–300	–	
Mixing speed	rpm	15–45	25–35	–	

9.5.8 CONTAINER AND CLOSURE SYSTEM

The obtained microemulsion-based gel would be filled into lacquered aluminium tubes enclosed with a screw-capped plastic closure. The aluminium tube would then be enclosed in a cardboard carton along with a separate applicator nozzle to facilitate vaginal application.

TNF was successfully formulated as a microemulsion-based vaginal gel using QbD and risk management principles.

9.6 PARENTERAL

Parenteral routes of administration include intramuscular, intravenous, intrathecal, intradermal, subcutaneous, intraarterial and intraarticular routes. Different dosage forms that may be utilised for parenteral product development include solutions, suspensions, powders for reconstitution, emulsions and irrigation solutions. There are a few considerations applicable specifically to parenteral product development. These include the requirement of a device such as a syringe or a pen for injection, the need for preservatives in multiple-dose products and different end users such as physicians, patients, nurses, pharmacists and so on (Lambert 2010). Sterility, pyrogen-free nature, freedom from hemolysis, freedom from particulate matter and tonicity are some of the key requirements specific for the parenteral route of administration.

The representative process of parenteral product development is elucidated in the following case study, which can vary from case to case on the basis of the type of dosage form, intended route, immediate release or depot and other specifications.

9.6.1   Case Study – Fast-Acting Intravenous Atovaquone Nanosuspension

In the present case study, development of a novel pharmaceutical product – fast-acting intravenous atovaquone nanosuspension for the treatment of severe malaria – is described.

Clinical need: Severe malaria is a medical emergency that can result in mortality if left untreated. Death occurs within a few hours of admission to the hospital and hence there is an urgent need to develop a rapid-acting intravenous formulation for its treatment. Many of the patients with severe malaria are unable to take oral medication because of nausea and vomiting, thus necessitating parenteral therapy. Currently used intravenous antimalarial medications have some drawbacks. Intravenous administration of quinine causes hypotension and cardiac arrest if given by rapid injection, thus necessitating rate-controlled intravenous infusion or thrice-a-day intramuscular injection. Intramuscular injection of quinine causes pain and irritation because of its low pH. Another artemisinin derivative, artesunate, in the injectable form is dispensed as a two-component product: anhydrous powder of artesunic acid with a separate ampoule of 5% sodium bicarbonate solution. Artesunic acid is dissolved in sodium bicarbonate to form sodium artesunate immediately before injection (WHO 2010). However, the product has poor dissolution properties, which froths and cakes upon addition of the dissolution medium (Ellis et al. 2010). The solution is then required to be diluted in 5 mL of 5% dextrose for intravenous or intramuscular injection. The solution needs to be made freshly for each administration and must not be stored because of its poor stability in aqueous solutions at neutral or acid pH (WHO 2010). Hence, the aim was to develop a rapid-acting intravenous formulation of atovaquone for the treatment of severe malaria, which is stable and convenient to administer. The product development process is summarised as follows.

9.6.2   Establishment of TPP

The TPP for the product development is depicted in Table 9.27.

9.6.3   Proposed Formulation and Manufacturing Process

The product contains two parts, namely, sterile lyophilised powder and Sterile Water for Injection. The manufacturing of Sterile Water for Injection was done in-house by the routine process, which was well controlled and hence will not be discussed in the forthcoming sections.

TABLE 9.27

TPP of the Atovaquone Nanosuspension for Intravenous Malaria Therapy

TPP Element	Target	Justification
Dosage form	Dry powder for reconstitution with Sterile Water for Injection (lyophilised nanosuspension)	Dry powder would avoid instability problems with aqueous solution of drug
Route of administration	IV bolus after reconstitution	Intravenous route was proposed for rapid onset of action and because of inability of severe malaria patients to take oral medications
Injection volume	2–10 mL	Since route of administration is intravenous bolus, 2–10 mL was decided as the injection volume
Identity	Positive for atovaquone	To ensure safety and efficacy of product
Strength	75 mg/vial	Dose required for antimalarial activity
Assay	97.5%–101.5% of label claim	To ensure safety and efficacy of product and as per USP
pH	5.5–7.5	Acceptable for parenteral administration to avoid pain and irritation
Reconstitution time	Easy reconstitution within 1–2 min	To ensure quick administration
Insoluble particulate matter	Meets pharmacopoeial requirements	To ensure safety and efficacy of product
Impurities and related substances	Meets ICH requirements	To ensure safety and efficacy of product
Content uniformity	Meets pharmacopoeial requirements	To ensure accurate dosing
In vitro release	Immediate release: greater than 80% release in 30 min	To ensure rapid onset of action
Bacterial endotoxin	Meets pharmacopoeial requirements	To ensure patient safety
Sterility	Meets pharmacopoeial requirements	To ensure patient safety

(*Continued*)

TABLE 9.27 (CONTINUED)

TPP of the Atovaquone Nanosuspension for Intravenous Malaria Therapy

TPP Element	Target	Justification
Excipients	Acceptable at levels used for the proposed market	To ensure patient safety
Shelf life	24 months	To maintain safety and efficacy of product
Container closure system	Single-dose vial with rubber closure sealed with aluminium caps	To maintain the sterility and therapeutic potential of drug product during shelf life

9.6.3.1   Nanosuspensions as Formulation Strategy

Atovaquone is a lipophilic drug with extremely poor aqueous solubility. Intravenous formulation development for atovaquone is challenging because of its poor solubility in hydrophilic cosolvents, surfactants, oils and lipids used/approved for intravenous administration (Scholer et al. 2001). Nanosuspensions provide a feasible alternative to formulate such lipophilic drugs having a high melting point and low solubility in water and oils (Rabinow 2004).

9.6.3.2   High-Pressure Homogenisation as Manufacturing Process

High-pressure homogenisation (HPH) was chosen as the manufacturing process owing to its amenability to scale up, faster production compared to media milling and extensive experience and expertise with the technique (Patravale et al. 2004). The manufacturing process flow is depicted in Figure 9.22.

The formulation of the nanosuspension would be carried out in a controlled area (Class 10,000). The sterilisation of the nanosuspension would be done by filtration through a 0.22-μm membrane filter to avoid risk of particle aggregation, which can occur if autoclaving is used for sterilisation. The sterile nanosuspension obtained would then be lyophilised to obtain a powder ready for reconstitution to avoid any stability issues. The entire processing of the nanosuspension after its sterilisation through membrane filtration would be carried out in an aseptic area (Class 100). These steps included lyophilisation and sealing of the lyophilised product inside vials. The packaging of vials would then be done in a carton with Sterile Water for Injection (which would be manufactured separately), in a controlled area (Class 10,000).

9.6.4   IDENTIFICATION OF CQAs BASED ON TPP AND PROPOSED MANUFACTURING PROCESS

On the basis of the proposed formulation and manufacturing process, CQAs of the atovaquone nanosuspension were identified and categorised as those of high, medium and low risk, as shown in Figure 9.23.

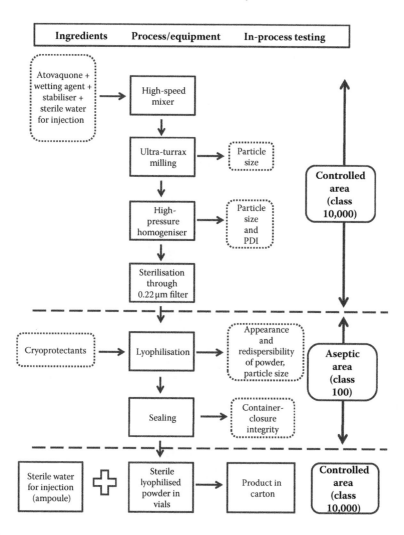

FIGURE 9.22 Process flow for manufacture of the atovaquone nanosuspension.

The product attributes identified to be highly critical were assay, impurities, particle size of premilled suspension for HPH, particle size of the nanosuspension, redispersibility of the lyophilised powder/cake and release profile. These were subjected to further risk analysis.

9.6.5 Experimental Work

9.6.5.1 Procurement of API

Parenteral-grade atovaquone was obtained from the selected API source. The API source was selected on the basis of compliance with pharmacopoeial specifications,

Quality attribute	Target/acceptance criteria	Criticality	Justification
Appearance	Yellow lyophilised powder/cake	Low	Not directly linked to safety and efficacy of product, so low criticality.
Identity	UV spectrum and chromatogram match the reference standard	Low	This will affect the clinical safety and efficacy but can be controlled with appropriate API source selection.
Assay	97.5%–101.5%	High	This will affect the safety and efficacy. Process variables may influence the assay and hence assay should be evaluated at each step, which makes it critical.
pH	5.5–7.5	Medium	This is required to avoid pain and irritation but can be controlled through preliminary evaluation and monitoring at product release, thus considered to be medium critical.
Redispersibility of lyophilised powder/cake	Easy dispersion within 1–2 min	High	Critical to obtain quick and complete redispersion and nanosuspension of the desired particle size and could be affected by material attributes and process parameters and hence would be studied in detail during manufacture.
Insoluble particulate matter	Meets pharmacopoeial requirements	Medium	Critical to ensure safety and efficacy of product but would be controlled through proper selection of API, excipients and control of manufacturing facility and monitored at product release.
Impurities and related substances	Meets ICH requirements	High	To ensure safety and efficacy of product.
Content uniformity	Meets pharmacopoeial requirements	Medium	Critical to ensure accurate dosing and would be monitored after each unit operation and at product release and hence medium critical.
Bacterial endotoxin	Meets pharmacopoeial requirements	Medium	Critical to safety but would be controlled through proper control of raw materials used in the manufacture and stringent environmental control of the manufacturing facility.

FIGURE 9.23 CQAs of the atovaquone nanosuspension.   (*Continued*)

Quality attribute	Target/acceptance criteria	Criticality	Justification
Sterility	Meets pharmacopoeial requirements	Medium	Critical to safety but would be controlled through proper control of raw materials used in the manufacture and stringent environmental control of the manufacturing facility.
Excipients	Meets pharmacopoeial requirements	Medium	Critical but can be controlled through proper excipient source selection.
Particle size of premilled suspension for HPH	Less than 25 μm	High	Critical to prevent clogging of HPH valve and needs to be monitored before proceeding for HPH.
Particle size of the nanosuspension	100–200 nm	High	Is critical since release profile is directly related to the particle size of the nanosuspension, and material attributes and process parameters can affect particle size, and hence will be monitored throughout the manufacture.
Viscosity of the nanosuspension	Acceptable syringeability and easy injectability	Medium	Critical to ensure syringeability and injectability but will be controlled and monitored at release; hence, medium criticality.
Release profile	Greater than 80% release in 30 min	High	Any variation in release profile will lead to batch-to-batch variation and affect the pharmacokinetic profile and so is highly critical.

FIGURE 9.23 (CONTINUED) CQAs of the atovaquone nanosuspension.

lowest particle size and cost. Different API sources were screened, and amongst the available sources, the lowest particle size of API ($D_{0.9}$ of less than 100 μm) complying with pharmacopoeial standards was selected. API having the lowest particle size was selected as the lower is the initial size of the API, the higher is the possibility of achieving the desired particle size reduction. Also, prior experience with Ultra-Turrax milling and HPH technique confirmed the feasibility of working with API of particle size ($D_{0.9}$ of less than 100 μm) for nanosuspension formulation. The procured API was tested for compliance with USP specifications.

9.6.5.2 Selection of Stabiliser and Wetting Agent

Various excipients were screened for their role as a wetting agent and stabiliser on the basis of the following criteria: compendial status, acceptability for parenteral administration, Inactive Ingredients Database limits and low microbial load. Contact angle studies were conducted to select the optimum wetting agent, as shown in Table 9.28. Pure atovaquone showed a contact angle of 120°.

TABLE 9.28

Contact Angles Obtained for Atovaquone with Water and 1% Wetting Agent Solution

Wetting Agent	Contact Angle (°)
Tween 80	35.14
Solutol HS 15	42.06
Cremophor EL	50.12
Poloxamer 188	59.01
Poloxamer 407	65.41
Kollidon 12 PF	68.74
Kollidon 17 PF	70.01
Polyvinyl alcohol	77.81

Tween 80 was chosen as the wetting agent since it showed the lowest contact angle when used at 1% concentration. Particle size measurements of nanosuspensions containing 0.5% w/v atovaquone, 1% w/v Tween 80 and various stabilisers at 1% w/v (processed through HPH) were done to select the best stabiliser, as shown in Table 9.29.

Kollidon 12 PF was chosen as the stabiliser since it showed the lowest particle size when used at 1%, and the particle size remained unchanged after 24 h. Chemical compatibility of the drug with Tween 80 and Kollidon 12 PF was studied and found satisfactory.

9.6.5.3 Optimisation of the Premilling Process Step Using Ultra-Turrax

Premilling of the atovaquone suspension was done using Ultra-Turrax milling before proceeding for HPH because of two reasons. First, it is well established that the lower the particle size of the presuspension loaded into HPH, the better is the size reduction achieved. Second, the homogeneous gap in the homogenising valve can get clogged owing to higher particle size of the input material; hence, an in-house specification of $D_{0.9}$ less than 25 µm was set for the particle size of the atovaquone suspension to be loaded into the homogeniser. Premilling optimisation with respect to speed and time was done based on a series of experiments from which the lowest speed and least processing time at which the desired particle size of $D_{0.9}$ less than 25 µm was obtained (which remained stable for 12 h) were finalised. It was found to be 15,000 rpm for 5 min. The HPH process is usually initiated within 2–3 h after

TABLE 9.29

Effect of Various Stabilisers on the Particle Size of the Nanosuspensions

Stabiliser	Particle Size (nm)	Particle Size after 24 h (nm)
Tween 80	250.10 ± 5.42	265.21 ± 5.04
Solutol HS 15	200.32 ± 6.54	240.44 ± 6.42
Cremophor EL	230.16 ± 8.09	250.03 ± 4.87
Poloxamer 188	280.32 ± 4.32	340.12 ± 7.32
Poloxamer 407	300.17 ± 5.01	370.12 ± 4.98
Kollidon 12 PF	150.21 ± 4.98	160.14 ± 5.04
Kollidon 17 PF	180.20 ± 4.02	200.14 ± 4.12
Polyvinyl alcohol	260.32 ± 6.51	279.32 ± 5.90

premilling. However, the physical stability of the premilled suspension was studied for a period of 12 h to support the process lag time interval between premilling and HPH processing. The premilled suspension was stored at 25°C before HPH processing.

9.6.6 RISK ASSESSMENT

A risk assessment was done as per ICH Q9 to identify the CMAs of excipients and CPPs of manufacturing operations likely to have a major impact on the product CQAs. The risk assessment was done in two steps for the two operations, namely, formulation of the nanosuspension by HPH and lyophilisation of the nanosuspension.

9.6.6.1 Initial Risk Assessment for Formulation of the Nanosuspension by HPH

Initial risk assessment for formulation of the nanosuspension is shown in Figure 9.24, wherein the CMAs and CPPs that were identified to be of high risk for formulation of the nanosuspension were studied further during product development. Previous knowledge was utilised to identify the variables having a low potential for causing risk to product quality.

The API obtained was of high purity and its purity was verified by chromatography; hence, it would be a low-risk factor to the assay and impurities of the final product. Also, the particle size of the API was considered to be of low risk since an API having a constant particle size, $D_{0.9}$, of less than 100 μm was obtained from the

| Product CQA | API purity | Critical material attributes | | | | Critical processing parameters | |
		API particle size	Concentration of Tween 80	Concentration of Kollidon 12 PF		Homogenisation pressure	Number of HPH cycles
Assay	Low	Low	Low	Low		Low	Low
Impurities	Low	Low	Low	Low		Low	Low
Particle size	Low	Low	High	High		High	High
Release profile	Low	Low	Medium	Medium		Medium	Medium

FIGURE 9.24   Risk assessment for formulation of the nanosuspension.

API supplier. Similarly, on the basis of previous experience with the HPH technique for nanosuspension manufacture, the concentration of Tween 80, concentration of Kollidon 12 PF, homogenisation pressure and number of HPH cycles were reported to be of high risk to the particle size and hence were optimised further for product development. Drug–excipient compatibility studies revealed the concentration of Tween 80 and Kollidon 12 PF to be of low risk to the assay and impurities of the product. Similarly, the drug was found to be stable to HPH processing and hence homogenisation pressure and number of HPH cycles were expected to be of low risk to the assay and impurities of the product. The concentration of Tween 80, Kollidon 12 PF, homogenisation pressure and number of HPH cycles were identified to be of medium risk to the release profile since the excipients chosen were for promoting solubility and wettability and the HPH process would result in particle size reduction. Also, the release profile of the nanosuspension is directly linked to its particle size, and particle size and the effect of all these variables on particle size was studied further by DoE.

9.6.6.1.1 DoE Study for Optimising Tween 80 and Kollidon 12 PF Concentration

Design Expert was utilised to study the effect of Kollidon 12 PF (at 0.5%, 0.75% and 1% levels) and Tween 80 concentration (at 0.5%, 0.75% and 1% levels) on the particle size of the nanosuspension and particle size after 24 h, as shown in Figure 9.25.

The lyophilisation process is usually initiated within 4–6 h after HPH. However, the physical stability of the nanosuspension was studied for a period of 24 h to support the process lag time interval between HPH and lyophilisation. The nanosuspension was stored at 25°C before lyophilisation. Figure 9.26 depicts the impact of Kollidon 12 PF and Tween 80 concentration on the particle size of the nanosuspension and particle size of the nanosuspension after 24 h at a constant homogenisation pressure (1200 bars) and number of HPH cycles (25 cycles). The particle size of the nanosuspension decreased with an increase in the concentration of Tween 80 and Kollidon 12 PF. From Figure 9.26, it was observed that the desired response of a particle size, $D_{0.9}$, of 100–200 nm, which remained stable for 24 h, was obtained in the range of 0.7%–0.8% of both Tween 80 and Kollidon 12 PF.

Factor	Range investigated (% w/v)	Design space (% w/v)
Concentration of Tween 80	0.5–1	0.7–0.8
Concentration of Kollidon 12 PF	0.5–1	0.7–0.8

Response	Goal	Acceptance
Particle size of the nanosuspension	$D_{0.9}$ of 100–200 nm	$D_{0.9}$ of 100–200 nm
Particle size of the nanosuspension after 24 h	$D_{0.9}$ of 100–200 nm	$D_{0.9}$ of 100–200 nm

FIGURE 9.25 Impact of Kollidon 12 PF and Tween 80 concentration on the particle size of the nanosuspension.

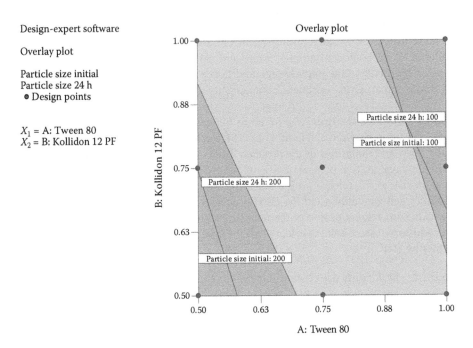

Design-expert software

Overlay plot

Particle size initial
Particle size 24 h
● Design points

X_1 = A: Tween 80
X_2 = B: Kollidon 12 PF

FIGURE 9.26 **(See colour insert.)** Impact of stabiliser and wetting agent concentration on the particle size of the nanosuspension and the particle size of the nanosuspension after 24 h at a constant homogenisation pressure (1200 bars) and number of HPH cycles (25 cycles).

9.6.6.1.2 DoE Study for Optimising Homogenisation Pressure and Number of HPH Cycles

Design Expert was utilised to study the effect of homogenisation pressure (at 900, 1200 and 1500 bars) and number of HPH cycles (at 20, 25 and 30 cycles) on the particle size of the nanosuspension at a constant concentration of Tween 80 and Kollidon 12 PF (0.75%) and 0.5% w/v atovaquone, as shown in Figure 9.27.

Figure 9.28 depicts the impact of homogenisation pressure and number of HPH cycles on the particle size of the nanosuspension and its polydispersity index (PDI) at a constant ratio of drug/stabiliser/wetting agent. From Figure 9.28, it was observed

Factor	Range investigated	Design space
Homogenization pressure	900–1500 bars	1140–1260 bars
Number of HPH cycles	20–30 cycles	23–27 cycles

Response	Goal	Acceptance
Particle size of the nanosuspension	$D_{0.9}$ of 100–200 nm	$D_{0.9}$ of 100–200 nm
PDI	0.1–0.2	0.1–0.2

FIGURE 9.27 Impact of homogenisation pressure and number of HPH cycles on the particle size and PDI of the nanosuspension.

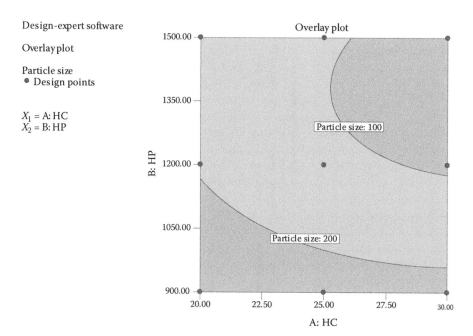

FIGURE 9.28 (See colour insert.) Impact of homogenisation pressure (HP) and number of HPH cycles (HC) on the particle size of the nanosuspension at a constant ratio of drug/stabiliser/wetting agent.

that an increase in homogenisation pressure and number of HPH cycles resulted in a decrease in the particle size and PDI. The design space obtained is depicted in Figure 9.28.

9.6.6.2 Updated Risk Assessment for Formulation of the Nanosuspension and Their Justification

After successful product and HPH process optimisation, all the variables and unit operations were reassessed towards the risk and it was observed that the risk was minimised at each stage, as depicted in Table 9.30.

Based on experiments, it was concluded that by operating within the defined design space, as shown in Figures 9.26 and 9.28, the product having the desired particle size and stability was obtained. Thus, the risk associated with the concentration of Kollidon 12 PF, concentration of Tween 80, homogenisation pressure and number of HPH cycles on particle size was minimised. Also, the risk associated with release profile was minimised since release profile is directly related to the particle size and the risk associated with particle size was minimised.

9.6.6.3 Risk Assessment for Lyophilisation Process

Similarly, a risk assessment was done for the lyophilisation process. As shown in Figure 9.29, the CMAs and CPPs that were identified to be of high risk to the lyophilisation operation were studied further during product development.

TABLE 9.30

Updated Risk Assessment for Formulation of the Nanosuspension

Product CQA	Critical Material Attributes				Critical Processing Parameters	
	API Purity	API Particle Size	Concentration of Tween 80	Concentration of Kollidon 12 PF	Homogenisation Pressure	Number of HPH Cycles
Assay	Low	Low	Low	Low	Low	Low
Impurities	Low	Low	Low	Low	Low	Low
Particle size	Low	Low	Low (0.7%–0.8%)	Low (0.7%–0.8%)	Low (1140–1260 bars)	Low (23–27 cycles)
Release profile	Low	Low	Low	Low	Low	Low

Product CQA	Critical material attributes		Critical processing parameters		
	Type of cryoprotectant	Concentration of cryoprotectant	Temperature and time for freezing	Temperature and time for primary drying	Temperature and time for secondary drying
Assay	Low	Low	Low	Low	Low
Impurities	Low	Low	Low	Low	Low
Particle size	High	High	Low	Low	Low
Release profile	Medium	Medium	Low	Low	Low
Redispersibility	High	High	Low	Low	Low

FIGURE 9.29 Risk assessment of the lyophilisation process.

Previous knowledge was utilised to identify the variables having low potential for causing risk to product quality. The nature of the cryoprotectant and its concentration are known to have a major impact on the particle size of the nanosuspension and the redispersibility of the lyophilised cake/powder. Drug–cryoprotectant compatibility studies were done to ensure that the nature and concentration of the cryoprotectant are of low risk to the assay and impurities. The commonly used temperatures employed for different stages of lyophilisation, namely, freezing, primary drying and secondary drying, were found to be of low risk to the assay, impurities, particle size and redispersibility of powder/cake on the basis of preliminary experiments. The release profile of the nanosuspension is directly linked to its particle size, and the particle size and the effect of type and concentration of the cryoprotectant on particle size were studied. Hence, these variables were of medium risk to the release profile.

9.6.6.3.1 Optimisation of the Concentration of the Cryoprotectant Based on Empirical Design

Five commonly used cryoprotectants, namely, trehalose, sucrose, dextrose, lactose and mannitol, were screened at three concentrations – 1%, 2% and 5% w/v – using empirical design. Trehalose at 5% w/v was chosen as the final cryoprotectant based on the S_f/S_i ratio (1.34) (ratio of the particle size after lyophilisation to the particle size before lyophilisation), ease of reconstitution (within 1 min) and the appearance of the lyophilised powder. Sucrose and mannitol showed particle aggregation resulting in high S_f/S_i ratio. Dextrose and lactose gave a lyophilised powder, which was difficult to redisperse. For the freeze drying process, preliminary experiments established the freezing temperature, primary drying temperature and secondary drying temperature to be of low risk and freezing was done at −40°C for 24 h, primary drying was done at 0°C for 5 h followed by 10°C for 2.5 h and 15°C for 2 h and secondary drying was done at 25°C for 2.5 h. The chamber pressure was kept at 20 Pa and the cold trap temperature was kept at −50°C.

9.6.6.4 Updated Risk Assessment for Lyophilisation of the Nanosuspension and Their Justification

After successful optimisation of cryoprotectant concentration, all the variables were reassessed towards the risk and it was observed that the risk was minimised at each stage, as depicted in Table 9.31.

9.6.7 Container and Closure System

The sterile nanosuspension was then lyophilised in an aseptic area (Class 100) using trehalose as a cryoprotectant and sealed. Packaging of the vials containing sterile lyophilised powder with ampoule containing Sterile Water for Injection in a carton was done in a controlled area (Class 10,000).

TABLE 9.31
Updated Risk Assessment for Lyophilisation of the Nanosuspension and Their Justification

| Product CQA | Critical Material Attributes | | Critical Processing Parameters | | |
	Type of Cryoprotectant	Concentration of Cryoprotectant	Temperature and Time for Freezing	Temperature and Time for Primary Drying	Temperature and Time for Secondary Drying
Assay	Low	Low	Low	Low	Low
Impurities	Low	Low	Low	Low	Low
Particle size	Low (trehalose at 5% w/v)		Low	Low	Low
Release profile	Low	Low	Low	Low	Low
Redispersibility	Low (trehalose at 5% w/v)		Low	Low	Low

TABLE 9.32
Composition of the Atovaquone Nanosuspension

Ingredient	Concentration (% w/v)
Atovaquone	0.5
Kollidon 12 PF	0.75
Tween 80	0.75
Trehalose	5
Sterile Water for Injection	q.s. 100 mL

9.6.8 Control Strategy for Manufacturing of the Atovaquone Nanosuspension

The final composition of the atovaquone nanosuspension is detailed in Table 9.32.

A presuspension of atovaquone and the selected stabilisers was prepared by Ultra-Turrax milling followed by HPH to obtain a nanosuspension having $D_{0.9}$ in the range of 100–200 nm. The nanosuspension was sterilised by membrane filtration. The sterile nanosuspension was then lyophilised in an aseptic area (Class 100) using trehalose as a cryoprotectant and sealed. Packaging of the vials containing sterile lyophilised powder with Sterile Water for Injection was done in a controlled area (Class 10,000). Overall control strategy for the lyophilised atovaquone nanosuspension is depicted in Table 9.33.

The atovaquone nanosuspension was successfully developed for intravenous therapy of severe malaria by applying QbD and risk management principles.

9.7 INHALATION

Inhalation therapy remains the mainstay for the treatment of pulmonary ailments such as obstructive airways diseases. The main advantages are site-specific delivery, rapid onset of action, bypassing first-pass metabolism, dose minimisation and subsequently reduction of drug side effects. The major dosage form devices used for the pulmonary administration of drugs include pressurised inhalers, nebulisers and the newly developed dry powder inhalers. The major anatomical and dosage form-related factors and variables that can affect the efficacy of the inhaled drug and thereby the performance of the product are illustrated in Figure 9.30.

The conventional pulmonary drug delivery devices suffer from disadvantages such as drug loss in the device and oropharynx, which leads to unwanted side effects, reduced drug effectiveness and increased costs of inhaled medications. This deposition arises from aerosol 'ballistic effects' and turbulent dispersion. It has been suggested that by reducing the ballistic effect, the performance of spray inhalers can be

TABLE 9.33

Overall Control Strategy for the Lyophilised Atovaquone Nanosuspension

Parameter	Range Studied	Proposed Range	Fixed Point	Control Criteria
Atovaquone concentration	–	0.5% w/v	0.5% w/v	
Ultra-Turrax speed	10,000–20,000 rpm	14,500–15,500 rpm	15,000 rpm	To ensure desired particle size ($D_{0.9}$ of less than 25 μm) for the HPH, process which remained stable for 12 h
Ultra-Turrax milling time	5–15 min	4–6 min	5 min	To ensure desired particle size ($D_{0.9}$ of less than 25 μm) for the HPH process, which remained stable for 12 h
Concentration of Tween 80	0.5%–1.5%	0.7%–0.8%	0.75% w/v	To ensure desired particle size ($D_{0.9}$ of 100–200 nm) and stability after 24 h
Concentration of Kollidon 12 PF	0.5%–1.5%	0.7%–0.8%	0.75% w/v	To ensure desired particle size ($D_{0.9}$ of 100–200 nm) and stability after 24 h
Homogenisation pressure	900–1500 bars	1140–1260 bars	1200 bars	To ensure desired particle size ($D_{0.9}$ of 100–200 nm) and PDI (0.1–0.2)
Number of HPH cycles	20–30 cycles	23–27 cycles	25 cycles	To ensure desired particle size ($D_{0.9}$ of 100–200 nm) and PDI (0.1–0.2)
Sterilisation process	Filtration through a 0.22-μm membrane filter			To ensure product safety
Concentration of trehalose	1%–5%	5% w/v	5% w/v	To ensure desired particle size, appearance of lyophilised powder and easy reconstitution
Temperature and time for freeze drying	Freezing at −40°C for 24 h, primary drying at 0°C for 5 h, followed by 10°C for 2.5 h and 15°C for 2 h and secondary drying at 25°C for 2.5 h			To ensure preservation of product efficacy, desired particle size, appearance of lyophilised powder and easy reconstitution

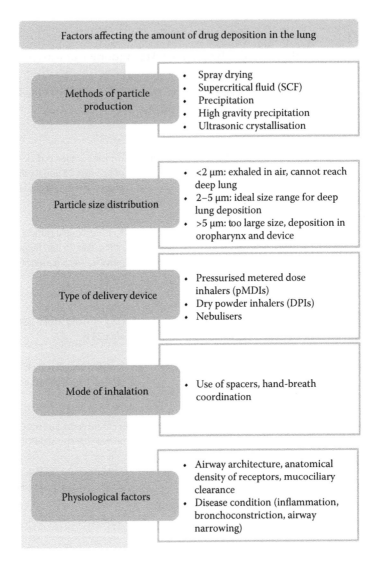

FIGURE 9.30   Factors influencing the performance of an inhalation product.

significantly improved. This can be achieved by reduction of propellant (completely flushing it out) and optimisation of the speed of emerging particles. This has been achieved to some extent by utilising spacers and valved holding chambers (VHCs). The development of new-generation devices that eliminate the use of propellant and spacers/VHCs is of interest nowadays mainly because of intrapatient and interpatient variability of dose delivered and loss of drug in the spacers.

A regular product development starts from characterisation of the drug substance to formulation development, analytical development, scale-up and finally technology transfer fulfilling regulatory requirements for final commercialisation. QbD has been applied as a USFDA initiative for development of many dosage forms, and it must be applied to inhalation products for their effective development. At the initial stages of inhalation product development, many factors have to be considered as the final performance will be dependent upon both the formulation and the device design. This can be defined in the QTPP. QTPP can be determined by understanding the patient need and the typical disease condition in which the final product performance is required. The major parameters and variables that can affect the CQAs that need to be considered during inhalation product development are presented in Figure 9.31.

As a subset of QTPP, various CQAs are determined, which can be altered potentially by formulation or process variables, for example, aerodynamic properties. There are many factors involved and they must be considered on a case-to-case basis. As in the case of dry powder inhalers, CQAs can be particle size of the raw material, blending process variables (shear rate and time), filling process variables such as powder flow and adhesion properties of powder. Metered dose inhaler CQAs can be drug/surfactant/cosolvent concentration, propellant ratio, container closure system variables, addition order of drug and surfactant/cosolvent, process temperature/filling to exhaustion and duration of process. In the case of nebulisers, pH, tonicity and viscosity of the solution can play an important role as CQAs. Major CQAs correlating patient requirements are presented in Table 9.34.

9.7.1 Case Study – Respimat® Soft Mist Inhaler

The present case study describes the development of Respimat Soft Mist Inhaler (SMI) by Boehringer Ingelheim (Ingelheim, Germany). The main goal of development of the device was to eliminate the use of propellant and generate the spray pattern that allows deep lung deposition of the drug. This was achieved by using mechanical power from a spring instead of a liquid–gas propellant for atomisation. Here, how the concept was developed into a successful delivery device (Respimat SMI) is presented.

The Respimat SMI forms an aerosol by forcing a drug solution through two micronozzles that create a converging jet system. The advantages of the Respimat SMI include a high proportion of respirable particles, a slow aerosol plume velocity and an extended aerosol generation time (approximately 1.5 s), which effectively increases the efficacy of the delivered drug, results to low device or mouth–throat deposition and reduces breath coordination, respectively. The fine particle fraction (FPF; percentage of mass with a droplet diameter less than 5.8 μm) in the case of Respimat SMI is 65%–80%, while current hydrofluoroalkane-based pressurised metered dose inhalers (pMDIs) have a FPF in the range of 20% to 50%. Figure 9.32 depicts the stage-wise development of Respimat.

9.7.2 The Concept of a Soft Mist Inhaler

The generation of an inhalable aerosol from a drug solution requires the metered dose of liquid to be converted into appropriately sized droplets without the use of propellants. Techniques to generate soft mist are illustrated in Table 9.35.

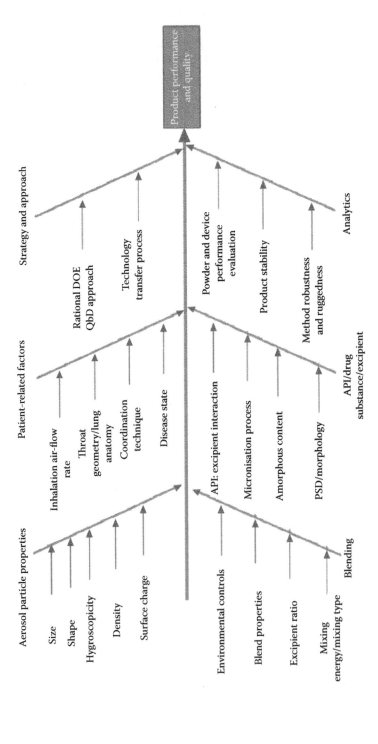

FIGURE 9.31 Understanding CQAs.

TABLE 9.34

Patient Requirements and Related CQAs for Pulmonary Product Development

Patient Requirements	QTPP (Product Requirements)	CQAs	
		Formulation and Process	Device
Accurate, precise and reproducible dose delivery	Uniform aerodynamic particle size distribution	Fine particle fraction Physicochemical properties of the drug and carrier particles	Nozzle diameter Mechanism of torque generation for spray production
	Drug content	Method of filling, agglomeration properties of drug and excipients	–
	Reproducible dispensing of powder or spray	Production method controls, viscosity	Consistent action of device components
Safety of the product	Product stability and purity	Particle interactions Microbial limits	Device material and formulation interactions Protection from moisture ingression
Product acceptability	Taste	Drug and excipient characteristics	–
	Low frequency of dosing	Drug content per actuation	–
	Compact device size	–	Device component selection depending on mechanism of spray generation

9.7.3 Prototype Development and Demonstration of the Concept of Providing Mechanical Energy

In the early 1990s, the prototype was made up of a metal pump body and a syringe, which served as a solution reservoir. The device was operated by means of a lever arm, which simultaneously tightened the mainspring and withdrew a metered volume of drug solution (approximately 13.5 μL) from the reservoir. A button was pressed, which released the spring, and the liquid was forced through a two-channel nozzle, which resulted in aerosol generation. Nozzle openings were made through a stainless steel disk having tiny holes. Although the droplet size distribution in the aerosol was demonstrated to be in the range suitable for inhalation, a nozzle design better suited for mass production was needed.

9.7.4 Development of an Easy-to-Use Model

The major targets of device development are depicted in Table 9.36.

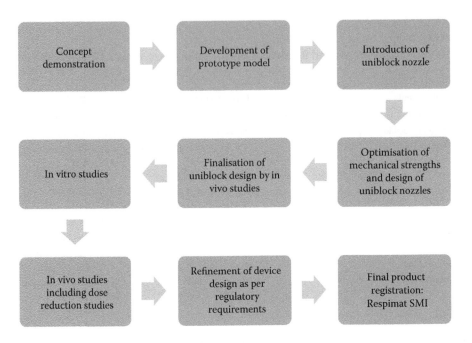

FIGURE 9.32 Stages of development of Respimat SMI.

TABLE 9.35
Techniques to Generate Soft Mist

Technique	Device/Mechanism
Electrical energy	Ultrasonic and piezo-electric devices, electrohydrodynamic effect
Mechanical energy	Compressible spring, extrusion through micron-sized holes

TABLE 9.36
Targets of Development of Respimat

Targets	Strategy Employed
Eliminate propellant use	Use of mechanical energy
Optimum spray velocity	Optimised uniblock design and spray nozzle placement angle
Dose reduction	Optimised spray velocity gives high-dose deposition in deep lung, thereby reduction in dose

9.7.5 INTRODUCTION OF THE 'UNIBLOCK' NOZZLE

The first feasible prototype was obtained when a uniblock with two channels was constructed. Its liquid jets impacted at a distance of 25 mm from the nozzle outlet to produce the aerosol of the desired particle size range of 1–5 μm.

This was achieved by developing a miniature 'sandwich' concept, the uniblock, consisting of a rectangle (2 × 2.5 mm) borosilicate glass plate bonded to a structured silicon wafer. Structures that prevent the nozzle from becoming blocked (inlet, outlet and filter channels) were etched into the silicon wafer using a microchip production-derived technology. This technique allowed a large-scale production of uniblocks with high precision and accuracy. The configuration of the inlet and outlet channels was engineered to produce a high FPF. Having tested several different principles, the impinging jet nozzle design was selected because of reproducible production of the nozzle assembly.

9.7.6 OPTIMISATION OF THE MECHANICAL STRENGTH OF THE UNIBLOCK

Finite-element simulation techniques were employed to ensure that the device has sufficient strength.

9.7.7 REFINEMENT OF DEVELOPED MODEL

For the further development of this first model, all metal components were made from polymers whenever possible. To easily generate the energy for the aerosol production by hand, the torque required for loading the dose was minimised (approximately 40 cNm).

Additional refinement of the device occurred after patient focus groups evaluated four different design prototypes. The preferred version of the device was used for phase II and phase III clinical trials. Several additional aesthetic modifications have been made to the device in advance of its launch onto the market. These include a hinged cap, colour coding to identify specific drug classes contained in the device and a transparent base to allow easy identification of the drug product.

After initial stability and technical performance tests, the device was successfully used in the first lung deposition study, in comparison with a pMDI containing chlorofluorocarbons, carried out on healthy volunteers. The experience accumulated with the various prototypes (I–IV, not shown), and the need to make a functionally practical device with a smaller number of individual components was combined with the results of device-handling studies, in which patients evaluated the four different design concepts. The resultant device, prototype IV, incorporated a radical change in design. Compared to the prototype I design, which released the dose at an angle of 90° relative to the axis of the pump chamber (i.e. the mouthpiece was at right angles to the drug cartridge), the final design releases the dose in a direction parallel to the axis. The device has been equipped with a dose indicator.

9.7.8 WORKING OF RESPIMAT

The principal parts of Respimat are shown in a schematic diagram (Figure 9.33a). After removal of the transparent base, the cartridge containing the drug solution is inserted and the transparent base is mounted again. This connects the cartridge to

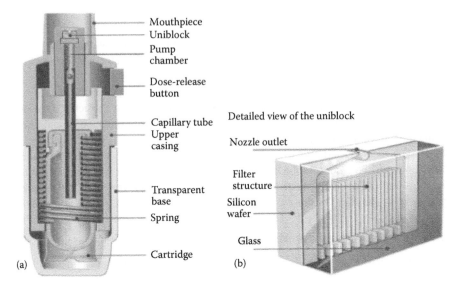

FIGURE 9.33 (a) Schematic cross section of the Respimat Soft Mist Inhaler. (b) 3D representation of the uniblock nozzle assembly. (Reprinted from Dalby, R., Spallek, M., Voshaar, T., *International Journal of Pharmaceutics*, 283 (1–2), pp. 1–9. Copyright 2004, with permission from Elsevier.)

the uniblock by a capillary tube, which contains a nonreturn valve. In its initial use, the device has to be primed to expel air from its inner parts. The device is then ready for use. The transparent base is turned 180° to load a dose.

This rotation is transformed into a linear movement by the helical cam gear. This leads to tightening of the spring and movement of the capillary with the nonreturn valve to a defined lower position. Through the capillary, the drug solution is drawn into a pump chamber, as shown in Figure 9.33a during this movement. When the patient presses the dose release button, the mechanical power of the spring pushes the capillary with the now closed nonreturn valve to the upper position. This operation drives the metered volume of the drug solution through the nozzle in the uniblock. At the nozzle outlets of the uniblock, two fine jets of liquid are produced. The resulting impact of the two jets of liquid generates a slow-moving aerosol cloud, or 'soft mist'. It is then the task of the patient to inhale slowly and deeply in order to achieve optimum delivery to the lungs.

9.7.9 In Vitro Testing

The particle size distribution at different relative humidity conditions (RH = 50% ± 10% and > 95%) was checked, as shown in Table 9.37. It showed that the droplets of both ethanol- and water-based formulations remained small enough in order to reach their targets in the lungs.

9.7.10 In Vivo Testing

Improved lung delivery by Respimat SMI in in vitro studies was verified in various scintigraphic deposition studies in patients as well as volunteers.

TABLE 9.37

In Vitro Testing

Formulation	Solvent	RH%	FPF%	MMAD (µM)	GSD
A	Water	50	63	2.0	2.6
B	Ethanol	50	83	1.1	2.2
C	Water	>95	60	4.2	2.0
D	Ethanol	>95	85	1.1	1.9

Note: GSD, geometric standard deviation; MMAD, mass median aerodynamic diameter.

In these studies, the radionuclide 99mTc was physically associated with the micronised drug particles in a suspension formulation (e.g. of a pMDI) or was incorporated into the droplets of a solution formulation. The topographical deposition of the aerosol in the lungs was visualised using a gamma camera and quantified in terms of the percentage of lung or oropharyngeal deposition with reference to the metered dose.

The deposition data showed that the soft mist generated by Respimat SMI, both from an aqueous solution of fenoterol and from an ethanolic solution of flunisolide, resulted in a twofold to threefold increase in lung deposition compared to the corresponding pMDI. In parallel, the oropharyngeal deposition was significantly reduced for the aerosol administered by Respimat SMI. Finally, the ratio of deposition in peripheral regions to deposition in the central lung zone was similar for Respimat SMI and a pMDI.

9.7.11 CONFIRMATORY STUDIES ON DOSE REDUCTION VIA RESPIMAT SMI

Since Respimat SMI increases the amount of drug deposited in the lungs, Boehringer Ingelheim expected that inhaled drugs would maintain their efficacy even when smaller doses than those normally be delivered by a pMDI would be administered.

In two clinical studies for aqueous drug solutions with fenoterol (Berotec) and the combination of fenoterol/ipratropium bromide (Berodual), this expected result was confirmed. For Berotec, 12.5 and 25 mg administered by Respimat SMI were therapeutically equivalent to either 100 or 200 mg administered via pMDI. For Berodual, the bronchodilatory effects of 25/10 or 50/20 mg (fenoterol/ipratropium bromide) doses administered via Respimat SMI were equal or slightly superior to the recommended dose of 100/40 mg given via pMDI. These results suggest that the improved lung deposition observed with Respimat SMI allows lower absolute doses to be administered for a similar clinical effect in the local treatment of lung diseases.

Respimat fulfills the main objectives of the development, that is, avoids the use of propellant and reduces oropharynx and device deposition of the dose. It also demonstrates the dose reduction of the inhaled drug. Scintigraphic studies show that Respimat improves the delivery of drug into the lungs, as depicted in Figure 9.34. The graphs depicted in Figure 9.34 have been designed based on the data obtained from studies by Newman et al. (1998) and Pitcairn et al. (2008).

Respimat device development has been translated into the development of three successful marketed products, which are presented in Figure 9.35. On May 26, 2015, Boehringer Ingelheim announced the USFDA approval for once-daily Stiolto Respimat (tiotropium bromide and olodaterol) Inhalation Spray (see Figure 9.35d).

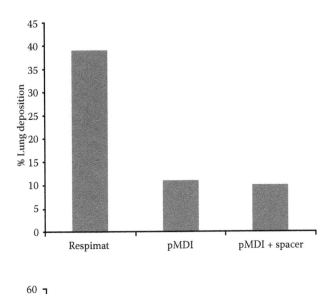

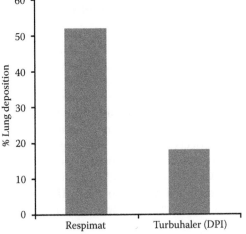

FIGURE 9.34   Comparison of lung deposition of Respimat versus pMDI and DPI.

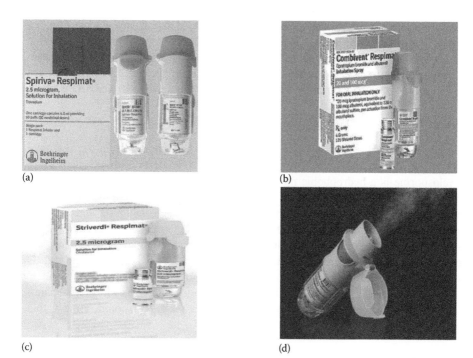

FIGURE 9.35 **(See colour insert.)** Marketed products that use Respimat as a delivery device for COPD treatment: (a) SPIRIVA – tiotropium bromide, (b) COMBIVENT – ipratropium bromide and albuterol sulfate spray, (c) STRIVERDI – olodaterol and (d) STIOLTO – tiotropium bromide and olodaterol. (From Combivent Respimat Inhalation 2015; SPIRIVA Respimat Inhaler 2015; STIOLTO Respimat 2015; Striverdi Respimat 2015.)

9.8 OPHTHALMIC

Eye drops are conventional dosage forms that account for 90% of currently accessible ophthalmic formulations. Despite excellent acceptance by the patients, one of the major problems encountered is rapid precorneal drug loss. To improve ocular drug bioavailability, there is a significant effort directed towards new drug delivery systems for ophthalmic administration.

An ophthalmic formulation could be a solution, suspension, ointment, gel, emulsion or an insert. A typical eye care product is sterile, is nearly isotonic, has some buffering capacity, contains preservatives (unless the active itself is bacteriostatic) and is packaged into a suitable tamperproof, multidose dispensing system. Nowadays, unit dose eye drops are preferred as they are preservative free. However, they are expensive.

During formulation development, the choice of raw materials must be based upon physiological comfort and product stability, and preferably with a proven track record of prior use in ophthalmic products. The ideal pH for an ophthalmic formulation is 7.4, equivalent to tear fluid. However, most drugs are chemically unstable at this pH. Therefore, a buffer, if included, must facilitate pH as close as possible to the physiological pH, while not causing chemical instability. Thickening agents such as methyl cellulose or hydroxypropyl methylcellulose may be added to prolong the contact time of formulation with the eye surface. Colouring agents are not recommended for ophthalmic products. Once the formulation profile is identified, the first step in product development is establishing its physical and chemical attributes such as appearance, viscosity, surface tension, osmolarity, zeta potential, globule size, assay, refractive index and pH.

9.8.1   Case Study – Cyclosporine Ophthalmic Emulsion

Topical cyclosporine (cyclosporine A; Restasis) has been approved for the treatment of chronic dry eye associated with inflammation (Henderer and Rapuano 2011). Cyclosporine is a hydrophobic molecule with a log P value of 8.2 (Mondon et al. 2011). Because of its poor solubility, it is very difficult to formulate it in the form of an aqueous solution. Immense literature is available on the solubility and bioavailability enhancement of cyclosporine. Researchers have developed nanoparticles, oily solutions, cationic emulsions and liposomes (Mosallaei et al. 2012) to enhance the penetrability of cyclosporine into the eye. Limited aqueous solubility of the drug substance is the most common rationale for developing an ophthalmic emulsion.

The present study deals with the formulation development of a cyclosporine nanoemulsion having the same qualitative (Q1) and quantitative (Q2) composition as that of Restasis.

9.8.2   Elements of Pharmaceutical Development

9.8.2.1   Quality Target Product Profile

The objective of the present work is to develop a product equivalent to Restasis. To begin with, as per the QbD concept, QTPP is defined keeping in view patient safety. The QTPP for this formulation is explained in Table 9.38.

9.8.2.2   Critical Quality Attributes

Potential drug product CQAs derived from the QTPP and prior knowledge are used to guide the product and process development. Relevant CQAs are identified by an iterative process of quality risk management and experimentation that assesses the extent to which their variation can have an impact on the quality of the drug product. Various CQAs were identified in order to meet the QTPP relevant to patient safety and efficacy. These CQAs are listed in Table 9.39.

9.8.2.3   Risk Assessment: Linking Material Attributes
and Process Parameters to Drug Product CQAs

9.8.2.3.1   Preparation of the Cyclosporine Nanoemulsion

The preparation scheme for formulation of the cyclosporine nanoemulsion is illustrated in Figure 9.36.

An Ishikawa fishbone diagram was built to identify a potential cause–effect relationship amongst the probable material attributes, processing parameters and CQAs of the formulation. Figure 9.37 pictorially presents the resultant fishbone diagram depicting the effect of key formulation attributes and process parameters to be considered as variables for the development of a cyclosporine emulsion.

TABLE 9.38

QTPP for the Cyclosporine Nanoemulsion

QTPP Element	Target	Justification
Dosage form	Emulsion	Emulsion helps in stabilisation and enhanced ophthalmic delivery of cyclosporine
Route of administration	Ophthalmic	Same as for Restasis
Dosage strength	0.05% w/w	Pharmaceutically equivalent product and ensures efficacy
Microbial quality	Similar to Restasis	Opthalmic formulation should be sterile and free from pathogenic organisms
Dosage administration	Similar to Restasis	To achieve the requisite pharmacodynamic action and efficacy
Indications and usage	Indicated to increase tear production in patients suffering from keratoconjunctivitis sicca	Same as for Restasis
PD target	Equivalent to Restasis	To ensure safety and efficacy
Assay	95%–105%	For therapeutic activity
Osmolarity	303.7 ± 22.9 mOsm/kg	Ophthalmic product should be isotonic with eye in order to prevent the irritation so that it ensures patient comfort and safety
Impurities	As per ICH guidelines	For stability and safety of formulation
Viscosity	20–50 cps	To maintain the residence of formulation in the eye for sufficient time
Sterility	Should pass sterility test	Sterility directly related to safety of patients
pH	6.0–7.0	To prevent irritation on application
Zeta potential	Less than -30 mV and more than $+30$ mV	To prevent the coalescence of globules
Stability	At least 24 months at RT	To maintain the therapeutic potential during storage

(*Continued*)

TABLE 9.38 (CONTINUED)
QTPP for the Cyclosporine Nanoemulsion

QTPP Element	Target	Justification
Container closure system	Sterile single-use vials. Each vial contains 0.4 mL fill in 0.9 mL LDPE vial	For compatibility and stability of product
Use in specific population	Similar to Restasis	Use with precaution in nursing mothers
Storage condition	Store at 15°C–25°C	To maintain efficacy and stability .

Further, the criticality analysis for selecting the high-risk factors was carried out by constructing the REM, showing the potential risk(s) associated with each of the formulation parameters (CFPs) and process parameters (CPPs) of the cyclosporine nanoemulsion on its potential CQAs by assigning low, medium and high values to each of the variables. The criticality of potential factors is depicted in Figure 9.38.

9.8.2.3.2 Assessment of Low-Risk Factors

Low levels were decided on the basis of prior process/formulation-related knowledge and available literature.

9.8.2.3.3 Assessment of Medium-Risk Factors

Initial empirical experimentations were performed for assessment of medium-risk factors. Phase volume ratio was a medium-risk parameter for assay, cumulative drug release and viscosity. Initial experimental study suggested that a phase volume ratio of 1:8 converts these medium-risk parameters to low-risk parameters. pH adjustment method II (pH adjusted before formation of primary emulsion) was found to form stable and monodisperse emulsion as compared to pH adjustment method I (pH adjustment of final emulsion). The remaining critical parameters as mentioned in Figure 9.38, which had a medium risk on various CQAs, were subjected to experimental studies. The ranges that are mentioned in Table 9.40 were taken for optimisation with the aim of changing the criticality of studied parameters to low level.

9.8.2.3.4 Assessment of High-Risk Factors

The factors associated with high criticality were finally subjected to factor screening studies employing central composite design.

9.8.3 SELECTION OF VARIABLES

The risk assessment and process development experiments described in Section 9.8.2.3 can lead to an understanding of the linkage and effect of process parameters and formulation attributes on product CQAs and also help identify the variables and their ranges within which consistent quality can be achieved.

TABLE 9.39

CQAs for the Cyclosporine Nanoemulsion

Quality Attribute		Target	Is This a CQA?	Justification
Physical attribute	Colour	Clear	No	Physical attributes of the formulation were not considered as critical, as these are not directly linked to the efficacy and safety
	Odour	No unpleasant odour		
	Appearance	Acceptable to end user		
Globule size		200–400 nm	Yes	To achieve efficacy and demonstrate bioequivalence with Restasis
Polydispersity index		0.1–0.3	Yes	To maintain stability of emulsion and ensure patient safety
Assay		95%–105%	Yes	To achieve therapeutic equivalence and efficacy
Cumulative drug release		More than 80% in 60 min	Yes	Equivalent to Restasis and critical for efficacy
Zeta potential		Less than −30 mV and more than +30 mV	Yes	Higher zeta potential prevents the coalescence of globules and thus enhances the stability
Osmolarity		303.7 ± 22.9 mOsm/kg	Yes	Ophthalmic product should be isotonic with tear fluid in order to prevent irritation
Viscosity		30–50 cps	Yes	Viscosity determines the residence time in the eye and should be similar to Restasis for efficacy
Sterility		Should pass sterility test	Yes	Sterility directly related to safety of patients
pH		6.5–8.0	Yes	To prevent irritation on application of the formulation
Impurities		As per ICH guidelines	Yes	To ensure the quality of the product

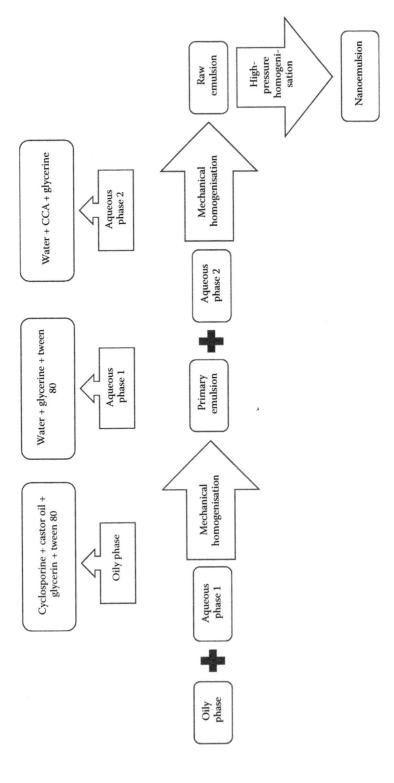

FIGURE 9.36 Process flow for formulation development.

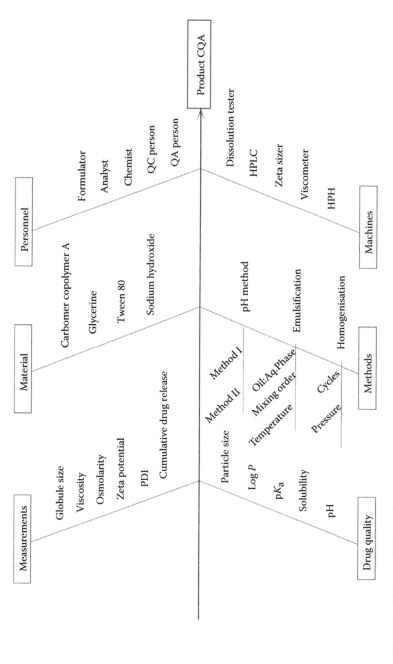

FIGURE 9.37 Ishikawa fishbone diagram for the cyclosporine emulsion.

Drug product CQAs	Critical formulation parameters		Critical process parameters		
	Phase volume ratio	pH Adjustment method	Temperature of primary emulsion	HPH pressure	Number of cycles
Globule size	High	Medium	Medium	High	High
Polydispersity index	Medium	Medium	Medium	Medium	Medium
Assay	Medium	Low	Low	Low	Medium
Cumulative drug release	Medium	Low	Low	Low	Medium
Zeta potential	High	Low	Low	High	High
Osmolarity	High	Medium	Medium	Low	Low
Viscosity	Medium	Low	Medium	Medium	Medium

FIGURE 9.38   Initial risk assessment matrix for the cyclosporine emulsion.

TABLE 9.40
Independent Formulation and Process Variables and Their Levels

Factors	Low Level (−)	High Level (+)
Temperature of primary emulsion (°C) X_1	40	70
Pressure of HPH (bar) X_2	300	600
Number of cycles X_3	5	10

9.8.3.1   Design of Experiments

The effect of formulation and processing variables on the CQA of the Cyclosporine emulsion was studied using DoE experimentation. The formulation and process variables likely to affect the CQA of the product were identified on the basis of product/process understanding and preliminary results. The process variables were temperature of primary emulsion (X_1), HPH pressure (X_2) and number of cycles (X_3). The ranges over which these variables were studied are mentioned in Table 9.40.

Central composite design (Design Expert version 9.0.4.1, State-Ease, Inc.) was used to study the effect of formulation and process variables (factors) with four centre points. Experimental matrix involves the preparation of 28 formulations. Responses evaluated against experiments were globule size (Y_1), zeta potential (Y_2) and osmolarity (Y_3). Results of the highly critical responses are explained in Section 9.8.4.

9.8.4   Nanoemulsion Characterisation

9.8.4.1   Globule Size

The globule size of the nanoemulsion is critically important as it relates to physical stability as well as clinical outcome. The smaller globule size of the nanoemulsion has lesser tendency to cream and coalesce and will be physically more stable in comparison to bigger size emulsion. The Z-average of nanoemulsion droplets was found to vary from 225.3 ± 0.84 to 400.5 ± 3.64 nm.

The factors significantly ($p < 0.05$) affecting the Z-average were phase volume ratio, HPH pressure and number of HPH pressure cycles. The influence of HPH pressure and its pressure cycles on nanoemulsion globule size is shown in Figure 9.39. Its value decreased with an increase in HPH pressure, and a reverse trend was observed at low process pressure. The higher the pressure, the more intense the impact of shear forces responsible for globule size reduction, and vice versa.

9.8.4.2 Zeta Potential

Zeta potential is one of the physical properties of the dispersed system describing its stability. The dispersed particles or droplets with large zeta potential (either positive or negative) would repel each other and prevent the occurrence of droplet fusion (coalescence). As per the literature, the optimum value of zeta potential should be greater than +30 mV and less than −30 mV to confer stability to a dispersed system. In this study, the negative charge is imparted by the carboxylate group of carbomer copolymer type A polymer. The zeta potential of the prepared cyclosporine nanoemulsion varied from −16.25 to −70.24 mV. None of the studied independent factors significantly ($p > 0.05$) affected the zeta potential except the HPH pressure (X_2) and number of pressure cycles (X_3), as shown in Figure 9.39b. Higher HPH pressure and a greater number of pressure cycles generate smaller droplets owing to the intense and repetitive effect of shear forces. Since the zeta potential is a surface phenomenon, decreasing the particle size yields a greater number of smaller droplets and thereby increases their zeta potential.

9.8.4.3 Osmolarity

The osmolarity of the human eye is 303.7 ± 22.9 mOsm/kg. Ophthalmic products should be isotonic in order to prevent irritation to the eye. Hypertonic or hypotonic solution/dispersion damages the epithelia of the eye. Glycerine is the primary ingredient of the nanoemulsion formulation responsible for its osmolarity. The osmolarity of the formulation in the absence of glycerine was 12–15 mOsm/kg. With the investigated formulation and process variables, the osmolarity values ranged from 262 to 295 mOsm/kg. There is no significant impact of the HPH pressure (X_2) and number of pressure cycles (X_3) on osmolarity, as shown in Figure 9.39c.

9.8.5 Validation of Optimised Formulations

One checkpoint formulation from the overlay plot (Figure 9.39d) was evaluated to validate the design. Based on numerical optimisation and target response, the software gave numerous solutions. From obtained solutions, one set of the data points from control design space having desirability close to 1 was selected and further experiments were carried out in replicates ($n = 5$). The predicted and experimental values of all the response variables and percentage error in prognosis are depicted in Table 9.41. Comparison of the observed response with that of the anticipated responses reveals that all response values between 95% prediction interval and hence the selected model were found to be significant ($p < 0.05$).

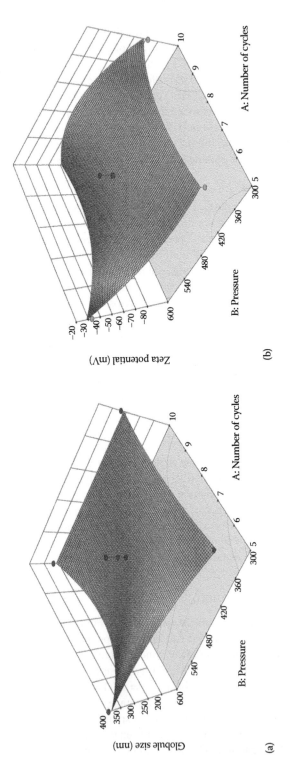

FIGURE 9.39 (See colour insert.) Effect of product and process variables on nanoemulsion: (a) globule size and (b) zeta potential.   (*Continued*)

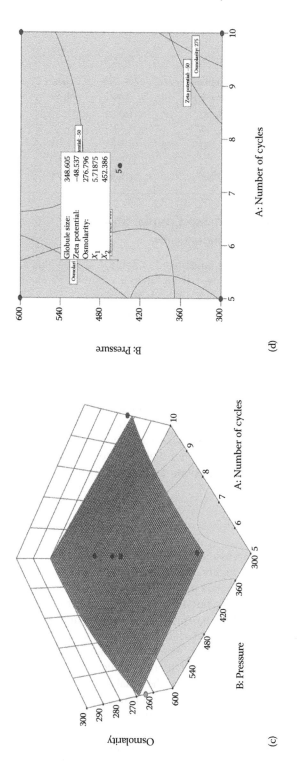

FIGURE 9.39 (CONTINUED) **(See colour insert.)** Effect of product and process variables on nanoemulsion: (c) osmolarity and (d) overlay plot depicting controlled design space.

TABLE 9.41

Validation of Predicted Response

Response	Mean	Standard Deviation	n	95% PI Low	Data Mean	95% PI High
Globule size (nm)	314.40	54.53	5	160.80	335.20	467.96
Zeta potential (mV)	−47.20	11.41	5	−79.33	−49.80	−15.06
Osmolarity (mOsm/kg)	279.60	10.34	5	252.80	284.50	306.39

9.8.6 UPDATED RISK ASSESSMENT AND CONTROL STRATEGY OF THE CYCLOSPORINE NANOEMULSION

The overall risk assessment data as mentioned in Table 9.42 show that all the high-risk variables can be converted to low-risk variables and thus QTPP can be achieved by keeping all the critical variables within the control range.

Table 9.43 depicts the control strategy for the development of cyclosporine nanosuspension.

The present study describes the implementation of the QbD approach in the systematic development of the optimised cyclosporine nanoemulsion employing simple, effectual and cost-effectual techniques.

The process and formulation variables affecting the CQAs of the cyclosporine nanoemulsion were studied. The investigated formulation- and process-related factors were found to have a significant impact ($p < 0.05$) on the globule size, zeta potential and osmolarity of the cyclosporine nanoemulsion. Criticality of studied factors was reassessed to convert it from a high level to a low level, and the final optimised product having the composition as given in Table 9.44 was developed using a control strategy.

9.9 NASAL

The nasal route is an easily accessible, convenient and reliable route with a porous endothelial membrane and a highly vascularised epithelium that provides rapid absorption into the systemic circulation, avoiding the hepatic first-pass elimination. In addition, intranasal delivery enables dose reduction, quicker onset of action and fewer side effects (Ugwoke et al. 2001). Nasal administration minimises the lag time associated with oral drug delivery and offers noninvasiveness, self-administration, patient comfort and patient compliance, which are the hurdles in intravenous drug therapy (Marttin et al. 1998). There are reports available stating that hydrophobic drugs are generally well absorbed from the nasal cavity with pharmacokinetic profiles, which are generally alike to those obtained after an intravenous injection with a bioavailability close to 100% (Hicke 2004). The absorption of hydrophilic drugs, on the other hand, can be increased by means of absorption enhancers. Drugs ranging from small chemicals to large macromolecules including peptide/protein

TABLE 9.42
Updated Risk Assessment of the Cyclosporine Nanoemulsion

	Critical Formulation Parameters		Critical Process Parameters		
Drug Product CQAs	**Phase Volume Ratio**	**pH Adjustment Method**	**Temperature of Primary Emulsion**	**HPH Pressure**	**Number of Cycles**
Globule size	Low Control on phase volume ratio	Low	Low	Low Control on pressure	Low Optimum number of cycles
Polydispersity index	Low	Low	Low	Low	Low
Assay	Low	Low	Low	Low	Low
Cumulative drug release	Low	Low	Low	Low	Low
Zeta potential	Low	Low	Low	Low Control on pressure	Low Optimum number of cycles
Osmolarity	Low Controlled addition of glycerin keeps osmolarity higher	Low	Low	Low	Low
Viscosity	Low	Low	Low	Low	Low

TABLE 9.43
Control Strategy for the Cyclosporine Nanoemulsion

Factor	Unit	Range Studied	Design Space	Purpose of Control
Temperature of primary emulsion X_1	°C	25–85	60–70	To ensure the development of a stable, efficacious cyclosporine ophthalmic nanoemulsion. It should have minimal globule size, isotonic to lacrimal fluids, and a zeta potential higher than +30 mV and less than −30 mV
Pressure of HPH X_2	bar	300–600	370–540	
Number of cycles X_3	n	3–15	5–7	

TABLE 9.44
Final Formulation and Process Parameters

Formulation Composition

S. No.	Ingredients	Final Composition
1	Cyclosporin	0.05% w/w
2	Carbomer copolymer A	0.2% w/w
3	Glycerin	2.2% w/w
4	Tween 80	1% w/w
5	Castor oil	1.25% w/w
6	Sodium hydroxide	q.s. to adjust the pH 7.0
7	Distilled water	q.s
Process Parameters		
1	Temperature of primary emulsion (°C) X_1	68
2	Pressure of HPH (bar) X_2	380
3	Number of cycles X_3	6

therapeutics, hormones and vaccines are being delivered through the nasal cavity. The nasal delivery seems to be a favourable way to circumvent the obstacles for the blood–brain barrier, allowing the direct drug delivery in the biophase of the central nervous system (CNS). During the past several decades, the feasibility of drug delivery via the nasal route has received increasing attention from pharmaceutical scientists and clinicians.

9.9.1 CASE STUDY – APOMORPHINE IN SITU GEL

Parkinsonian patients on long-term management with levodopa frequently develop sudden and unpredictable motor response fluctuations and other *off* phase symptoms such as depression, anxiety, pain, panic, delusions and dystonia. Apomorphine, a dopamine agonist, is effective in the management of parkinsonism, especially the *on–off* response fluctuations. It rapidly and reproducibly reverses the *off* periods (Chaudhuri and Clough 1998). Further advantages over other dopamine agonists such as bromocriptine and pergolide include fast onset of action, potent decrease in *off* periods and lower psychological complications. Unfortunately, as a result of the short duration of the therapeutic *on-phase* effect (90 min), high injection frequencies (up to 10–15 times a day) could be required (Sam et al. 1995). This justifies a need for a sustained-release apomorphine delivery system. Prolonged delivery of apomorphine by subcutaneous administration with portable pumps was attempted, but it caused local ulcerations (Stibe et al. 1988). Apomorphine is well absorbed after nasal administration of an aqueous solution. The time of onset and duration of action are equivalent to the subcutaneous injection (Van Laar et al. 1992). However, administration of an intranasal solution faces several problems such as drainage from the site of application and irregular deposition within the nasal cavity. Other disadvantages include rapid clearance owing to ciliary movement and instability of solutions. A sustained release nasal drug delivery system may be an ideal approach to solve this problem of short therapeutic half-life (Ugwoke et al. 2001). In this study, we report the formulation of mucoadhesive in situ gel dosage forms of apomorphine using sodium CMC.

Furthermore, the Box–Behnken design (BBD) was used to optimise the formulation as per ICH Q8(R2) pharmaceutical development guidelines.

9.9.2 Elements of Pharmaceutical Development

9.9.2.1 Quality Target Product Profile

To begin with, as per the QbD concept, QTPP was defined enlisting the summary of quality characteristics for the development of an apomorphine in situ intranasal gel (Table 9.45). Various CQAs were identified in order to meet various QTPPs that are related to patient safety.

9.9.2.2 Critical Quality Attributes

Potential drug product CQAs derived from the QTPP and prior knowledge are used to guide the product and process development. Relevant CQAs are identified by an iterative process of quality risk management and experimentation that assesses the extent to which their variation can have an impact on the quality of the drug product. Various CQAs were identified in order to meet various QTPPs related to patient safety. These CQAs are listed in Table 9.46.

9.9.2.3 Preparation of the Apomorphine In Situ Gel

Water-soluble additives such as sodium CMC (0.5%, 1.0% and 1.5% w/v), PEG 6000, sodium chloride and benzalkonium chloride (0.02% v/v) were dissolved in distilled water using SS316 tank with agitator. Sodium CMC was dispersed uniformly in this mixture followed by the addition of Pluronic F127 (20% w/v) at a temperature of 4°C–8°C. The solution was filled in an intranasal device and stored in a cool place.

9.9.2.4 Risk Assessment: Linking Material Attributes and Process Parameters to Drug Product CQAs

The prioritisation for selecting the high-risk factors was carried out by constructing the REM, showing the potential risk(s) associated with each of the formulation parameters (CFPs) and process parameters (CPPs) of the apomorphine in situ gel on its potential CQAs by assigning low, medium and high values to each of the studied factors (Figure 9.40). The factors associated with high criticality were finally subjected to factor screening studies employing the BBD.

TABLE 9.45

QTPP for the Apomorphine In Situ Gel

QTPP Element	Target	Justification
Dosage form	Solution	Ease of administration
Route of administration	Intranasal	For superior therapeutic action
Dosage strength	3.5 mg	To attain the desired therapeutic action
Microbial quality	<100 cfu	As per pharmacopoeial limits
Indications and usage	Indicated in depression, anxiety, local ulcerations, pain, panic, delusions and dystonia	To treat the *off* phase symptoms caused by levodopa administration
PK target safety/ efficacy	C_{max} and T_{max} should be achieved to get therapeutic action	Efficacy and safety
Impurities and degradation products	Total impurities should not be more than 0.1%	Stability of the formulation
Microbial load	As per pharmacopoeial limit	Product should be free from microorganisms
Assay	95%–105%	To attain therapeutic equivalence
Cumulative drug release	More than 80% release in 12 h	To achieve C_{max} in vivo
Consistency index	Maximum	To get mucoadhesion for prolonged release
Viscosity	Minimum	For easy dose dispensing from device
pH	6.0–7.0	For effective nasal delivery
Osmolarity	302 ± 22.9 mOsm/kg	To maintain tonicity
Stability	At least 24 months at RT	To maintain therapeutic potential during storage
Gelation temperature	32°C–37°C	To form an in situ gel, gelation temperature should be close to body temperature
Container closure system	Intranasal device	For application to nasal mucosa as nasal drops and for stability during shelf life
Storage condition	Store at 15°C–25°C	To maintain efficacy and stability

TABLE 9.46

CQAs for the Apomorphine In Situ Gel

Quality Attribute		Target	Is This a CQA?	Justification
Physical attribute	Colour	Clear	No	Physical attributes of the formulation were not considered as critical, as they were not directly linked to the efficacy and safety
	Odour	No unpleasant odour		
	Appearance	Acceptable to end user		
Assay		95%–105%	Yes	To achieve therapeutic equivalence
Cumulative drug release		More than 80% release in 12 h	Yes	To achieve C_{max} in vivo
Gelation temperature		32°C–37°C	Yes	To form an in situ gel, gelation temperature should be close to body temperature
Consistency index		Maximum	Yes	To get mucoadhesion for prolonged release
Sterility		Should pass sterility test	Yes	Sterility directly related to safety of patients
pH		6.0–7.0	Yes	For effective nasal delivery
Osmolarity		302 ± 22.9 mOsm/kg	Yes	To maintain tonicity
Viscosity		20–50 cps	Yes	Minimum viscosity for effective delivery from the device

9.9.2.4.1 Assessment of Low-Risk Factors

Low levels were decided on the basis of prior process/formulation-related knowledge and available literature.

9.9.2.4.2 Assessment of Medium-Risk Factors

Initial empirical experimentations were performed for assessment of formulation- and process-related medium-risk factors. Concentration of sodium chloride was a medium-risk factor for osmolarity, which was converted to low risk by fixing the range 1%–2% w/v. Similar studies revealed that sodium CMC when used in the concentration of 0.5%–1.5% w/v gave the desired viscosity. From preliminary screening,

Drug product CQAs	Critical formulation parameters				Critical process parameters		
	PEG 6000	Sodium CMC	Sodium chloride	Pluronic F 127	Stirring speed	Stirring time	Processing temperature
Gelation temperature	High	High	High	Medium	Medium	Low	Low
Consistency index	High	High	Low	Medium	Medium	Medium	Medium
Assay	Low	Low	Low	Low	Low	Low	Medium
Cumulative drug release	High	High	Low	Low	Low	Low	Medium
Osmolarity	Low	Low	Medium	Low	Medium	Low	Low
Viscosity	Low	Medium	Low	Medium	Medium	Medium	Medium

FIGURE 9.40 Initial risk assessment matrix for the apomorphine in situ gel.

TABLE 9.47
Optimised Parameters of Medium-Risk Factors

Concentration of Pluronic F127	Stirring Speed	Stirring Time	Processing Temperature
20% w/v	400–500 rpm	45–60 min	20°C–25°C

20% w/v of Pluronic F127 was selected for further study. Empirical experimental studies changed the criticality of all process-related medium-risk factors into low-risk factors. Table 9.47 depicts the final parameters.

9.9.2.4.3 Assessment of High-Risk Factors
The factors associated with high criticality were finally subjected to factor screening studies employing the BBD.

9.9.2.5 Selection of Variables
Risk assessment (Section 9.9.2.4) describes the effect of material attributes on product CQAs. It also helps in identifying the variables and their levels within which the desired quality can be achieved.

9.9.3 DESIGN OF EXPERIMENTS

A three-level BBD was used to study the effects of process parameters or independent variables on their responses. Sodium CMC, PEG 6000 and sodium chloride concentration were the three independent factors. The factors and their levels are mentioned in Table 9.48. All these factors had a high and a low level, whereas a midpoint value was by default provided by the software. In addition to these, five centre points were selected. A three-factor BBD with 17 randomised experimental runs (including five centre points) was used to optimise the formulation. The measured dependent responses were time for 100% cumulative drug release (Y_1), gelation temperature (Y_2) and consistency index (Y_3).

TABLE 9.48
Independent Formulation and Process Variables and Their Levels

Factors (% w/v)	Low Level (−1)	High Level (+1)
PEG 6000 (X_1)	0.5	1.5
Sodium CMC (X_2)	0.5	1.5
Sodium chloride (X_3)	1	2

9.9.4   In Situ Gel Characterisation

9.9.4.1   Time Required for 100% Drug Release

From the best-fit quadratic equation obtained, it was marked that the quadratic model was significant. PEG 6000 (factor X_1) decreases the time of 100% drug release in a concentration-dependent manner. This may apparently be attributed to increased hydrophilicity of Pluronic F127 in the presence of PEG 6000 leading to a decrease in viscosity. As expected, a concentration-dependent increase in mucoadhesive sodium CMC (factor X_2) prolonged drug release. The n for all experimental runs was found to be $0.49 < n < 0.92$, indicating that release from sodium CMC polymer followed non-Fickian diffusion modulating release by both diffusion and erosion mechanism. The 3D plot (Figure 9.41a) generated for the X_1X_2 interaction for PEG 6000 and sodium CMC illustrates that both in combination increase the time required for release.

9.9.4.2   Gelation Temperature

The interactive effect of PEG 6000 and sodium CMC had an antagonistic effect on the gelation temperature; that is, the gelation temperature decreased. This may be attributed to the fact that increased concentration of PEG 6000 and sodium CMC together causes the enhancement in the entanglement, which weakens the hydrogen bonding between the hydrophilic polymeric part of sodium CMC and Pluronic F127, causing dehydration at lower temperature. Furthermore, the presence of a tonicity agent (sodium chloride) along with sodium CMC lowers the gelation temperature by counteracting the increased gelation temperature that was achieved by the interaction between X_1 and X_2 (PEG 6000 and sodium CMC) (Figure 9.41b). Factor X_3 (tonicity agent concentration) has an antagonistic effect on the gelation temperature.

9.9.4.3   Consistency Index

Figure 9.41c illustrates that as the concentration of PEG 6000 increases, the consistency index decreases. A lower consistency index corresponds to lower viscosity. The factor X_2 (sodium CMC) was found to have a synergistic effect with PEG 6000, suggesting that the increased concentration of the mucoadhesive polymer increased the viscosity of the system without altering shear thinning behaviour. It was evident that as the concentration of PEG 6000 increases, the consistency index decreases.

9.9.5   Validation of Optimised Formulations

An overlay plot (Figure 9.41) gives the controlled design space. From the design space, one checkpoint formulation was taken up to validate the model.

Based on numerical optimisation and target response, the software gave numerous solutions. From the obtained solutions, one set of data points from the control design space was selected and further experiments were carried out ($n = 5$). The predicted and experimental values of all the response variables and percentage error are depicted in Table 9.49. Comparison of the observed response with that of the

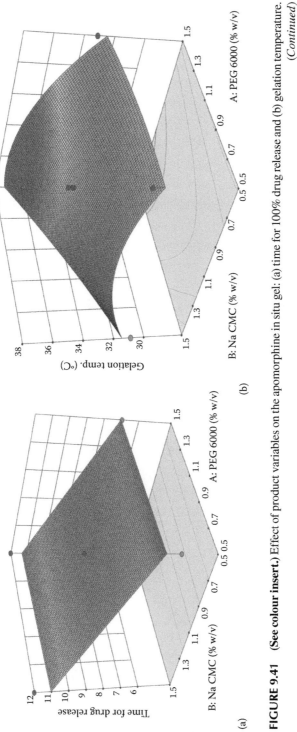

FIGURE 9.41 **(See colour insert.)** Effect of product variables on the apomorphine in situ gel: (a) time for 100% drug release and (b) gelation temperature.

(Continued)

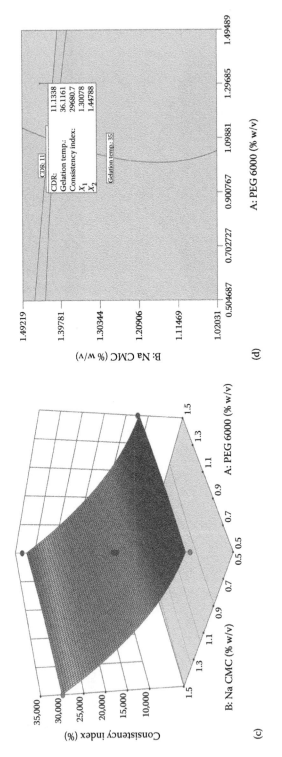

FIGURE 9.41 (CONTINUED) **(See colour insert.)** Effect of product variables on the apomorphine in situ gel: (c) consistency index and (d) overlay plot.

TABLE 9.49

Validation of Predicted Response

Response	Mean	Observed	Standard Deviation	95% PI Low	Data Mean	95% PI High
Time for 100% drug release (h) (Y_1)	11.56	12.1	0.16	10.5	6.87	14.5
Gelation temperature (°C) (Y_2)	36.60	36.84	0.78	34.26	36.84	38.93
Consistency index (Y_3)	31,669.64	33,324	1045.83	28,567.62	33,324	34,771.65

anticipated responses reveals that all response values lie between 95% prediction range and hence the selected model is significant.

9.9.6 UPDATED RISK ASSESSMENT AND CONTROL STRATEGY OF THE APOMORPHINE IN SITU GEL

An overall risk assessment of the drug product identified based on optimisation of formulation and process parameters is depicted in Table 9.50. It shows that all the variables exhibit low criticality and the QTPP can be achieved by keeping all the critical variables within the control range (Table 9.50). The control strategy for the apomorphine in situ gel is depicted in Table 9.51.

The process and formulation variables affecting the CQAs of the apomorphine in situ nasal gel were studied. The investigated formulation- and process-related factors were found to have a significant impact ($p < 0.05$) on the time required for 100% drug release, gelation temperature and consistency index of the apomorphine in situ nasal gel.

Final composition of the apomorphine in situ gel, after reassessment of risk and application of control strategy, is depicted in Table 9.52.

TABLE 9.50

Updated Risk Assessment of the Apomorphine In Situ Gel

Drug Product CQAs	Critical Formulation Parameters				Critical Process Parameters		
	PEG 6000	Sodium CMC	Sodium Chloride	Pluronic F127	Stirring Speed	Stirring Time	Processing Temperature
Gelation temperature	Low Control on addition of PEG 6000 in the range of 0.95%–1.5% w/v	Low Control on addition of 1.38%–1.5% w/w	Low Controlled addition in the range of 1.2%–1.7% w/v	Low	Low	Low	Low
Consistency index	Low Control on addition of PEG 6000 in the range of 0.95%–1.5% w/v	Low Na CMC acts by synergism with PEG. Hence, concentration in the range of 1.38%–1.5% w/v could be used.	Low	Low	Low	Low	Low
Assay	Low	Low	Low	Low	Low	Low	Low
Time for 100% drug release	Low PEG decreases time for 100% drug release, which can be compensated by addition of Na CMC	Low Control on addition of 1.38%–1.5% w/w	Low	Low	Low	Low	Low
Osmolarity	Low	Low	Low	Low	Low	Low	Low
Viscosity	Low	Low	Low	Low	Low	Low	Low

TABLE 9.51

Control Strategy for the Apomorphine In Situ Gel

Factor	Range Studied	Design Space	Purpose of Control
PEG 6000 (% w/v) (X_1)	0.5–1.5	0.95–1.5	To obtain a product having a higher safety, efficacy and stability
Sodium CMC (% w/v) (X_2)	0.5–1.5	1.39–1.5	
Sodium chloride (% w/v) (X_3)	1–2	1.2–1.7	

TABLE 9.52

Final Formulation and Process Parameters

Formulation Composition	
Ingredients	**Final Composition**
Apomorphine	3.5 mg/actuation
Sodium CMC	1.39%–1.5% w/v
PEG 6000	0.95%–1.5% w/v
Pluronic F127	20% w/v
Sodium chloride	1.2%–1.7% w/v
Benzalkonium chloride	0.02% w/v
Water	q.s.
Process Parameters	
Stirring time (X_2)	45–60 min
Processing temperature (X_3)	15°C–25°C
Stirring Speed (X_1)	400–500 rpm

REFERENCES

Bash, K. L. 2000. Review of vaginal pessaries. *Obstetrical and Gynaecological Survey* 55(7): 455–60.

Chaudhuri, K. R., Clough, C. 1998. Subcutaneous apomorphine in Parkinson's disease: Effective yet underused. *Br Med J* 316:641.

CMC-IM. 2008. Pharmaceutical development case study: 'Ace tablets'. https://www.google.com .in/?gfe_rd=cr&ei=JMukVcvKIcqM8QeIm6OgAw#q=Pharmaceutical+Development +Case+Study:+%E2%80%9CACE+Tablets%E2%80%9D (accessed June 19, 2015).

Combivent Respimat Inhalation. 2015. http://www.webmd.com/drugs/2/drug-161259 -6247/combivent-respimat-inhl/ipratropium-albuterolsalbutamolinhaler-oral/details (accessed June 12, 2015).

Dalby, R., Spallek, M., Voshaar, T. 2004. A review of the development of Respimat Soft Mist Inhaler. *Int J Pharm* 283(1–2):1–9.

Desai, P., Date, A., Patravale, V. 2011. Overcoming poor oral bioavailability using nanoparticle formulations – Opportunities and limitations. *Drug Discov Today Technol* 9(2):e-87–95.

Dhiman, S., Singh, T. J., Rehni, A. K. 2011. Transdermal patches: A recent approach to new drug delivery system. *Int Jo Pharm Pharm Sci* 3(5):26–34.

Drug Bank: Tenofovir (DB00300). 2015. http://www.drugbank.ca/drugs/DB00300 (accessed May 31, 2015).

Ellis, W. Y., Lim, P., Maniar, M. 2010. Methods for the formulation and manufacture of artesunic acid for injection. Patent No.: US 7678828 B2.

Gattefosse. 2015. http://www.gattefosse.com/en/applications/suppocire-ap-pellets.html?admin istration-route,rectal-vaginal (accessed June 19, 2015).

Henderer, J. D., Rapuano, C. J. 2011. Ocular pharmacology. In *Goodman & Gilman's The Pharmacological Basis of Therapeutics*, 12th ed., L. L. Brunton, B. A. Chabner, B. C. Knollmann, eds. McGraw Hill: New York.

Hicke, A. J. 2004. *Pharmaceutical Inhalation Aerosol Technology*. Marcel Dekker, Inc: New York.

Hussain, A., Ahsan, F. 2005. The vagina as a route for systemic drug delivery. *J Control Rel* 103(2):301–13.

Joshi, M., Pathak, S., Sharma, S., Patravale, V. 2008. Solid microemulsion preconcentrate (NanOsorb) of artemether for effective treatment of malaria. *Int J Pharm* 362:172–8.

Kurz, A., Farlow, M., Leferve, G. 2009. Pharmacokinetics of a novel transdermal rivastigmine patch for the treatment of Alzheimer's disease: A review. *Int J Clin Pract* 63(5):799–805.

Lambert, W. J. 2010. Considerations in developing a target product profile for parenteral pharmaceutical products. *AAPS PharmSciTech* 11:1476–81.

Laxmi, P. J., Deepthi, B., Rama, R. N. 2012. Rectal drug delivery: A promising route for enhancing drug absorption. *Asian J Res Pharm* 2(4):143–9.

Marttin, E., Nicolaas, G. M., Schipper, J. et al. 1998. Nasal mucociliary clearance as a factor in nasal drug delivery. *Adv Drug Del Rev* 29:13–38.

Mondon, K., Zeisser-Labouebe, M., Gurny, R., Moller, M. 2011. Novel cyclosporin A formulations using MPEG-hexyl-substituted polylactide micelles: A suitability study. *Eur J Pharm Biopharm* 77(1):56–65.

Mosallaei, N., Banaee, T., Farzadnia, M. et al. 2012. Safety evaluation of nanoliposomes containing cyclosporine A after ocular administration. *Curr Eye Res* 37(6):453–6.

Newman, S., Brown, J., Steed, K. et al. 1998. Lung deposition of fenoterol and flunisolide delivered using a novel device for inhaled medicines: Comparison of RESPIMAT with conventional metered-dose inhalers with and without spacer devices. *Chest J* 113:957–63.

Patravale, V. B., Date, A. A., Kulkarni, R. M. 2004. Nanosuspensions: A promising drug delivery strategy. *J Pharm Pharmacol* 56:827–40.

Paudel, S. K., Milewski, M., Swadley, C. L. et al. 2011. Challenges and opportunities in dermal/transdermal delivery. *Ther Deliv* 1(1):109–31.

Pitcairn, G., Reader, S., Pavia, D., Newman, S. 2008. Deposition of corticosteroid aerosol in the human lung by Respimat® Soft Mist™ Inhaler compared to deposition by metered dose inhaler or by Turbuhaler® dry powder inhaler. *J Aerosol Med* 18:264–272.

Prausnitz, M. R., Langer, R. 2009. Transdermal drug delivery. *Nat Biotechnol* 26(11):1261–8.

Rabinow, B. E. 2004. Nanosuspensions in drug delivery. *Nat Rev Drug Discov* 3:785–96.

Roumen, F. 2008. Review of the combined contraceptive vaginal ring, NuvaRing®. *Ther Clin Risk Manag* 4(2):441–51.

Sam, E., Jeanjean, A. P., Maloteaux, J. M., Verbeke, N. 1995. Apomorphine pharmacokinetics after intranasal and subcutaneous application. *Eur J Drug Metab Pharmacokinet* 20:27–33.

Scholer, N., Krause, K., Kayser, O. et al. 2001. Atovaquone nanosuspensions show excellent therapeutic effect in a new murine model of reactivated toxoplasmosis. *Antimicrob Agents Chemother* 45:1771–9.

Sharma, R., Sanodiya, B. S., Bagrodia, D. 2012. Efficacy and potential of lactic acid bacteria modulating human health. *Int J Pharm Bio Sci* 3(4):935–48.

SPIRIVA® Respimat Inhaler. 2015. http://www.boehringer-ingelheim.com/products/prescription_medicines/respiratory/copd/spiriva/respimat (accessed June 12, 2015).

Stibe, C. M., Kempster, P. A., Lees, A. J., Stern, G. M. 1988. Subcutaneous apomorphine in parkinsonian on-off oscillations. *Lancet* 1:403–6.

STIOLTO Respimat. 2015. http://respiratorytherapycave.blogspot.in/2015/07/stiolto-new-option-for-copd.html (accessed June 12, 2015).

Striverdi Respimat. 2015. https://www.medicines.org.uk/emc/medicine/28992 (accessed June 12, 2015).

Ugwoke, M. I., Verbek, N., Kinget, R. 2001. The biopharmaceutical aspects of nasal mucoadhesion drug delivery. *J Pharm Pharmacol* 59:3–22.

Van Laar, T., Jansen, E. N. H., Essink, A. W. G., Neef, C. 1992. Intranasal apomorphine in parkinsonian on-off fluctuations. *Arch Neurol* 49:482–4.

World Health Organization (WHO). 2010. *Guidelines for the Treatment of Malaria*, 2nd ed. WHO: Geneva, Switzerland.

10 Packaging of Pharmaceuticals

Manasi M. Chogale and Maharukh T. Rustomjee

CONTENTS

10.1 PHARMACEUTICAL PACKAGING: INTRODUCTION AND GENERAL CONSIDERATIONS

Development of a pharmaceutical pack, which was earlier considered as the last step of product development, is fast becoming a core element of the pharmaceutical product development process. Pharmaceutical packaging development is currently being considered not only as an expedient for the development of an efficient product but also as an indispensible component of the product marketing landscape, which assists the manufacturers to singularise their products from those of their rivals. A pharmaceutical packaging system generally implies a combination of a suitable container along with a well-fitting closure enclosed in an outer cover, suitably complimented with an appropriate label. Pharmaceutical companies are now increasingly looking at packaging and labelling as a medium to advocate and protect their products, thereby improving the patient acceptability and regulatory compliance of the products. Hence, the need for an impeccable packaging system is at an all-time high.

Stability data of a drug product in the final pack over the proposed shelf life is a prerequisite to obtain approval from regulatory agencies to market any drug product. Although an appropriate formulation design is of paramount importance in the development of a stable drug product, it is crucial to note that an efficient formulation strategy itself may not always be adequate to provide the stability of the product during storage. The various drug molecules incorporated into the drug product possessing diverse properties are susceptible to interactions with surrounding environmental conditions. Hence, such drug products may demand protection against ambient degrading factors, such as humidity, photo exposure and oxygen, which can be ensured by employing a suitable pack. Hence, an appropriate pack profile coupled with a proper formulation design alone can impart stability to the product over its shelf life. The screening process to identify the appropriate packaging system for a product to ensure the required protection is therefore of chief significance in product development (Dean et al. 2000).

Compliance with regulatory and compendial requirements, providing sufficient protection to products against degrading factors, performing the desired function, compatibility with the drug products, its safety to the user and assisting in the shipping and transport of the drug product are some of the fundamental requirements of a successful packing material. Moreover, the packaging should be dignified for marketing, easy to use and economical for the benefit of patients and product manufacturers.

Based on its function, packing systems may be classified as 'primary' packaging, 'secondary' packaging and 'tertiary' packaging systems. Primary packaging components comes in direct contact with the dosage form, for example, bottles, liners, desiccant contained in bottles with dosage forms, blister films and so on. Secondary packaging components enclose the primary pack containing the drug product, for example, cartons and overwraps for blister cards. They serve as a medium for labels and leaflets and are of prime importance from the marketing perspective of the product. Tertiary packing materials comprising of larger cartons and pallets used for shipping and transport are responsible for ensuring safe and damage-free transport of the products. Packing materials can also be classified in other ways as illustrated in Figure 10.1.

Pack requirements differ for different types of products depending on their susceptibility to degradation and their intended applications. Concerns for drug–container interaction are the highest for liquid dosage forms, for example, injectables, ophthalmics and liquid oral products, and lowest for topical and solid oral dosage forms. Therefore, a pack found appropriate for one type of product cannot be assumed to be suitable for another type of product (Bauer 2009).

An efficient container closure system is expected to provide protection to the products from moisture, light, oxygen or other gases, microbial contamination and mechanical stress, depending on the susceptibility of the product to degradation. Protection from microbial contamination can be accomplished by employing well-sealed containers and by implementing stringent control on the microbial load of materials, manufacturing and packaging processes. Drug products can be protected from physical stress by the use of packs with satisfactory durability, limited

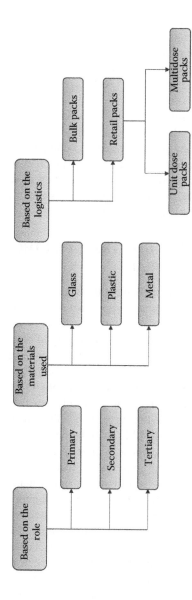

FIGURE 10.1 Classification of packaging materials.

headspace and use of fillers (e.g. cotton), which prevents rattling of products during shipping and handling. Containers are generally incapable of offering protection from thermal stress. Thermolabile products are typically handled in refrigeration during storage, shipping and end use. Hermetically sealed containers are very commonly used for the protection of the contained products from external degrading factors, whereas child-resistant packs are used to protect the paediatric population from accidental usage. Hence, protection against moisture, light and oxygen are predominantly dependent on the primary pack selected. The pharmaceutical pack also bears the identification and correct, up-to-date information of the product. It is also responsible for providing proof for tamper evidence and protection against counterfeiting of the product (Chen 2009).

Primary pack is generally selected based on the knowledge of the physicochemical properties of the drug product, its susceptibility to degradation, the protective properties of available packing materials and its compatibility with each other to enable the desired ease of use by the target patient population. A rational selection of packs will minimise the risk of failure in stability studies, increase the efficiency of product development and reduce the cost for the patients and also the pharmaceutical industry by reducing the risk of commercial recalls.

10.2 PERMEABILITY CONSIDERATIONS FOR THE SELECTION OF A PACKING MATERIAL

As mentioned in Section 10.1, the fundamental function of a primary pack is to provide maximum protection to the contained product over its intended shelf life, and thus the permeability of the primary packing material becomes the most crucial factor to be considered along with the product's sensitivity to surrounding permeants.

Permeability of a packing material to water vapour and oxygen are the most vital elements to be deliberated and are reported as moisture/water vapour transmission rate (MVTR/WVTR) and oxygen transmission rate (OTR), respectively (Chen 2009). Table 10.1 lists the MVTR and OTR values of a few commonly used packaging materials (Kamal and Jinnah 1984; Weeren and Gibboni 2002).

On the basis of the data in Table 10.1, it can be noted that glass and aluminium foil are the only two materials completely impermeable to moisture and oxygen and hence are capable of rendering maximal protection to the contained product. However, these materials are being substituted by newer materials bearing MVTR and OTR values almost comparable to glass and aluminium foil.

Packaging materials should therefore be screened based on the specific requirement of each individual product. Since it is observed that materials possessing low MVTR values do not necessarily have low OTR values, a combination of materials can be employed based on the specific requirements of the drug product.

Although aluminium foil may be regarded as the ideal barrier to gases, thinner foils (thickness, 25 μm) may possess 'pinholes', which may be responsible for increasing the permeability of the foil. Pinholes may be the result of either the presence of organic impurities in the molten aluminium or the stress of the rolling process endured by the metal during foil production. The number of pinholes per specific area of the foil is inversely related to the thickness of the foil. This factor is

TABLE 10.1

Moisture and Oxygen Transmission Rates of Some Packaging Materials

Materials	MVTR (g·mil[a]/100 in²/day, at 100°F, 90% RH)	OTR (cc·mil/100 in²/day at 77°F)
Cold form aluminium foil	0.00	0.00
Glass	0.00	0.00
Polychlorotrifluoroethylene (PCTFE)	0.016	7.00
Polyvinyl dichloride	0.1–0.2	0.15–0.90
High-density polyethylene (HDPE)	0.3–0.4	139–150
Polypropylene (PE)	0.69–1.0	182
Low-density polyethylene (LDPE)	1.0–1.5	463–500
Polyethylene terephthalate (PET)	1.2–2.0	3–5
Polyvinyl chloride (PVC)	0.9–5.1	5–20
Ethylene vinyl alcohol (EVOH)	1.4–5.4	0.05–0.90
Polystyrene (PS)	7–10	350–400
Nylon	16–20	1.0

[a] The unit 'mil' is equal to the thousandth of an inch.

critical in the selection and designing of foil materials as a barrier. Aluminium foils with thickness above 35–40 μm are considered to be almost pinhole free and hence are commonly used for the packaging of specialised dosage forms such as effervescent granules or effervescent tablets (Dean et al. 2000).

Thus, the selection of a packaging material/system for a product is based on the following:

- Type of dosage form; for example, solid orals are packed in high-density polyethylene (HDPE) bottles, semisolids in collapsible/noncollapsible tubes, liquid orals in glass or low-density polyethylene bottles, sterile parenterals in ampoules, vials or prefilled injectors and so on.
- Susceptibility of the drug product; for example, products highly sensitive to moisture/oxygen require a packing material with low MVTR and OTR values.

- Regulatory aspects; in the United States, solid dosage forms are required to be supplied either in HDPE pack/blister pack with child-resistant features only.
- Intellectual property aspect; this issue is of major consideration while selecting the pack and its material especially when the innovator has filed a patent or design for a particular type of packaging.

10.3 DEVELOPMENT OF A PHARMACEUTICAL PACK

The first step in the development of a pharmaceutical pack is establishing the type of development intended. The type of development may either be a new pack for a new product, a major pack change (new pack for an existing or established product), a major pack change associated with reformulation of a product or the pack either acting as an administrative aid or being used in a device.

The principal stage of pack development is a stability test in the final packaging system in which the product is to be marketed. However, adequate care needs to be taken when performing stability testing at accelerated conditions; since at such extremes of temperature and humidity the pack properties may undergo some changes leading to incorrect stability results. For example, in conditions where the samples are prepared at 18°C and then stored for a prolonged period at higher or lower temperatures, differential expansion or contraction between pack and closure may cause loosening or tightening of the closure, thereby altering properties of the contained product (Dean et al. 2000). It is also crucial to note that a regular stability test does not encompass the factors likely to be associated with bulk warehousing, handling and shipping. Hence, the routine stability studies should be supplemented with other tests that cover these aspects to ensure that stability data are relevant at all conditions. Laboratory-simulated tests or actual 'field trials' may be performed to suffice the requirement. Another significant aspect to be considered is that the product pack may have a limited shelf life once the pack is opened and is in use. This modification is normally found in products such as powders for reconstitution or ophthalmic/otic drops for multiple use. In such cases, it makes sense to establish two shelf-life periods: for the unopened pack (more than 3 years) and for the opened pack ('to be used within 4 weeks' or 'to be discarded after 1 month').

Occasionally, there may be differences in the product pack used for stability testing and the pack eventually finalised for the product. This is contrary to the ideal situation that dictates performing the stability study in the final 'to be marketed' container itself. This may occur in scenarios where financial limitations do not allow the installation of the fabrication moulds for a novel pack or a characteristic tooling is required for the production of the pack wherein the situation is constrained by the costly tooling or where a pack either acts as or incorporates a delivery system involving complicated and expensive tooling (Bauer 2009).

The materials used for a primary or secondary pack should be nonmigratory, nontoxic and nonirritant, and this information is corroborated by performing a thorough analysis of the constituents of the materials in terms of their toxicological data and availability of suitable and sensitive analytical methods to detect extraction and migration of these constituents into the contained product.

It is important to record and maintain systematic details pertinent to the fabrication and processing of the product pack. Details relevant to the ambient conditions, machine types, speeds and settings, fill volumes and so on should also be documented and retained for further reference.

To summarise, the development of a pharmaceutical pack is a scientific art incorporated as an integral part of product development. The primary pack is screened purely on the basis of the physicochemical properties of the drug product and a thorough investigation of the packaging materials available that may suit the drug product along with their compatibility under normal and accelerated conditions. Selection of the secondary and tertiary packing material is more or less driven by the cost, transport conditions and marketing requirements. The basic objectives of secondary and tertiary packs are to protect the product during transit and serve as a marketing tool for the product.

10.4 REGULATORY CONSIDERATIONS FOR DEVELOPMENT OF A PHARMACEUTICAL PACK

The Food and Drug Administration (FDA) guideline on 'Container Closure Systems for Packaging Human Drugs and Biologics' is an important document to provide guidance on general principles for submitting information on packaging materials for human drugs and biologics. Current good manufacturing practices for control of drug product containers and closures are included in 21 CFR parts 210 and 211. The FDA requirement for tamper-resistant closures is included in 21 CFR 211.132 and the Consumer Product Safety Commission requirements for child-resistant closures are included in 16 CFR 1700 (USFDA 1999).

The United States Pharmacopoeial Convention has established requirements for containers that are described in many of the drug product monographs in the 'General Notices and Requirements (Preservation, Packaging, Storage and Labelling)' section of the USP/NF for capsules and tablets. These requirements relate to the design characteristics of the container (e.g. tight, well closed or light resistant), and for injectables, materials of construction to be used are also addressed.

The Centre for Drug Evaluation and Research (CDER) and Centre for Biologics Evaluation and Research (CBER) approve a container closure system to be used in the packaging of a human drug or biologic as part of the application (new drug application or abbreviated new drug application) for the drug or the biologic. Since the specifications of a container closure system change from one product to another, each application should contain sufficient information to show that each proposed container closure system and its components are suitable for its intended use. The type and extent of the information to be provided in an application will depend on the dosage form and the route of administration. More detailed information usually should be provided for a liquid-based dosage form than for a powder or solid, since the former is more vulnerable to interaction with the packaging components. The guideline classifies dosage forms on the basis of their likelihood for interaction with the packaging components and the degree of concern associated with each dosage form (Table 10.2).

According to this guideline, the data to be submitted to the regulatory agency for the approval of a packaging system for a dosage form is categorised under four

TABLE 10.2
Packaging Concerns for Common Classes of Drug Products

Degree of Concern with Route of Administration	Likelihood of Packaging Component–Dosage Form Interaction		
	High	Medium	Low
Highest	Inhalation aerosols and solutions, injectables	Sterile powders and powders for injection, inhalation powders	–
High	Ophthalmic solutions, transdermal ointments/ patches, nasal sprays	–	–
Low	Topical and oral solutions	Topical and oral powders	Oral tablets and capsules

sections, namely, *description, suitability, quality control* and *stability*. The *suitability* of a container closure system is associated with four parameters: *protection, safety, compatibility* and *performance* of the packaging system.

Specialised dosage forms such as inhalations and ophthalmics require additional information apart from those mentioned in Table 10.3. For example, the submission for injectables and ophthalmics mandates the mention of sterilisation procedures, sterility and seal integrity testing for containers along with an extensive data on safety testing of the elastomeric closures and glass containers.

The European Medicines Agency (EMEA) has a similar guideline on 'Plastic Immediate Packaging Materials', which covers the specific requirements for the approval of plastics to be used as an immediate (primary) packaging material (EMEA 2009).

10.5   LABELS AND LEAFLETS

The term *label* can imply label on the immediate (primary) pack, the label on the secondary pack (carton) or the outer label on the case or pallet (tertiary pack). The function of the label and leaflet is to convey information to the pharmacist/wholesaler/ manufacturer, to control the product in terms of its distribution and medical aspects and to curtail the risk of product liability claims. The key EU legislations for labels and leaflets are as follows:

- Directive EC 65/65 provides the particulars required for labels in outline.
- Directive 75/319 provides the particulars required for leaflets in outline.
- Directive 89/341 makes a leaflet compulsory if all the details required are not displayed on the label.
- Directive 92/27 consolidates and provides greater detail on the requirements for labels and leaflets.

  The directive lists the details to be mentioned on the immediate packaging material (i.e. name, common name, strength of the product, etc.). For

TABLE 10.3

Data for Submission as per FDA Guideline on 'Container Closure Systems for Packaging Human Drugs and Biologics'

Parameter	Description
Description	Overall general description of the container closure system, and for each packaging component: • Name, product code, manufacturer, physical description • Materials of construction (for each: name, manufacturer, product code) • Description of any additional treatments or preparations
Suitability	*Protection* (by each component and/or the container closure system, as appropriate) • Light exposure, reactive gases (e.g. oxygen), moisture permeation • Solvent loss or leakage • Microbial contamination (sterility/container integrity, increased bioburden, microbial limits) • Filth *Safety* (for each material of construction, as appropriate) • Chemical composition of all plastics, elastomers, adhesives, etc. • Extractables, as appropriate for the material • Extraction/toxicological evaluation studies, as appropriate • Appropriate USP testing • Appropriate reference to the indirect food additive regulations (21 CFR 174-186) • Other studies as appropriate *Compatibility* (for each component and/or the packaging system, as appropriate) • Component/dosage form interaction, USP methods are typically accepted • May also be addressed in postapproval stability studies *Performance* (for the assembled packaging system) • Functionality and/or drug delivery, as appropriate
Quality control	For each packaging component received by the applicant: • Applicant's tests and acceptance criteria • Dimensional (drawing) and performance criteria • Method to monitor consistency in composition, as appropriate For each packaging component provided by the supplier: • Manufacturer's acceptance criteria for release, as appropriate • Brief description of the manufacturing process
Stability	Stability studies to be carried out in the proposed packaging system

primary packaging systems that are too small to allow displaying all the labelling requirements (e.g. blister packs), it is possible to reduce the number of details to be displayed. But such packs must at least display the name of the product, method of administration, expiry date, batch number and contents by weight, volume or unit (Dean et al. 2000).

• Directive 92/73 makes similar provisions for homeopathic drugs.

10.5.1 LEAFLETS

All products must contain a patient information leaflet (PIL) unless all the information can be conveyed on the outer packaging label. The PIL is based on the Summary of Product Characteristics, the only difference being that the language in a PIL must be suitably simplified so that it is easily comprehensible and readable to the patient population. The content of the leaflet and order of presentation are defined in Directive 92/27 and depicted in the following (Dean et al. 2000; MHRA 2015):

- Identification of product
- Dosage, method and route of administration, frequency of administration, duration of treatment where limited, action to be taken in the case of an overdose or lack of dosing and risk of withdrawal effects where possible
- Name of product, statement of active ingredients, pharmaceutical form, pharmacotherapeutic group, name and address of authorisation holder
- Undesirable effects
- Therapeutic indications
- Effects that can occur under normal use of the product and action to be taken
- Information needed for taking the product
- Expiry date
- Contraindications, precautions for use, interactions, special warnings, including use in pregnancy, elderly, effect on ability to drive vehicles and details of any excipients that may be important for the safe and effective use of the product
- Warning against use of the product after the date, appropriate storage precautions and warning against visible signs of deterioration
- Instructions for use
- Date on which package leaflet was last revised

10.6 MATERIALS FOR FABRICATION OF PACKAGING SYSTEMS

As illustrated in Figure 10.1, pharmaceutical packaging systems can be classified based on the material used for their fabrication. The selection of a packaging material is based on numerous factors such as compatibility with the contained drug product, special protection requirements of the product, ease of handling and patient convenience. Glass, plastics and metals are some of the commonly used materials for the fabrication of a pharmaceutical pack and are discussed in Sections 10.6.1 through 10.6.3.

10.6.1 GLASS AS A PACKAGING MATERIAL

Glass has served the pharmaceutical and cosmetic industries as an effective packing material for many decades. With the advent of plastics, glass was virtually destined as an archaic material and a rapid decline in its use was anticipated. However, it is important to note that glass is the safest option as a packaging material owing to the absence of any contaminants that are normally found in plastics and also its ability to offer maximum protection to the contained material as is evident from its MVTR and OTR values mentioned in Table 10.1. Type I (neutral, borosilicate glass) and Type II (surface-treated soda lime glass) are mentioned in pharmacopoeias as being commonly used for fabrication of packaging materials. Surface lubricity of glass may be improved by various surface treatments such as silicone coatings. These treatments may help in reducing damage by impact or giving additional decorative effects. More significantly, such surface treatment ensures uniformity of dosage, specifically in case of dosage forms such as suspensions.

The property of glass that drives its extensive use is its high degree of chemical inertness. Neutral glass is far more resistant to chemical attack. The use of neutral glass becomes mandatory in certain instances; ordinary insulin has been referred to as a prime example of application of neutral glass where pH control is crucial. Glass is completely impermeable to all gases, solutions and solvents. Clear glass also has a high degree of transparency. However, when fabricated in either amber or certain green shades, glass offers significant ultraviolet resistance to the contained product. It is lighter than most metals. Its rigidity and ability to endure stacking are considerable, which makes it capable of withstanding internal as well as external pressures. Heat resistance and a high melting point make it suitable for both moist and dry sterilisation, resulting in extensive applications in the packaging of parenterals. Borosilicate glass is particularly resistant to thermal shock. Soda glass is quite suitable for hot filling operations, provided that due attention is paid to design, pack size and the temperature difference.

Glass is therefore a rigid, inert pack that can withstand the rigors of handling, filling, closuring and use, and if required for reuse, it can also be easily recleaned. Its weight and stability when correctly designed enable extremely high-speed processing. Although soda glass is less expensive compared to plastics in terms of basic cost of raw materials, the energy processing cost is high. Glass is preferable in comparison to other materials in terms of total energy required for processing. Apart from its weight and breakage feature, glass rates as an almost ideal packaging material (Bauer 2009; Dean et al. 2000).

However, glass packs are more expensive than most competitive materials for warehousing. Whereas plastic packs can be made immediately before filling or consecutively by a form–fill–seal (FFS) process, this type of processing is literally impossible with glass because of the cost and the kind of processing conditions employed. Hence, glass containers are on the verge of being entirely replaced by plastic packs for the packaging of sterile and nonsterile products.

10.6.2 Plastics as a Pharmaceutical Packaging Material

The numerous disadvantages associated with the use of glass containers led to the development of plastics as a packaging material. Plastics are classified as 'thermoset' plastics, which are used for fabrication of closures, small cases (at one time used for menthol cones), protective lacquers and enamels (as applied internally and externally to metal packs) and a range of adhesive systems. 'Thermoplastics' are used for the fabrication of the main body of the primary pack. Examples of thermoplastics include polyethylene, polyvinyl chloride, polystyrene, polypropylene, nylon, polyester, polyvinylidene chloride (PVDC) and polycarbonate.

Unlike glass, plastics are not innocuous materials owing to the presence of various additives within them. Hence, before being selected as a pharmaceutical packaging material, the plastic materials undergo various rigorous tests, few of which are listed below:

Appearance and optical properties and general physical properties (tensile properties, shear and stress)

Distortion and creep, impact strength, tear properties, torsion stiffness test, hardness, abrasion resistance, coefficient of friction

Heat distortion temperature, ball indentation, melting point, specific heat of latent fusion, Vicat softening point, brittleness temperature, thermal conductivity, coefficient of linear expansion, shrinkage

Blocking, flammability

Gas permeation, water vapour permeation, water absorption, total immersion test, resistance to weak acids and alkalis, environmental stress cracking

Melt flow index

Extractive study, interaction study, migration study

10.6.2.1 Constituents in Plastics

Over the course of processing of plastics, various constituents are added to plastics to improve their efficacy and durability. From the toxicology and regulatory point of view, it is mandatory to identify and quantify the residues of these processing aids in the final product. A few of the commonly used additives are explained in Table 10.4.

10.6.2.2 Fabrication Processes for Plastic Films and Packs

According to the end use of the plastic, various fabrication or conversion processes may be employed. The type of fabrication process selected may depend on the type and grade of the plastic to be used, the constituents in it and the processing aids to be used. Plastic films intended to be used as bags or sachets can be prepared by a number of processes such as extrusion, solvent casting, calendaring and coextrusion. These processes can fabricate plastic films with a thickness below 0.010 inch. Moulding processes for fabrication of hollow packs include injection moulding, extrusion moulding, rotational moulding, soft moulding and so on. Weaknesses in barrier properties of plastics as mentioned earlier can be considerably lowered by the use of glass-based or metal coatings of silicon oxide, metal and metal oxides. These coatings can be performed by numerous methods such as dip coating, solvent, aqueous and aqueous dispersion coating, plasma enhancement and vacuum deposition.

TABLE 10.4

Constituents in Plastics

Constituent	Uses
Plasticisers	• Improves flow properties of the materials and increases softness and flexibility However, they tend to migrate into the contained product For example, phthalate esters, phosphate esters, sebacate and adipate esters, polymeric plasticisers and citrates
Filler	• Inert solid substance to reduce the degradation of plastic or to reduce their cost • The addition of fillers to thermosetting plastics generally reduces physical properties and, depending on the actual filler(s) used, is likely to reduce chemical resistance and increase permeability For example, carbon black (up to 5%), chalk or calcium carbonate, talc, china clays, silica and magnesium carbonate
Toughening agents/impact modifiers	• Used to improve the drop strength and top loading strength of naturally brittle plastics For example, 15% of methyl methacrylate butadiene styrene copolymer, chlorinated polyethylene, vinyl acetate
Lubricants	• As a processing aid to prevent adhesion with the metal parts For example, waxes, stearates
Antioxidants	• Prevent or retard oxidative degradation of plastic products For example, BHA, BHT, hindered phenols and cresols, secondary amines

Apart from these coatings, barrier enhancers can also be assimilated into the plastic. These barrier enhancers may include various metal oxides, glass fiber, mica and so on. Barrier properties of the plastics can also be improved by enclosing a foil ply between layers of plastic (Dean et al. 2000).

Plastics are commonly used for the packaging of oral solids and oral liquids. Currently, few of these materials are also used for the packaging of sterile/nonsterile parenteral and ophthalmic formulations. The commonly used plastics for oral solids include PET and HDPE. PET bottles are used owing to their advantages of low weight and nonfragile nature. HDPE bottles offer better barrier properties compared to PET but still suffer from issues such as migration of components, which may result in product instability. Table 10.5 summarises the various materials commonly used for fabrication of plastic bottles (Shams 2011).

One of the early restrictions with the use of plastics as a packaging material was with its inability to withstand the sterilisation process. However, the currently available plastics can be sterilised by various processes as listed in Table 10.6. The use of plastics for packaging of sterile products is discussed in detail in Section 10.8.

TABLE 10.5
Plastics Used for Fabrication of Bottles

Name	Properties	Density (g/mL)	MVTR (g/mil/100 in²)	Stiffness (jpsi*1000)	Hot Fill Max
PET	Clarity, strength	1.35	3.5	400	160
HDPE	Toughness, barrier to moisture and gas	0.96	0.5	150	160
PVC	Versatility, ease of blending, strength	1.35	3.0	300	160
LDPE	Versatility	0.920	1.0–1.5	120	180
PP	Strength, toughness, resistance to heat, chemical and grease	0.91	0.3	200	200

TABLE 10.6
Methods Used for Sterilisation of Plastic Containers

Method of Sterilisation	Comments
Autoclaving by steam sterilisation (moist heat)	• May be used for both filled (water-based) and empty packs • Can be used for HDPE, PP, PC, PA and, under selected conditions, plasticised PVC
Gaseous treatment using ethylene oxide	• Can be operated as 100% ethylene oxide under negative pressure to ensure that any leakage is internal, thereby reducing the risks of explosion or using ethylene oxide at a concentration of 10%–15% with an inert gas diluent • Testing for ethylene oxide retention is advised for each different grade of plastic with extended periods of extraction until total extraction of ethylene oxide is proven
Accelerated electrons or beta irradiation	• Similar to gamma irradiation; but milder in effect

10.6.3 Metals for Pharmaceutical Packaging

The application of metals for packaging of pharmaceuticals is mainly restricted to the use of tin plate containers used for the packaging of aerosols, aluminium and associated alloys for the fabrication of collapsible tubes and aluminium foils used as laminations for blister, strip and sachet packaging. Metal containers are strong, relatively unbreakable, sturdy and opaque. They are impermeable to moisture vapour, gases and microbial contamination provided they are pinhole free. They are also resistant to the extremes of temperatures. However, use of metal containers necessitates application of coatings and lacquers to prevent chemical reaction and corrosion. The lacquers including acrylics, phenolics, polyesters, epoxy and vinyl resins are normally applied by organic or aqueous based or powder-based coatings.

Collapsible metal tubes are used for packaging a wide range of pharmaceuticals including creams, pastes, ointments, jellies and semiliquids. Apart from their high degree of impermeability, such metal tubes are convenient for patient use. Since the contents from a metal tube are expelled by squeezing the tube, there is no tendency for the walls to revert back to their original shape once the pressure is released, thereby ruling out the risk of air entering the pack and reacting with the products or drying it out. A modification of a conventional metal tube is in the form of a laminated tube, which consists of a polyethylene/aluminium foil/polyethylene-laminated body fitted with a polyethylene nozzle. A majority of the collapsible tubes are made from aluminium, which is a cheap and 'soft' metal. Its susceptibility to attack by acidic and alkaline chemicals has been overcome by widespread use of coating by epoxy, phenolic and vinyl resins. Certain pharmaceutical semisolid products require a special applicator nozzle, which either can be permanently attached to the tube or may be provided as a separate applicator. The nozzle and orifice size are dictated by the viscosity of the product and the amount to be dispensed per application. For cases where the contact of the product with a metal applicator is not desired, elongated plastic nozzles are designed. Such plastic nozzles are widely used for ophthalmic products. There is an FDA mandate on the particulate contamination of eye ointment tubes by both metal particles and foreign matter. Eye ointment tubes are commonly sterilised by irradiation or exposure to ethylene oxide before filling. Collapsible tubes are commonly provided with a double-ended cap incorporating a piercing device.

Aerosol containers are commonly fabricated from tin-plated steel or aluminium. The tin-plated steel container is most preferred for manufacturing an aerosol container. Aluminium is used to fabricate extruded or seamless aerosol containers. Aluminium is commonly used as an aerosol container because of its lower potential for incompatibility, seamless nature and resistance to corrosion. Stainless steel containers are restricted to smaller sizes owing to production issues and high cost. They are extremely sturdy, are resistant to most chemicals and require no internal organic coating. They are usually used for inhalation aerosols (Sciarra and Cutie 2009). Nonaqueous products are generally filled in an unlined tin-plated metal container. Products of low pH and those containing water require containers with organic lining of epoxy and/or vinyl resins. A commonly used organic coat is composed of an undercoat of vinyl and a top coat of epoxy resin. Special precautions are to be taken for products containing alcohol that are packaged in aluminium containers. Stability testing of an aerosol formulation involves minute examination of the empty container for the presence of pinholes, or softening or blistering of the internal lacquer.

10.7   BLISTER AND STRIP PACKAGING

Blister and strip packs have been used since the early 20th century and are now well-established forms of pharmaceutical 'unit dose' packaging. Unit dose packs like

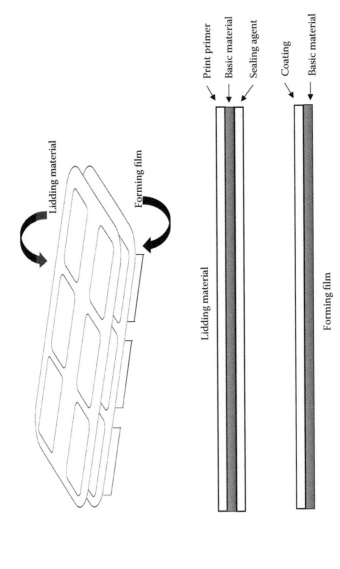

FIGURE 10.2   Composition of a blister pack.

these offer numerous advantages such as individual protection until the dose therein is removed, personal dosage, child safety, no cross-contamination risks, no opening and reclosing problems and so on. This has curbed the growth of their multipack equivalents. However, elder patient compliance in terms of handling is a point of concern. A very significant advantage of blister packs is that they are tamper-proof.

The four basic components of pharmaceutical blister packages include the forming film, the lidding film, the heat-seal coating and the printing ink. The forming film and the lidding film should match accurately to form an integrated package (Pilchik 2000). Figure 10.2 illustrates the composition of a blister pack.

The forming film is that part of the blister pack that houses the product in deep moulded pockets. It is generally composed of transparent plastics. The factors to be considered for screening the appropriate forming film for the blisters include its property type, grade and thickness. Other factors to be considered include the height and weight of the product, presence of pointed edges on the product and the impact resistance, aging, migration and cost of the film (Bhat 2015; Shams 2011).

Table 10.7 lists the various materials commonly used as forming films along with a comparison of WVTR and its characteristics (Bhat 2015).

TABLE 10.7
Characteristics of Commonly Used Forming Films of Blister Packs

Type and Thickness of Forming Film (mil[a])	WVTR[b] (g/m²/day)	Characteristics
PVC (10)	1.1	Inexpensive, clear, stiff, with low WVTR, excellent thermoformability, a high flexural strength, good chemical resistance, easy tintability, low cost
PVC/PVDC (10/1.2)	0.17	PVDC helps reduce permeability of PVC; PVDC helps provide oxygen barrier
PVC/CTFE (8/0.76)	0.07	Lowest WVTR of all films. PCTFE has the longest shelf life and does not turn yellow with use. May cause environmental concerns
Polypropylene (12)	0.20	Comparable WVTR, recyclable, no release of toxins
PET (10)	2.6	In combination with PVDC, can be used as a replacement to PVC
Polystyrene (12)	6	Compatible with thermoforming
OPA/aluminium/PVC	0	Excellent moisture barrier

[a] The unit 'mil' is equal to the thousandth of an inch.
[b] Measured at 20°C/85% RH.

TABLE 10.8
List of Commonly Used Lidding Materials

Lidding Material	Weight[a]
0.8 mil Aluminium, hard, push through	1
0.8 mil Aluminium, hard, heat seal-coated, side-printed, push through	1.25
1 mil Aluminium, soft, child resistant	1.15
45 g/m²/1 mil Paper/aluminium, peel off	1.55
45 g/m²/0.48 mil Paper/PET/aluminium peel off–push through	2.00

[a] Where 1 represents the weight per unit area of 0.8 mil aluminium, hard, push through.

Cyclic olefin copolymer (COC) films are now being seen as viable alternatives for PVDC or polychlorotrifluoroethelyne (PCTFE) high-moisture barrier films.

The lidding film or the lidding material provides the structural base onto which the final blister package is built. It must be selected according to the size, shape and weight of the product as well as the style of the package to be produced. Most commonly preferred lidding materials are aluminium foils in the thickness range of 0.46–0.61 mm. The material of the lidding film must be able to withstand the heat-seal coating process. Clay coatings are added to the lidding films to enhance printing. Heat sealing and printability are both important considerations in blister packaging, and the lidding material must offer the best workable compromise (Bhat 2015; Shams 2011). Table 10.8 lists various lidding materials commonly used.

Formpack with dessicant is a leading product from Amcor, specifically used for enhanced protection of susceptible dosage forms. The laminate consists of two principal layers with a primer and adhesive in between; a polyamide (25.0 μm), aluminium thin strip as barrier material (45.0 μm) and a three-layer polyolefin (50.0 μm) with an embedded dessicant agent. The embedded desiccant agent (moisture absorber) reacts irreversibly with water (moisture). It is tinted with blue pigments for unambiguous identification purposes of the polyolefin side. This product displayed lower water ingress into the blister cavity through the seal area at conditions of 40°C/75% relative humidity (RH) over a period of 2 years compared to other variants of Formpack containing PVC and HDPE. At normal testing conditions, Formpack with dessicant showed lower water ingress over a period of 20 years compared to other products (Amcor 2015).

10.8 PACKAGING OF STERILE PRODUCTS

The selection of an appropriate packaging material for sterile products presents a more complex challenge to the formulator than for nonsterile products. A sterile dosage form is required to be completely devoid of all forms of microbial and particulate contamination. The efficacy, stability and safety of a parenteral dosage form on storage depend solely on the type and performance of the packaging components. Ability of the packaging material to withstand the sterilisation process before its use is an important prerequisite for the selection of a packaging material for sterile products. Also, it is required of the primary pack to maintain its sterility along the total shelf life of the product.

Owing to its safety, lower reactivity and ability to visualise the contained product, glass is considered to be a prime candidate for the packaging of parenterals. Glass ampoules and vials are still the most commonly used pack for small-volume parenterals. Ampoules, which are commonly heat sealed after filling, are subjected to a dye immersion test to ensure 100% integrity of the ampoule after sealing. Though ampoules are convenient from a manufacturing perspective, their popularity has reduced owing to the issues of glass particulate contamination arising during opening of the ampoules. Irrespective of the style of the container, the quality of glass to be used for the packaging of sterile product must be Type I neutral glass (Contract Pharma 2012).

The use of plastic containers for packaging of parenterals is becoming increasingly popular owing to their advantages over glass such as elimination of the risk of broken glass and glass leachables. They are also preferred for protein pharmaceuticals because of their lower tendency for protein adsorption. Moreover, the newer grades of plastics introduced recently are able to endure the various sterilisation processes (see Table 10.6). Blow–fill–seal (BFS) and FFS are utilised for the filling of numerous ophthalmic and other injectable/noninjectable sterile formulations. Low-density polyethylene (LDPE) containers are most commonly used for this technique. The BFS or FFS system involves moulding, aseptic filling and sealing of a plastic ampoule in a continuous process under aseptic environment. This technology is being exceedingly preferred for the packaging of parenteral products owing to its advantages like unparalleled flexibility that it offers in container design, improved product quality, high level of output and lower operational costs. However, caution needs to be exercised when using plastics because of their potential for interaction with the formulation and permeability to the external environment. Traditional plastics such as polyvinyl chloride and polyethylene are not as transparent as glass, thus making visual examination of the contents difficult. Also, these materials cannot withstand the thermal sterilisation process, hence mandating the use of ethylene oxide and radiation sterilisation followed by aseptic filling.

While ampoules and BFS/FFS systems are self-sealing systems, an elastomeric closure system is also necessary to seal containers such as vials. It is mandatory for the formulator to ensure that there is compatibility between the product and stopper by conducting suitable testing since the materials used as closures tend to exhibit higher reactivity and leachability. For lyophilised products, it is preferable to select a stopper with a lower 'moisture absorption capacity'. This is so because the moisture absorbed during the autoclaving process can subsequently be transferred to the product during storage, leading to deterioration of the product. Specialised stoppers such as Teflon-coated stoppers are available, which can suffice the need for an inert stopper. Dessicant-embedded stoppers are also under development for the maintenance of moisture-sensitive products (Chemtech Foundation 2012).

Stoppers are unable to endure the depyrogenation process and hence need to be preautoclaved. However, this presterilisation treatment should be validated to ensure its efficacy for endotoxin elimination. Stoppers usually require some degree of siliconisation to grant them easy processing in automatic filling lines. The silicone used for this siliconisation does not chemically interact with the rubber surface and hence may interact with the formulation components. Recent developments have led to the replacement of silicone with specialised silicone polymers that, apart from improving surface lubricity, ensure no possibility of an interaction with the contained formulation.

It is also critical to ensure that the stopper or any component of the primary pack do not leach out an excessive quantity of extractables into the product. Glass leachables are managed by using 'treated glass' or using specially coated glass that reduces the amount of leachables and enhances the chemical resistance of the glass surface. Plastic leachables are minimised by using plastic materials containing a low level of extractable materials. For example, polyvinyl chloride possessing a high level of extractable plasticiser (diethylhexylphthalate) can be replaced by a combination of polyethylene, polypropylene or other polyolefinic materials with a lower potential of extractables.

Sterilisation of glass containers can be achieved using either a continuous tunnel or a dry heat oven. The former technique is used for larger volumes and high-speed lines while the latter is predominantly used for smaller batches. Electron beam sterilisation may also be employed for the sterilisation of glass containers. More recently, the trend is shifting towards the procurement of depyrogenated glass vials and partially assembled syringes in sealed packages that can be directly filled at the customer's site. In such cases, the responsibility of the preparation, depyrogenation and aseptic packaging rests with the supplier.

Plastic packaging components can be sterilised using numerous methods as listed in Table 10.6. The BFS and FFS systems involve fabrication of the final sterile container just before the aseptic filling step. Strict monitoring of the endotoxin content of the LDPE and other polymeric beads used to fabricate the containers as well as the melting conditions employed to form them is necessary with the BFS. The FFS employs in-line sterilisation/drying of the film before shaping of the containers.

Intricacy in the packaging of parenteral products has been recently introduced by introduction of prefilled syringes. These reduce the possibility of needle-stick injuries and facilitate administration in an emergency situation. Beckton-Dickinson and Vetter are a few companies that specialise in this technology. Safety needles, autoinjectors, pens, needle guards and other needle free injector technologies have been developed for noninvasive delivery and to overcome needle-related issues.

10.9   ACCESSORY PACKAGING MATERIALS

It is expected that the selection of a suitable container closure system alone can provide the contained product with the highest degree of protection over its shelf life. However, this is seldom possible owing to the limitations associated with each kind of packing material. Hence, in order to ensure absolute protection of the contained product, selection of an appropriate packing material is often complemented with the use of different accessory materials such as dessicants and oxygen busters to offer improved protection to the product, thereby enhancing its shelf life.

10.9.1   Gas Absorbers: Desiccants and Fillers

Desiccants are intended to be placed inside the primary/secondary pack to keep the contained products dry and stable. They function by absorbing moisture from the surrounding either by physical adsorption (e.g. silica gel) or by chemical reaction (e.g. calcium oxide), thereby reducing the humidity in the headspace of the sealed packs. Silica gel, clay and molecular sieves are the commonly used desiccants. Table 10.9 depicts in brief the characteristics of these materials.

The suitability of a material to be used as a desiccant depends on its 'moisture sorption capacity'. It is important to note that the moisture sorption capacities of the same type of desiccants may vary depending on the grade used. This principle holds true for clay and zeolite desiccants. Therefore, moisture sorption isotherms of desiccants must be established before the selection of materials. Moreover, the moisture sorption capacity of a desiccant is dependent on the temperature of use. This property of the desiccant should be taken into account during the screening process, particularly when the packaging system is designed for moisture protection at variable temperatures (Chen 2009).

The common methodology for using desiccants is to place a predetermined quantity of the 'activated' desiccant into the primary container containing the product and

---

TABLE 10.9
Characteristics of Commonly Used Desiccants

Silica Gel	Clay	Molecular Sieves (Zeolite)
• Highly porous, amorphous form of silica • Encloses a vast network of interconnecting microscopic pores or capillaries that draw and hold water • Possess a larger pore size with broad distribution • Functions on the mechanisms of surface adsorption and capillary condensation in the porous network • Functions well at ambient temperatures, reduced efficiency at higher temperatures	• Economical and efficient dessicant, especially at lower temperatures • Composed of silica aluminium oxide, magnesium oxide, calcium oxide and ferric oxide • Functions well at ambient temperatures, low efficacy at higher temperatures	• Highly porous crystalline material with precise monodispersed pores into which molecules of certain sizes can fit. • Small pore size with narrow distribution. • A molecular sieve with an effective pore size of 3 Å can selectively adsorb water molecules, while a molecular sieve with 4 Å pore size can also adsorb molecular nitrogen and oxygen, in addition to water. • Can function efficiently even at low temperatures. However, at humidity higher than 50%, displays lower efficiency.

---

then to seal the containers. The quantity of desiccant to be used is a critical parameter, since an inadequate quantity will not ensure the required protection, while its use in excess amount may result in overdrying of the product and a superfluous increase in the product cost. It is therefore necessary to understand the characteristics and the degradation mechanisms of the product as well as the humidity range suitable for the products before deciding upon the quantity of desiccant to be used. Once the desired range of humidity is determined, a suitable quantity of desiccant to be used can be calculated, based on the moisture sorption properties of both the desiccant and the drug product using a modelling method (Dobson 1987).

As much importance is given to the selection of the correct quantity of desiccant, it is equally crucial to ensure that the ambient conditions and the packaging process for placing desiccants into product packs are well controlled. This is to ensure that the efficacy of the desiccant is maintained during the packaging process. The desiccant used may rapidly absorb water vapour from the surrounding when exposed to ambient conditions, thereby hampering its efficacy. For example, activated clay, silica gel and molecular sieves when exposed to 25°C/75% RH for 1 h absorb approximately 10% of water, leading to a significant loss of the protection effect for products. Hence, the surrounding humidity and exposure time for placing desiccant into packs must be minimised, in order to minimise the loss of their efficacy. Care should also be taken when desiccants are used for gelatin capsules to prevent brittleness of the capsule shells from the loss of moisture (Chang et al. 1998; Kontny and Mulski 1989).

Fillers such as cotton and rayon are also commonly copackaged with solid pharmaceutical dosage forms to restrain product movement during shipping. However, cotton and rayon have been sometimes associated with instability problems for certain drug products. The oxidative agents used for the processing of cotton and rayon may be retained in these fillers and these residues can unintentionally be a source of degradation for some active pharmaceutical ingredient or result in cross-linking of gelatin capsules or gelatin-coated tablets. Residual formaldehyde retained in the filler may result into dissolution problems for the drug products.

The moisture sorption capacity of cotton when used as a filler is another factor to be considered. Cotton can absorb more than 6% of water at humidity higher than 60% RH at 25°C. When dry cotton is used as a filler, it can act as a dessicant and absorb water from the copackaged product. On the other hand, if equilibrated at high humidity before use, cotton can be a source of water for the copackaged dry products, if the products are hygroscopic. Although the amount of cotton used for pharmaceutical products is small, in general, the effects of cotton on moisture transfer may still be significant.

10.9.2 Oxygen Scavengers

For products that are sensitive to oxygen in the air, the various measures employed to counter this issue include the use of antioxidants in the formulations, packaging in glass or aluminium packs that are impermeable to oxygen, purging the pack headspace with an inert gas and so on. Enclosing oxygen scavengers along with the product is one such approach. The use of an appropriate amount of oxygen absorbers is sufficient to bring down the oxygen content in the headspace of a sealed pack to below 0.1%.

Oxygen scavengers, or oxygen absorbers, are materials that can react with oxygen from the environment, thereby eliminating the detrimental effects of oxygen when copackaged with products in closed packs. Oxygen scavengers are commonly used for food products but are less common for pharmaceutical products. The mechanism of product protection by oxygen scavengers is the rapid oxidation of the scavengers with oxygen from the headspace of closed packs and with the diffused oxygen during storage (Grattan and Gilberg 1994).

Successful oxygen scavenger candidates for products must be effective, safe to use, clean (so as not to contaminate products), convenient to use and of low cost. The commonly available oxygen scavengers also known as Oxygen Busters (O Busters) are capable of absorbing three times their weight in oxygen. O Busters are graded in 'cc', the number of cubic centimetres (cc) of oxygen that they will absorb. O Busters are available in a range of 20–1000 cc. The appropriate quantity of oxygen buster needs to be selected based on the oxygen absorbing capacity and the volume of the space to be protected (O-Busters 2015).

10.10 ANTICOUNTERFEITING TECHNOLOGIES

The pharmaceutical industry faces a constant threat from competitors that market counterfeit or fake goods. According to the World Health Organization (WHO), any medicine 'which is deliberately or fraudulently mislabelled with respect to identity or source' is counterfeit. WHO estimates the overall share of spurious drugs in the market to be approximately 10%.

Hence, the pharmaceutical companies are required to invariably employ extensive measures to protect patients from the adverse and sometimes fatal effects of such practices. Anticounterfeiting methods provide a means for product authentication, to distinguish genuine products from counterfeit ones. Ensuring application of these product authentication methods along with consumer and dealer awareness can ensure that such counterfeit or spurious products are completely eliminated from the supply chain. Anticounterfeiting technologies can be classified on the basis of the techniques used for authentication. The broad classification is as given in Table 10.10 (IMPACT 2015).

Development of a satisfactory pack profile for a pharmaceutical product is an exhaustive process and requires thorough study and understanding of the properties of the packaging material and the contained product and their interactions. Framing a proper and scientifically correct label/leaflet for a pharmaceutical product is as important as the selection of an appropriate packaging material for the success of the product.

Packaging has evolved as a science, and packaging expertise, combined with packaging technology, should receive utmost attention. This is essential if the most effective compromise is to be reached between those involved with such decisions, for example, marketing, medical, production and regulatory. For this purpose, a packaging coordinator who provides an overview is now becoming increasingly essential.

TABLE 10.10

Classification of Anticounterfeiting Technologies

Overt Technologies	Covert Technologies	Forensic Technologies	Serialisation Track and Trace System
• They are visible to the eye • Holograms • Optically variable devices • Colour shifting security inks and films • Security graphics • Sequential product numbering • On-product marking	• These are incorporated into the packaging and can only be identified with special equipment • Also include microscopic nanotext/images that have complex printing and hence are difficult to replicate • Invisible printing • Digital watermark • Laser coding	• They are built into the product and are well concealed • These features cannot be detected without the use of highly specialised equipment • Chemical taggants • Biological taggants • DNA taggants • Isotope ratios • Micro taggants	• They function by protecting the supply chain against breach and misuse

REFERENCES

Amcor. 2015. Formpack® Coldform Pharmaceutical Blister Packs. http://www.amcor.com/products_services/formpack/ (accessed June 20, 2015).

Bauer, E. 2009. *Pharmaceutical Packaging Handbook*. Boca Raton, FL: Taylor & Francis.

Bhat, J. M. 2015. Role of Packaging Material on Pharmaceutical Product Stability. http://www.ipapharma.org/events/stability/jbhat.pdf (accessed January 28, 2015).

Chang, R. K., Raghavan, K. S., Hussain, M. A. 1998. A Study on Gelatin Capsule Brittleness: Moisture Transfer between the Capsule Shell and Its Content. *Journal of Pharmaceutical Sciences* 87:556–558.

Chemtech Foundation. 2012. The Preferred Container Material for Parenteral Pharmaceutics. http://www.chemtech-online.com/P&B/Claudia_oct12.html (accessed January 19, 2015).

Chen, Y. 2009. Packaging Selection for Solid Oral Dosage Forms. In: *Developing Solid Oral Dosage Forms: Pharmaceutical Theory and Practice*, eds. Y. Qui, Y. Chen, G. G. Z. Zhang, 563–576. New York: Elsevier Inc.

Contract Pharma. 2012. Parenteral Packaging Considerations. http://www.contractpharma.com/issues/2012-04/view_features/parenteral-packaging-considerations/ (accessed January 16, 2015).

Dean, D. A., Evans, E. R., Hall, H. I. 2000. *Pharmaceutical Packaging Technology*. Boca Raton, FL: Taylor & Francis.

Dobson, R. L. 1987. Protection of Pharmaceutical and Diagnostic Products through Desiccant Technology. *Journal of Packaging Technology* 1:127–131.

EMEA. 2009. Committee for Medicinal Products for Human Use (CHMP) Committee for Medicinal Products for Veterinary Use (CVMP). Guideline on Plastic Immediate Packaging Materials. http://www.ema.europa.eu/docs/en_GB/document_library/Scientific_guideline/2009/09/WC500003448.pdf (accessed March 25, 2015).

Grattan, D. W., Gilberg, M. 1994. Ageless Oxygen Absorber: Chemical and Physical Properties. *Studies in Conservation* 39:210–214.

IMPACT. 2015. International Medical Products Anti-Counterfeiting Taskforce. Anti-Counterfeit Technologies for the Protection of Medicines. http://www.who.int/impact /events/IMPACT-ACTechnologiesv3LIS.pdf (accessed February 10, 2015).

Kamal, M. R., Jinnah, I. A. 1984. Permeability of Oxygen and Water Vapor through Polyethylene/Polyamide Films. *Polymer Engineering and Science* 24:1337–1347.

Kontny, M. J., Mulski, C. A. 1989. Gelatin Capsule Brittleness as a Function of Relative Humidity at Room Temperature. *International Journal of Pharmaceutics* 54:79–85.

MHRA. 2015. Medicines Information: SPC & PILs. http://www.mhra.gov.uk/spc-pil/ (accessed May 5, 2015).

O-Busters®. 2015. Oxygen Absorbing Packets and Strips. http://www.jamesdawson.com /pdf/o-buster-information-sheet.pdf (accessed June 20, 2015).

Pilchik, R. 2000. Pharmaceutical Blister Packaging, Part I Rationale and Materials. Pharmaceutical Technology. http://www.packagingconnections.com/sites/default/files /Pharma%20Blister%20Pack.pdf (accessed January 31, 2015).

Sciarra, J. J., Cutie, A. J. 2009. Pharmaceutical Aerosols. In: *The Theory and Practice of Industrial Pharmacy*, ed. L. Lachman, H. Lieberman, 589–618. Delhi, India: CBS Publishers and Distributors.

Shams, R. 2011. Critical Selection of Primary Packaging. http://www.interphex.com/rna /rna_interphex_v2/documents/2011/handouts/rustom_shams.pdf (accessed February 3, 2015).

USFDA. 1999. Guidance for Industry Container Closure Systems for Packaging Human Drugs and Biologics. http://www.fda.gov/downloads/Drugs/Guidances/ucm070551.pdf (accessed March 20, 2015).

Weeren, R. V., Gibboni, D. J. 2002. Barrier Packaging as an Integral Part of Drug Delivery. *Drug Delivery Technology* 2:48–53.

11 Quality Management Systems in Pharmaceutical Manufacturing

Abhinandan R. Patil and Maharukh T. Rustomjee

CONTENTS

11.1 INTRODUCTION

The pharmaceutical industry is responsible for making quality medicines, diagnostic agents and medical devices available to patients worldwide. Regulatory authorities all over the world have stringent rules and procedures for the review and approval of pharmaceutical products before granting permission for their marketing. The purpose of this stringent control on marketing approval for pharmaceutical products is to ensure their safety, efficacy and quality. During product development, correlation of the critical quality attributes to the safety, efficacy and quality of the product is well established (see Chapters 4, 7 and 9), and control measures during commercial manufacturing assure that the same consistent quality drug product is made available to patients. This is where the role of maintaining a state of control in the manufacture and supply chain of pharmaceutical products by applying pharmaceutical quality system (PQS), risk management, good manufacturing practices (GMP) and process validations is of utmost importance and the chapter will focus on the scope and implementation of PQS.

11.2 PHARMACEUTICAL QUALITY SYSTEM

In June 2008, the ICH harmonised tripartite guideline ICH Q10 on PQS was issued for implementation (ICH 2008). This guideline was necessary to bridge the differences in expectations on quality systems from the various regional regulations such as ICH Q7, ISO 9000, 9001:2000, 9004, 13485:2003, European Medicines Agency (EMA) guidelines and so on, thereby helping to achieve global harmonisation of quality systems and availability of medicines around the world in the interest of public health (EMA 2013; ICH 2000).

The ICH Q10 guideline describes the modern quality systems needed to establish and maintain a state of control that can ensure the realisation of a quality drug product and facilitate continual improvement over the life cycle of a drug product. The guideline applies to pharmaceutical drug substances and drug products including biotechnology products. However, for manufacture of investigational products, the regional GMP still apply.

The Q10 quality system augments the implementation of ICH Q8 (R2) and ICH Q9 guidelines, thus enabling the realisation of the full benefits of the concepts contained within these two guidelines, and in tandem, Q8 (R2), Q9 and Q10 guidelines assure product quality throughout its life cycle.

Using modern quality system elements for pharmaceutical manufacturing, the Q10 guideline enhances GMP and provides an opportunity for robust processes, resulting in drug substances and drug products that consistently meet their intended attributes. The aim of the PQS is to achieve product realisation, maintain a state of control at all times and facilitate continuous improvement in the product manufacturing.

Figure 11.1 illustrates the major elements of the ICH Q10 PQS model that serve as the major tools to meet the objectives (ICH 2008). These tools, if applied consistently

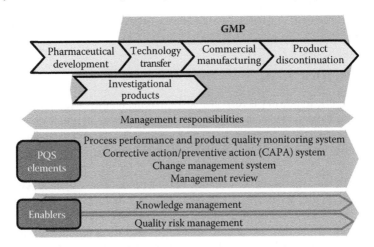

FIGURE 11.1 PQS as per ICH Q10. (Modified from International Conference on Harmonisation of Technical Requirements for Registration of Pharmaceuticals for Human Use, ICH harmonised tripartite guideline, pharmaceutical quality system Q10. Available at http://www.ich.org/fileadmin/Public_Web_Site/ICH_Products/Guidelines/Quality/Q10/Step4/Q10_Guideline.pdf.)

and appropriately to each life cycle stage, would help recognise and utilise opportunities for continual improvement in the product quality. These elements (particularly management review) drive the management and continual improvement of the pharmaceutical management system. Herein, the management responsibility is key to the success of the PQS. Leadership in establishing and maintaining a company-wide focus and commitment to quality is most essential. Senior management should provide inputs for establishing the quality policy, planning to achieve the quality objectives and suitable communication processes for the company. The management is also responsible for providing the requisite resources and periodic monitoring and review of the PQS to ensure proper governance, effectiveness and performance of the same. The figure also depicts the enablers for the PQS model that include knowledge management and quality risk management. These enablers support an effective and successful implementation of the PQS system. These tools assist in providing a mechanism for science- and risk-based decision making related to product quality matters.

Quality assurance (QA) and GMP are the backbone on which the PQS operates, and an understanding of these is essential to implement the PQS practically and effectively.

11.3 QUALITY ASSURANCE

The quality of a product is influenced individually or collectively by many factors, and the QA concept covers all of these. With regard to pharmaceuticals, QA covers the development, production, quality control (QC), distribution and inspection activities for any pharmaceutical company. Quality management is a management system to direct and control an organisation with regards to quality and is equivalent to the quality system as defined in ICH Q9. It is an assortment of business processes focussed on achieving the laid down quality policy and quality objectives of a company to meet their customer requirements. The organisational structure, policies, procedures, processes and resources needed to implement the quality policy and objectives perform the QA function for any organisation.

Quality management system (QMS) and QA are at the heart of quality initiatives in any company. Figure 11.2 depicts the various systems that are typically defined and controlled by the QA team in any organisation.

It is the responsibility of a manufacturing authorisation holder to manufacture and distribute medicinal products so as to assure that they are fit for their intended use, comply with the requirements of the product marketing approval and ensure adequate safety, efficacy and quality for the patient. QA requires the participation and commitment by staff in all the different departments and at all levels within the company and the quality objective is the responsibility of senior management. It also requires commitment from the company's suppliers and by the distributors in order to achieve it.

In a manufacturing set-up, a well-designed QA program is a way of preventing mistakes or defects in the manufactured product or services. The company should provide adequate resources in the form of competent personnel, appropriate premises, equipments and facilities for smooth working of the QA systems. All aspects of the QA systems should be well documented and regular monitoring of its effectiveness and evaluation of continual improvement opportunities is essential.

FIGURE 11.2 Quality assurance – an amalgamation of quality systems.

The QA of an organisation for the manufacture of medicinal products should ensure the basic minimum controls as listed below:

- Medicinal products are designed and developed in compliance to all good practices relating to manufacturing, clinical and laboratory
- Production and control operations are clearly specified and documented; qualified and trained personnel resources are available in all job roles and functions
- Managerial responsibilities are clearly specified in all departments; release of the product for market supply is certified by authorised personnel
- Adequate control on the supply and use of the correct starting and packaging materials
- Necessary controls on intermediate products, and any other in-process controls and validations of processes
- Facility, utility, equipment, environmental control and water systems are designed, qualified and validated
- The finished product is correctly processed and checked, according to the defined procedures
- Satisfactory arrangements are made for storage, distribution and handling of medicinal products such that its quality is maintained throughout the supply chain and during its shelf life
- Procedure for self-inspection or quality audit, which regularly appraises the effectiveness and applicability of the quality assurance system
- On the basis of the quality assessments, suitable remedial and precautionary actions or any revalidations with the proper documentation are undertaken to ensure the quality

The quality objective cannot be achieved consistently unless a comprehensively designed and correctly implemented system of QA incorporating GMP and appropriate QC along with quality risk management is implemented. At a broader level, QA includes all the good practices, that is, GMP, good clinical practices, good laboratory practices, good engineering practices and good distribution practices covering the activities in the pharmaceutical industry, and these together are referred to as the good '*something*' practices.

11.3.1 GOOD MANUFACTURING PRACTICES

In any pharmaceutical production system, there are inherent risks that cannot be eliminated by testing the final product. Some of the major risks are incorrect labels on containers, which could mean that patients receive the wrong medicine; insufficient or too much active ingredient, resulting in ineffective treatment or adverse effects; and cross-contamination of products, causing damage to health or even death. GMP is a systematic approach for ensuring that the occurrence of these risks is minimised and products are consistently produced and controlled according to quality standards. Correct procedures must be consistently followed at each step in the manufacturing process–every time a product is made, documented proof of the same needs to be generated.

Manufacturers must employ technologies and systems that are up-to-date in order to comply with the regulation and that is why GMP is also referred to as *cGMP*. The *c* stands for *current*, reminding us that technologies used to control manufacturing operations that were *state-of-the-art* 20 years ago may no longer be adequate today. Science and technology continue to evolve and so current application of these to the manufacturing operations to prevent contamination and errors is essential.

Current GMP refers to the Good Manufacturing Practice Regulations promulgated by the various regulatory bodies, some of which are listed in Table 11.1.

Detailed guidelines for GMP have also been published by the World Health Organization (WHO) and the Pharmaceutical Inspection Convention. GMP requirements have been established by many countries on the basis of these guidelines. Other countries have come together and harmonised their requirements, for example, the Association of Southeast Asian Nations, in the European Union and through the Pharmaceutical Inspection Convention.

These regulations can be legally enforced and they require that manufacturers, processors and packagers of drugs, medical devices and blood products take proactive steps to ensure that their products are safe, pure and effective. Failure of firms to comply with GMP regulations can result in very serious consequences with legal prosecution against the company and responsible individuals resulting in recall, seizure, fines and jail time. This is how the regulators protect the consumers from purchasing a substandard or dangerous product.

All issues related to manufacturing of pharmaceutical products such as records, personnel qualifications, sanitation, cleanliness, equipment verification, process validation, complaint handling and so on are addressed by GMP regulations. Most GMP requirements are recommendatory, general and open ended, and therefore it allows each manufacturer to decide individually how to best implement the necessary controls. This not only provides much flexibility and freedom in setting up internal quality systems but also requires that the manufacturer interpret the requirements in a manner that is suitable and appropriate for each individual business.

TABLE 11.1
Country-Wise GMP Regulations

Regulatory Agency	Regulations
United States Food and Drug Administration (USFDA, USA)	• 21 CFR Part 4 Current Good Manufacturing Practice Requirements for Combination Products • 21 CFR Part 210 Current Good Manufacturing Practice in Manufacturing, Processing, Packing or Holding of Drugs • 21 CFR Part 211 Current Good Manufacturing Practice for Finished Pharmaceuticals • 21 CFR Part 212 Current Good Manufacturing Practice for Positron Emission Tomography Drugs • 21 CFR Part 606 Current Good Manufacturing Practice for Blood and Blood Components
European Medicines Agency (EMA, European Union)	• Eudralex Volume 4 Good Manufacturing Practice Guidelines • Directive 2003/94/EC for medicinal products for human use and investigational medicinal products for human use
Pharmaceuticals and Medical Devices Agency (PMDA, Japan)	• Ministerial Ordinance on Standards for Manufacturing Control and Quality Control for Drugs and Quasidrugs
Therapeutic Goods Administration (TGA, Australia)	• Australian Codes of Good Manufacturing Practice • Australian Code of Good Manufacturing Practice for Medicinal Products (16 August 2002) • Australian Code of GMP for Human Blood and Tissues (24 August 2000)
Central Drugs Standard Control Organisation (CDSCO, India)	• Schedule M Good Manufacturing Practices and Requirements of Premises, Plant and Equipment for Pharmaceutical Products
Health Canada (Canada)	• Canadian Good Manufacturing Practices for Drugs (Part C Division 2 of Food and Drug Regulations)
China	• Regulations for Implementation of the Drug Administration Law of the People's Republic of China

11.3.2 QUALITY CONTROL

One of the most important aspects of the cGMP regulations pertains to the QC laboratory and product testing. The pharmaceutical QC laboratory serves a very significant function in pharmaceutical production and control. QC is that part of GMP that is concerned with sampling, specifications and testing of raw materials, packaging materials, intermediate products and finished products. QC includes laboratory management systems that organise the working practices, documentation and release procedures. This ensures that the necessary and relevant tests are actually carried out and that materials are not released for use, nor products released for sale or supply, until their quality has been tested and found to be satisfactory.

GMP regulations covering QC laboratory activities embody a set of principles that provides a structure and outline within which laboratory evaluation/studies are planned, performed, monitored, recorded, reported and archived. Product release decisions and risk/safety assessments are made on the evaluation of the testing results obtained during the product evaluation/study throughout its life cycle. During GMP audits and inspections by the regulatory authorities and company auditors, the QC laboratory is one of the focal points for the inspection as the quality of the products being released for human consumption is very much dependent upon the proper operations of the QC department. Historically, in the last few years, the regulators worldwide have been alarmed by observations related to lack of data integrity in many pharmaceutical companies and clinical study sites. These data integrity concerns raise many questions relating to the quality of medicinal products being released for human consumption and the United States Food and Drug Administration (USFDA), EMA and other regulators are hence pushing for adoption of the latest technologies such as automations in laboratories including data generation and traceability (USFDA 2011), process analytical technology (PAT) and continuous manufacturing so as to minimise human intervention in the manufacturing and testing of drug substances and drug products. This minimisation of human intervention is expected to lead to improved assurance of quality.

11.3.3 PROCESS VALIDATION

Process validation plays an important role in assuring the quality of medicinal products and is one of the key criteria to be considered before releasing drug products for human consumption. As per ICH Q7 – GMP guide for active pharmaceutical ingredients, process validation was defined as documented evidence that the process, operated within established parameters, can perform effectively and reproducibly to produce a medicinal product meeting its predetermined specifications and quality attributes. These and other guidelines used the prospective, concurrent and retrospective validation approach to perform the process validations.

However, the concept and practice of process validation have changed significantly over the last few years with the introduction of the product life cycle concept in quality management of medicinal products. The regulatory authorities (USFDA and EMA) have also released revised guidelines on process validation in compliance to ICH. Process validation now incorporates a life cycle approach linking product and process development, process performance qualification, validation of the commercial manufacturing process and maintenance of the process in a state of control during routine commercial production.

Process validation is now defined as the compilation and assessment of data, from the process design stage through commercial production, which establishes scientific evidence that a process is capable of consistently delivering quality products. As per this paradigm shift, the process validation takes place in three stages, namely, process design, process qualification and process verification (EMA 2014a). Chapter 8 provides an in-depth explanation on process validation as per the new paradigm and as applied throughout the product life cycle.

11.3.4 GMP INSPECTIONS

Inspections and audits of pharmaceutical manufacturing sites (these include manufacturing drug substances, drug products and medical devices) are one of the fundamental ways by which a regulatory authority can ensure that the sites are operating in compliance to GMP regulations and the medicinal products manufactured are of the requisite quality standard. Inspections are carried out by regulatory authorities for the reasons as listed below:

- Preapproval inspection, usually in relation to an application for marketing authorisation of a drug product
- Routine inspection of a regulated facility to ensure compliance to GMP and quality systems
- *For-cause* inspection to investigate a specific problem that has come to regulator's attention, for example, market complaints, recalls, reports of unexpected side effects

In recent times, the incidence of counterfeit medicines being found in the supply chain has increased and hence an important aim of pharmaceutical inspection is to eliminate this hazard by monitoring the quality of pharmaceutical products in distribution channels, from the point of manufacture up to delivery to the recipient.

Globalisation and outsourcing have resulted in a complex supply chain for medicinal products around the world that has extended the geographical sources for manufacturing of medicinal products and active pharmaceutical ingredients. This has led to an increase in the inspection burden on regulatory agencies to ensure the safety and effectiveness of drugs marketed in their country. There is an effort being made by all regulatory bodies to harmonise, mutually share and accept inspection findings across all agencies. Regulators are trying to move towards risk-based management for regulatory inspections to address the challenges of globalisation, and this is being supported by all stakeholders in the industry. To better control the risk to patients and to optimise the use of regulator resources, an appropriate level of regulatory oversight through continued international cooperation is the need of the hour. Quality risk management established by ICH Q9 applies to both industry and regulators. For quality risk-based inspections, key factors linked to the product (e.g. aseptic, toxic and paediatric) need to be focussed in addition to the inspection history of the site and the company.

Attempts are being made to develop a regulatory inspection framework approach at a global level (e.g. mutual recognition agreements between national/regional authorities) as part of an international collaboration on GMP inspections. Confidentiality agreements for sharing inspection reports, joint GMP inspections, acceptance of GMP certificates based on legal formats (e.g. WHO Certificate of Pharmaceutical Product) and recognition of harmonised approaches for GMP (e.g. WHO, the Pharmaceutical Inspection Convention and Pharmaceutical Inspection Co-operation

Scheme jointly referred to as PIC/S, ICH Q7, Q9 and Q10) are some of the attempts in this direction (ICH 2000, 2005; PIC/S 2007). The pharmaceutical industry also supports these collaborative approaches among the regulators and various programs such as the WHO prequalification program and ICH Global Cooperation Group, and it is expected that this collaborative inspection approach will also lead to capacity enhancement of regulatory authorities (IFPMA 2015; PIC/S 2014).

Despite all the inspection programs, it is pertinent to understand that the manufacturer is ultimately responsible for the quality of the products manufactured and is also accountable for the supply chain of the products that it controls.

11.4 FUTURE TRENDS IN QUALITY MANAGEMENT

As explained in Chapter 2, over the past two decades, pharmaceutical product manufacture has been plagued with critical drug shortages owing to product recalls and supply issues related to manufacturing problems. The increasing presence of counterfeit medicines in the supply chain has further complicated the matter for regulatory bodies. Shortage of drugs and biologics creates a considerable health hazard, delaying, and in some cases even denying, critically needed care for patients. Hence, preventing drug shortages and battling counterfeit medicines remain a top priority for drug regulators and the industry.

Over the last few years, various reviews and studies (EMA 2012; ISPE 2013) have been carried out by the regulators and industry bodies to understand the root causes of drug recalls and shortages. These surveys and studies have come to the conclusion that quality-related issues are the main root cause of the shortages, and this has led to an unprecedented *focus on quality* from various government and industry bodies, which is driving key changes and initiatives in the industry.

This drive towards quality is very apparent from initiatives worldwide and has resulted in many strategic actions/implementation plans for redressing this situation, notable among these are presented in Table 11.2.

11.4.1 CONTINUOUS MANUFACTURING

The above furious pace of interactions among the regulatory authorities and industry bodies has resulted in an acceptance of the need for the industry to absorb new technologies and QA approaches to make the manufacture and supply chain of pharmaceutical products more dependable. One way of doing this is to reduce human intervention as much as possible so that human errors and discretionary decision making are eliminated in the manufacturing of pharmaceutical products and hence the risk to quality is mitigated. The regulators are therefore encouraging the industry to adopt continuous manufacturing technologies by incorporating PAT with automation and feedback control loop.

Continuous manufacturing technology combines the traditional, independent steps of batch processing in pharmaceutical manufacturing into one cohesive process by continually verifying quality online and releasing products swiftly. This needs adoption of on-line and in-line screening of product attributes on a continuous basis during manufacture with application of PAT tools. Continuous manufacturing

TABLE 11.2
Strategic Action Plans for Preventing Drug Shortages

Reports	Key Recommendations
Strategic Plan for Preventing and Mitigating Drug Shortages (USFDA, October 2013) (USFDA 2013)	Short-term and long-term prevention strategies within USFDA and in collaboration with other stakeholders
Risk-Based Approach for Prevention and Management of Drug Shortages (PDA Technical Report No. 68 [TR 68]. Issued with input and support from EMA, several EU National Competent Authorities [NCAs] and USFDA, December 2014) (PDA 2014)	First practical approach to identify and mitigate drug shortage risks in a structured way
ISPE Drug Shortages Prevention Plan: A Holistic View from Root Cause to Prevention (Issued by ISPE on request by EMA, October 2014) (ISPE 2014)	Key recommendations include Corporate Quality Culture and a Robust Quality System, Metrics, Business Continuity Planning, Communications with Authorities and Building Capability
Prevention of Drug Shortages Based on Quality and Manufacturing Issues: Final Report (Issued by ISPE and PDA along with a multiassociation task force charged by EMA, December 2014) (EMA 2014b)	Single, collaborative action plan for the prevention of drug shortages resulting from manufacturing quality issues

hence leads to dramatically reduced throughput times, lower operating and investment costs and a smaller manufacturing footprint. The FDA has already approved many products using PAT tools such as Raman spectroscopy for drug substance identification and near-infrared spectroscopy for drying monitoring, blending monitoring and end point detection for these operations, identification of raw materials, assay and content uniformity.

The FDA supports the implementation of continuous manufacturing using a science- and risk-based approach considering the fact that the science for the same exists and there are no regulatory hurdles even though experience is limited. As with any new concepts, it is recommended that sponsors take up early and frequent discussion with the Agency before implementation (USFDA 2012).

11.4.2 QUALITY METRICS

In 2012, the Food and Drug Administration Safety and Innovation Act was passed; it authorised the FDA to collect manufacturing quality data from pharmaceutical companies and obtain certain records from a drug manufacturer in lieu of, or in advance of, an inspection. The previous requirement of a 2-year inspection frequency for drug facilities was replaced with a risk-based inspection approach, which meant that facilities that were at high risk relating to quality compliance would be inspected more often than those that were at low risk, which would be inspected at a lower frequency. This frequency determination would be done based on risk assessment by the authorities considering the history and status of the products and manufacturing facilities periodically. This would ensure that the resources of the regulator are focussed on the high-risk products and facilities and improve their overall effectiveness of quality surveillance resulting in reduction of drug recalls and shortages. Thus, the FDA launched an initiative to utilise well-defined quality metrics as an input to its risk-based inspection models, determine inspection schedules for a manufacturer, assess postmarket change reporting and restructure the format of inspection. A dialogue has since been initiated by the FDA with the industry to define and freeze the quality metrics to be used and find effective ones to assess processes and their risks. Companies routinely use metrics such as number of lot release tested, out-of-specification results and lots attempted, rejected, reworked and reprocessed. Each company collects data in its own way for its own internal use, using a range of definitions, sampling plans and periodicity. However, the ways these metrics are collected differ greatly from company to company and even from site to site within the same company. Hence, there is a need for standardisation in the way the metrics are defined, measured and generated so that for the FDA they are all assessed against a common yardstick using a standard scale and can help them benchmark the products and facilities from company to company and within different sites of the same company. The guiding principles for selection and definition of the metric are that they should be clearly defined to allow *consistent* reporting across sites. They should also be objective and meaningful, easy to capture, easy to report and 'normalised' based on factors (e.g. process differences and technical complexity) and should drive acceptable, not unwanted behaviours.

On behalf of the FDA, ISPE in cooperation with McKinsey has been running a pilot project (Wave 1) with 44 manufacturing sites/technologies of 18 participating companies to collect feedback from the stakeholders and to help define the metrics for the regulators. After a detailed evaluation of the various industry and regulatory considerations related to proposed pharmaceutical quality metrics, the ISPE pilot metric set was developed to include 14 measures reflecting a mix of leading and lagging indicators collected by site and product. Twelve of the metrics collect quantitative data responses and two of the metrics (for Process Capability and Quality Culture) collect qualitative data responses using a survey-based format. After the successful completion of the Wave 1 pilot, the Wave 2 pilot has been launched to continue to expand participation, assess a potentially adjusted metric set based on learning from Wave 1 and aligned with FDA technical regulatory agenda, incorporate input from subteams working on *Quality Culture* and *Process Capability* with more focus on current data collection and analysis and continue to expand cooperation with other groups and associations.

With the valuable inputs from the industry pilots, the quality metrics, when implemented in the next few years, are expected to cause a major change/transformation in the industry and the Agency. In this regards, a draft guideline 'Request for quality matrix' has been issued by USFDA in July 2015. Further, dialogue and involvement of all stakeholders is encouraged for a meaningful and successful implementation, which in the long run will help improve the overall quality of pharmaceutical products worldwide.

11.4.3   Office of Pharmaceutical Quality at FDA

In January 2015, the Office of Pharmaceutical Quality (OPQ) was set up as a new office within the Center for Drug Evaluation and Research (CDER) of the USFDA that creates a single unit dedicated to product quality. The new structure is expected to provide better alignment among all drug quality functions at CDER, including review, inspection and research. OPQ combines non-enforcement-related drug quality work into one superoffice, creating *one quality voice* and improving oversight of quality throughout the life cycle of a drug product. OPQ creates a uniform drug quality program across all sites of manufacture, whether domestic or foreign, and across all drug product areas – new drugs, generic drugs and over-the-counter drugs. Within the OPQ, many dedicated offices (as shown in Figure 11.3) have been set up to achieve the goals laid out in FDA's 21st Century Initiative.

The Office of Process and Facilities is one such section within the OPQ whose objective is to ensure that quality is built into the manufacturing processes and facilities over the product life cycle using risk-based approaches for efficient assessment of manufacturing facilities. Readers are referred to the white paper 'One Quality Voice' issued by the OPQ recently, which talks about the concept of one quality voice for all stakeholders – drug authorities, patients, industry, healthcare professionals and purchasers. Additionally, OPQ will

- Assure that all human drugs will consistently meet quality standards that safeguard clinical performance
- Utilise enhanced science- and risk-based regulatory approaches
- Transform product quality oversight from a qualitative to a quantitative and expertise-based assessment

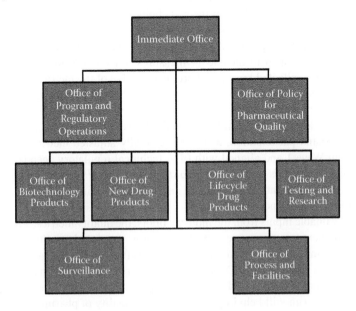

FIGURE 11.3   Office of Pharmaceutical Quality – USFDA.

- Provide seamless integration of review, inspection, surveillance, policy and research throughout the product life cycle
- Encourage the development and adoption of emerging pharmaceutical technology

The establishment of OPQ will result in enhanced transparency and communication between the Agency and industry related to manufacturing technologies, issues and capabilities, thereby preventing drug shortages and ensuring the availability of high-quality drugs.

In conclusion, in the years ahead, the industry and regulatory bodies are gearing up for major and constructive transformations towards which they are being propelled owing to the heightened focus on quality throughout the product life cycle for medicinal products. Adoption of these approaches and technologies is essential, and it is only a matter of time when these will become the normal way of working within the industry.

REFERENCES

EMA. 2012. Reflection paper on medicinal product supply shortages caused by manufacturing/ good manufacturing practice compliance problems. http://www.ema.europa.eu/docs /en_GB/document_library/Other/2012/11/WC500135113.pdf (accessed June 11, 2015).

EMA. 2013. *Guidelines for Good Manufacturing Practice for Medicinal Products for Human and Veterinary Use*, Volume 4, Chapter 1 Pharmaceutical Quality System. http:// ec.europa.eu/health/files/eudralex/vol-4/vol4-chap1_2013-01_en.pdf (accessed June 11, 2015).

EMA. 2014a. EMA finalizes guidance on process validation for manufacturers. http://www.in -pharmatechnologist.com/Processing/EMA-finalizes-guidance-on-process-validation -for-manufacturers (accessed June 11, 2015).

EMA. 2014b. Prevention of drug shortages based on quality and manufacturing issues: Final report. http://www.pda.org/docs/defaultsource/website-document-library/scientific-and -regulatory-affairs/drug-shortage/interrupted-supply-inter-associationsummary-final -report-2014.pdf?sfvrsn=4 (accessed June 11, 2015).

ICH. 2000. International conference on harmonisation of technical requirements for registration of pharmaceuticals for human use, ICH harmonised tripartite guideline, good manufacturing practices, Q7. http://www.ich.org/fileadmin/Public_Web_Site/ICH_Products /Guidelines/Quality/Q7/Step4/Q7_Guideline.pdf (accessed June 11, 2015).

ICH. 2005. International conference on harmonisation of technical requirements for registration of pharmaceuticals for human use, ICH harmonised tripartite guideline, quality risk management Q9. http://www.ich.org/fileadmin/Public_Web_Site/ICH_Products /Guidelines/Quality/Q9/Step4/Q9_Guideline.pdf (accessed June 19, 2015).

ICH. 2008. International conference on harmonisation of technical requirements for registration of pharmaceuticals for human use, ICH harmonised tripartite guideline, pharmaceutical quality system Q10. http://www.ich.org/fileadmin/Public_Web_Site /ICH_Products/Guidelines/Quality/Q10/Step4/Q10_Guideline.pdf (accessed June 19, 2015).

IFPMA. 2015. Inspections: International Federation of Pharmaceutical Manufacturers & Associations. http://www.ifpma.org/quality/inspections.html (accessed June 11, 2015).

ISPE. 2013. Report on the ISPE drug shortages survey. http://www.ispe.org/drugshortages /2013JuneReport (accessed June 11, 2015).

ISPE. 2014. Drug shortages prevention plan: A holistic view from root cause to prevention. http://www.ispe.org/drugshortagespreventionplan.pdf (accessed June 11, 2015).

PDA. 2014. Technical report 68: A risk-based approach for prevention and management of drug shortages. http://www.pda.org/docs/default-source/attendee-presentations /north-america/2014/2014-pda-fda-joint-regulatory-conference/emma-ramnarine .pdf?sfvrsn=4 (accessed June 11, 2015).

PIC/S. 2007. Pharmaceutical inspection convention, pharmaceutical inspection co-operation scheme, guidance to parametric release. http://www.picscheme.org/pdf/24_pi-005-3 -parametric-release.pdf (accessed June 11, 2015).

PIC/S. 2014. Pharmaceutical inspection convention, pharmaceutical inspection co-operation scheme, guide to good manufacturing practice for medicinal products. http://www .picscheme.org/publication.php?id=4 (accessed June 11, 2015).

USFDA. 2011. Guidance for industry part 11, electronic records; electronic signatures – Scope and application. http://www.fda.gov/downloads/RegulatoryInformation/.../ucm125125 .pdf (accessed June 11, 2015).

USFDA. 2012. FDA perspective on continuous manufacturing. http://www.fda.gov/downloads /AboutFDA/CentersOffices/OfficeofMedicalProductsandTobacco/CDER/UCM341197 .pdf (accessed June 11, 2015).

USFDA. 2013. Strategic plan for preventing and mitigating drug shortages. http://www.fda .gov/downloads/Drugs/DrugSafety/DrugShortages/UCM372566.pdf (accessed June 11, 2015).

12 A New Era of Drug Products

Opportunities and Challenges

*Preshita P. Desai, John I. Disouza
and Vandana B. Patravale*

CONTENTS

12.1 INTRODUCTION

The pharmaceutical industry focuses on design and development of novel drug delivery systems so as to leverage its advantages (patient compliance; controlled or site-specific drug release; improved pharmacokinetics; possible reduction in dose, dosing frequency, non-site specific side effects, etc.) to achieve either novel products with a desired target profile or to improvise existing drug products with a business proposition in terms of market entry and exclusivity (owing to novelty, efficacy and intellectual property). This changing paradigm is evident from the estimated significant increase in global revenue of advanced drug delivery systems from $151.3 billion in year 2013 to $173.8 billion by 2018, presenting a compound annual growth rate of 2.8% (BCC 2014). However, it must be noted that most of these advanced drug delivery systems demand extensive and specialised technical understanding, meticulous and stringent in-process quality controls, more intricate safety and efficacy studies and so on.

In this context, regulatory authorities are encouraging the development of such new delivery systems subject to the satisfactory fulfilment of their concern with respect to the quality, long-term safety and superiority in terms of efficacy. Some of these newer product trends are described in Sections 12.2 through 12.4.

12.2 NANOTECHNOLOGY BASED PRODUCTS

A wide range of nanoformulations (submicron size) are being explored for drug delivery (Figure 12.1). Many such products have been approved by regulatory authorities

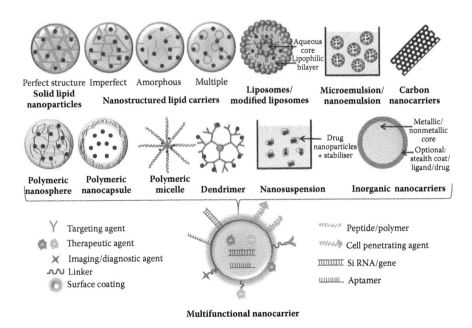

FIGURE 12.1 (See colour insert.) Nanopharmaceuticals: spectrum and advances.

(e.g. Doxil, Ambisome, Rapamune, TriCor, etc.) as they offer advantages such as enhanced dissolution, tissue permeability and bioavailability, protection to encapsulated drug, controlled release, reduced dose and dosing frequency, reduced toxicity and so on. Development of nanoformulations demands additional critical parameter controls and are discussed in Chapters 7 and 9. The newer technologies being explored in this area include surface modification (charge, targeting ligand, etc.) that allows targeted drug delivery, selective permeation, longer blood circulation time and so on, and are anticipated to roll out as next-generation medication for high-risk drugs (i.e. anticancer, antiretroviral, central nervous system ailments, etc.).

In this area, a new concept of *nanosimilars* is also emerging, which relates to generic versions of innovator nanoformulations and is being considered as a special category owing to the need for additional safety and efficacy data (Scott 2014).

Being recently developed, regulatory authorities are mostly treating their approval process in a case-to-case manner and are currently building up additional regulatory guidelines to streamline and expedite the development and approval process.

12.3 DRUG AND DEVICE COMBINATION PRODUCTS

These are widely being used in the management of various ailments, and the scope of applications of drug–device combinations is summarised in Figure 12.2. These can be identified as specialised devices (mainly class II and III as per the United States Food and Drug Administration [USFDA]) that are used for site-specific or controlled delivery of drug formulation. The continual progress in this area essentially moves towards the development of better-controlled, automated and patient-complaint

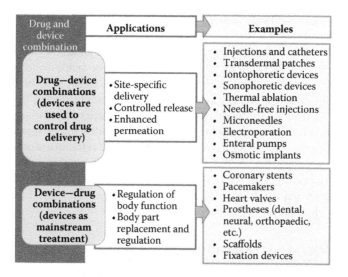

FIGURE 12.2 Drug and device combinations: scope and applications.

devices and can be exemplified with the help of the following example. Insulin, an antidiabetic peptide, was conventionally given as a subcutaneous/intravenous injection. Owing to its chronic therapy, it was thought that a noninvasive route will have a better market acceptability and thus Pfizer Ltd. introduced the inhaled insulin (EXUBERA) in early 2006, but it was withdrawn from the market by the end of 2007 as it did not receive the predicted market response (~1% market share). The failure of this drug–device combination was mainly attributed to the large device size, high cost, no additional efficacy advantage and physicians' concern about chronic pulmonary administration of insulin. With this withdrawal, other companies developing inhalable insulin (Eli Lilly and Company and Nectar Pharmaceuticals Ltd.) soon halted their development. However, after thorough understanding of market concerns, Sanofi India Ltd. introduced the inhalable insulin AFREZZA with a small whistle-shaped delivery device (a low-cost and patient-compliant product) in the US market in early 2015 and is being promisingly accepted by the end user (Afrezza 2015). On the other side, insulin pumps have now entered the market with a glucometer; hence, the dose is adjusted automatically as blood glucose levels change, making the therapy more accurate and compliant (MiniMed 530G with Enlite by Medtronic Pvt. Ltd.). This additionally indicates the impact of patient needs and market trends on possible development of new drug products (Medtronic 2015).

The new combination product trends are also being envisaged for implantable devices (coronary stents, pacemakers, central and peripheral nervous system, cochlear implants, etc.) wherein drugs are being introduced in combination to increase the foreign body acceptability of devices with reduced side effects (Aarsiwala et al. 2014). The illustrative example of this progressive development is advances in drug-eluting coronary stents. Stents are expandable tubular structures used to treat narrowed arteries specifically in coronary heart disease. In earlier times, bare metallic stents were used for this purpose, but their prime drawback was restenosis owing to uncontrolled neointimal cell growth. Thus, it was thought worthwhile to introduce antiproliferative drugs at a very low concentration and paclitaxel-loaded bioerodible polymer-coated stents came in the market (Conor). With an aim to have sustained release of these inhibitors up to 3 months to completely tackle restenosis and to have additional patent/market exclusivity, a paclitaxel-loaded, slow biodegradable polymer-coated stent (Infinium) was introduced. As an advancement, sirolimus (an antiproliferative drug that inhibits cell division at an earlier phase than paclitaxel) was identified to be a better candidate and sirolimus-loaded stents were introduced (Supralimus Core/SupraFLEX). The regulatory authorities (India) not only approved the advanced products but also increased the shelf life from the initial 1 year to 2.5 years for SupraFLEX in 2015 (SMPTL 2015). Thus, a continuous improvement in strategy on the basis of the identified market need is the major driver for advancements in new pharmaceutical products.

12.4 PERSONALISED MEDICINE

Personalised medicine is emerging as a newer therapy module with better regulatory acceptance worldwide in recent years. It refers to the specially designed/tailored medical regimen for each person based on individual characteristics (e.g. genetic make-up), needs and preferences. Knowing the fact that all patients do not respond to medication in a similar manner owing to variation in genetic set-up, personalised medication generally involves two steps, that is, assessment of the genetic make-up of the patient/group of population for the presence of specific gene/gene mutation/ representative biomarker to identify if the patient will respond to a particular drug and prescribing a drug based on that. This is a promising tool for critical diseases such as cancer, autoimmune disorders and so on wherein the drugs either turn ineffective or have major side effects. For example, the USFDA approved a drug, KALYDECO (ivacaftor), for treatment of cystic fibrosis, which is only effective in patients with a mutation in the G551D gene that controls the salt and water transport in the body (only ~4% of the overall cystic fibrosis patient population) (USFDA 2015). It must be noted that most of the personalised therapies are preceded by a diagnostic test to identify a specific genetic compliance (genetic mutation, specific gene presence, etc.). On account of this, personalised medicines are mostly available as a combination of diagnostic kit (companion diagnostic) and a drug product.

This field is further getting revolutionised with the emergence of biopharmaceuticals, biosimilars, gene therapy, stem cell therapy and regenerative medicine, which makes the therapy specific and individual oriented based on the patient's molecular set-up. In addition to the modified drug therapy (based on genetic, anatomical or physiological parameters), the scope of personalised medicine broadens to therapy modulation based on individual patient monitoring and is getting more advanced with the development of more accurate and sensitive diagnostic technologies. For example, a combination of drugs, Tafinlar (dabrafenib) and Mekinist (trametinib), with a diagnostic THxID BRAF test is approved for treatment of advanced unresectable melanoma. Herein, the patient is diagnosed for the presence of V600E or V600K mutation in the BRAF gene (found in 50% of the patient population). Here,

Tafinlar is prescribed for patients with the V600E mutation while Mekinist is prescribed for patients who present either of the mutations (USFDA 2015).

To summarise, the success of personalised medication depends on meticulous diagnostics that assures the patients' positive response to the drug, and thus, there is a trend towards development of more sensitive and accurate diagnostics along with advanced drug products and coupling thereof for a regulatory consent as a combination product (USFDA 2015).

On a drug development front, the paradigm is shifting from low-molecular-weight chemical entities to complex biomacromolecules (e.g. peptides, antibodies, genes, etc.). Regulatory authorities though have approved drug products based on these molecules for the treatment of various diseases such as cystic fibrosis, transplant rejection management, cancer, HIV and so on; they are very stringent regarding the safety and efficacy data of such products and mostly treat the approval on a case-to-case basis. It is predicted that this area has greater avenues for development of novel delivery systems in recent times as the challenge remains in the development of suitable delivery systems to ensure their delivery to the site of action (PWC 2015).

REFERENCES

Aarsiwala, A., Desai, P., Patravale, V. 2014. Recent advances in micro/nanoscale biomedical implants. *Journal of Controlled Release* 189:25–45.

Afrezza. 2015. Affrezza (insulin human) inhalation powder. https://www.afrezza.com/ (accessed June 19, 2015).

BCC. 2014. Global markets and technologies for advanced drug delivery systems. http://www.bccresearch.com/market-research/pharmaceuticals/advanced-drug-delivery-markets-phm006j.html (accessed June 19, 2015).

Medtronic. 2015. MiniMed 530G® with Enlite® by Medtronic Diabetes. http://www.medtronicdiabetes.com/treatment-and-products/minimed-530g-diabetes-system-with-enlite (accessed June 19, 2015).

PWC. 2015. Pharma 2020: Virtual R&D which path will you take? https://www.pwc.nl/nl/assets/documents/pwc-pharma2020.pdf (accessed June 19, 2015).

Scott, M. 2014. Nanosimilars. http://www.fda.gov/downloads/ForIndustry/UserFees/GenericDrugUserFees/UCM398891.pdf (accessed June 19, 2015).

SMPTL. 2015. Sahajanand Medical Technologies Pvt. Ltd. http://www.smptl.com (accessed June 19, 2015).

USFDA. 2015. Paving the way for personalized medicine. http://www.fda.gov/downloads/ScienceResearch/SpecialTopics/PersonalizedMedicine/UCM372421.pdf (accessed June 19, 2015).

Index

Page numbers followed by f and t indicate figures and tables, respectively.

Printed and bound by CPI Group (UK) Ltd, Croydon, CR0 4YY

17/10/2024

01775709-0013